33. Colloquium der Gesellschaft für Biologische Chemie
25.—27. März 1982 in Mosbach/Baden

Biochemistry of Differentiation and Morphogenesis

Edited by L. Jaenicke

With 158 Figures

Springer-Verlag Berlin Heidelberg GmbH 1982

Professor Dr. LOTHAR JAENICKE
Institut für Biochemie
der Universität zu Köln
An der Bottmühle 2
D-5000 Köln 1, FRG

ISBN 978-3-642-68835-5 ISBN 978-3-642-68833-1 (eBook)
DOI 10.1007/978-3-642-68833-1

Originally published by Springer-Verlag Berlin Heidelberg New York 1982
Softcover reprint of the hardcover 1st edition 1982

2131/3130-543210

Preface

The topic of this Mosbach Colloquium was meant as a question to begin with. When I started to study differentiation and morphogenesis in *Volvox* I hoped for a straightforward answer along prepared grooves - only to find out that also here things follow Murphy's Law: they were much more complicated than expected! Succour had to be sought. Thus, the idea arose to put this question before a board of experts.

Experience would have warned any ex-service man never to utter an idea or else you would be made responsible, and it came as it had to come: I was made impressario of this gremium; I had to assemble the experts. These Proceedings contain their expertise. I cannot even say that I biased it by my picking. In the beginning I aimed at setting different accents by inclination and force of habit. Then, by trial and error, by advice and declination, the programme shaped itself. It eventually gained momentum of which also the size of this volume is indicative.

In this volume are printed all the papers presented - with two regretted exceptions - but not the sometimes lively discussion, which clarified and pruned here and there. It would just have made the size too unwieldy.

Differentiation and morphogenesis start with the expression of genes. The development programme reels off the genome and is regulated by the position of the appropriate genes. Their structure is in the focus of gene biochemistry since the decisive tools have become available. Gene expression is under hormonal and environmental control in plants and animals. This offers possibilities of influencing developmental processes at the fundamental level at will.

Transfer of genes by plasmids or viruses changes natural differentiation programmes. This is studied at different levels of organization from plant crown gall tumours to animal neoplasias. Modification of deoxy and ribonucleic acids is part of the game.

Cellular differentiation gives rise to specific tissues. Thus, spores of bacteria are formed or somatic and gonidial cells are differentiated in flagellates up to the determination of function during embryogenesis in mammalian development. Insects offer particularly attractive objects for the study of gene activation.

Tissue is formed by adhesion of cells which recognize each other by means of surface-specific information. This may be chemical, such as glycoproteins or glycolipids, or physical, such as ion gradients. The cell surface changes with functional differentiation. By messages of this kind the ultimate network of functional organization in the most highly evolved forms of life is finally connected.

Messengers of morphogenesis have been found and identified widely in nature. The spectrum reaches from small molecules to complex proteins, the research objects from unicellular leucocytes and simply organized creatures to embryonic development of higher animals. Also here, thorough study of a relatively primitive system led to a break-through and demonstrated again the basic generality of information flow by peptide "hormones" through all classes of living systems.

Much had to be left out of the programme of this meeting, but, hopefully, the pointillistic treatment of the subject animates and attracts new interest in one of the oldest fields of general biochemistry since Spemann first postulated molecular organizers and inducers to explain differentiation and morphogenesis almost 80 years ago. At least we can state that much more detailed knowledge has accumulated through the effort of three generations, and the time may be ripe to apply it to uncover another layer of the intricate origins of life.

This issue has a rather toilsome history of birth: Gratefully acknowledged punctuality alternated with disenchanting delays and even refusals, sometimes at a last minute and after much nerve-racking correspondence.

Now, after the feast, I may thank all the active participants and patient advisors, the audience and the helpers, on the stage as well as behind the scene, for making the 33rd Mosbach Meeting a reasonable success. Springer-Verlag, as usual, took care to produce a fine book. May the reader be pleased with the contents, too, in finding up-to-date information on a timely borderline field of Biochemistry, presented from different and sometimes unusual angles. It was the aim to show the broad overlaps and indentations that make Biochemistry such a strong binding link between physical and biological sciences. Eight years ago, the same intention also underlaid the 25th Mosbach Colloquium on Biochemistry of Sensory Functions: Both volumes are like a *tema con variazioni* on the *leitmotiv*: Signal reception and processing - then to answer a message, to-day to direct function and shape.

Polite friends said that the Colloquium was a success. If this is true - which I hope - I have to thank the colleagues who helped to fill and fit the programme.

Oktober 1982

L. JAENICKE

Contents

Gene Expression

Transfer of Genes

Contributors

You will find the addresses at the beginning of the respective contribution

Gene Expression

Synchrony of Gene Expression and Regulation of the Developmental Program in Normal and Leukemic Cells

Leo Sachs[1]

"The described cultures thus seem to offer a useful system for a quantitative kinetic approach to hematopoietic cell formation and for experimental studies on the mechanism and regulation of hematopoietic cell differentiation" [1].

Cloning and Differentiation of Normal Hematopoietic Cells in Culture

The cloning and differentiation of normal hematopoietic cells in culture, has made it possible to study the controls that regulate growth (multiplication) and differentiation of various haematopoietic cell types. We first showed [1,2], as was then confirmed by others [3] that normal mouse myeloid precursor cells cultured with a feeder layer of other cell types can form colonies of granulocytes and macrophages in culture. We also found that the formation of these colonies is due to secretion, by cells of the feeder layer, of specific inducers that induce the formation of colonies and the differentiation of cells in these colonies to macrophages or granulocytes in mice (Fig. 1) [2,4,5] and in humans [6]. After we first detected their presence in culture supernatants [4,5], these protein inducers have been referred to by a number of names [7-14] and I shall use the name macrophage and granulocyte inducers (MGI) (Table 1) [12-14]. These inducers can be produced and secreted by various normal and malignant cells in culture and in vivo [15]. Their production can be induced by a variety of compounds [16-19], and some cells produce these proteins constitutively [5,11,12, 20,21]. MGI are a family of proteins that exist in a number of molecular forms that have different biological activities.

Growth-Inducing and Differentiation-Inducing MGI Proteins

The family of MGI proteins include some proteins that induce cell growth and others that induce differentiation. Those that induce growth, which are also required for normal cell viability, we now call MGI-1 [13,14]. They include different forms that can induce the formation of macrophage colonies (MGI-1M) [5,14,20], granulocyte colonies (MGI-1GM) [12,21,22]. MGI-1 has previously been referred to as mashran gm [9], colony stimulating factor [10], colony stimulating activity [11], and MGI [12] (Table 1). The existence of an antibody that

1 Department of Genetics, Weizmann Institute of Science, Rehovot 76100, Israel

33. Colloquium - Mosbach 1982
Biochemistry of Differentiation and Morphogenesis

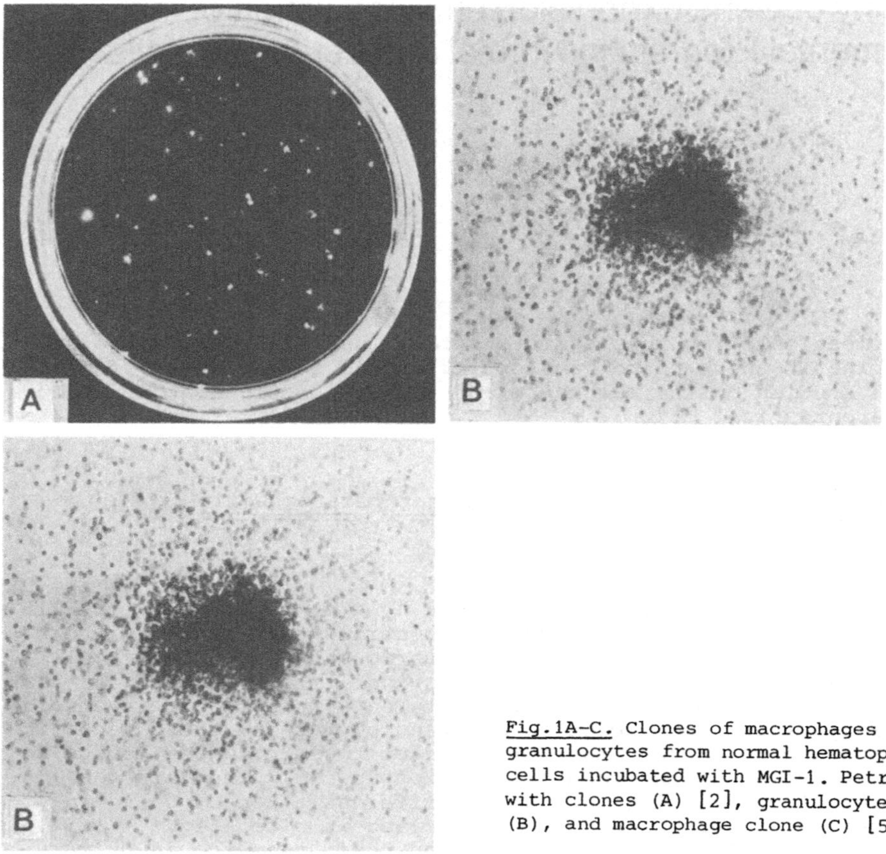

Fig.1A-C. Clones of macrophages and granulocytes from normal hematopoietic cells incubated with MGI-1. Petri dish with clones (A) [2], granulocyte clone (B), and macrophage clone (C) [5]

does not react with all forms of MGI-1M has shown, that there can be different antigenic sites on molecules that belong to the same form of MGI-1 [14]. The other main type of MGI, which we now call MGI-2 [13,14,23], induces the differentiation of myeloid precursor cells, either leukemic [24] or normal [13,23], without inducing colony formation. This differentiation-inducing protein [8] has also been referred to as MGI [24], D factor [25,26], and GM-DF [27]. It has been suggested that there are different forms of MGI-2 for differentiation to macrophages or granulocytes [23]. The regulation of MGI-1 and MGI-2 appears to be under the control of different genes [19].

These macrophage and granulocyte inducers can be proteins or glycoproteins, depending on the cells in which they are produced, and the presence of carbohydrates does not appear to be necessary for their biological activity [21]. Their molecular weights are mostly around 23,000 or multiples of this number [12,14,17,20,28] and MGI-2 activity is more sensitive to proteolytic enzymes and high temperature than MGI-1 activity [21], MGI-2 has a shorter half-life in serum than MGI-1 [29]. The ready separability of the different forms of MGI seems to depend on the cells from which they are derived [14]. Further studies should determine whether different forms of MGI are derived from a common precursor, and whether tumor cells with the appropriate gene

Table 1. Cloning and Clonal Differentiation of Normal Hematopoietic Cells in Culture

Cloning and differentiation in liquid medium (mast cells and granulocytes) [1]
Cloning and differentiation in agar (macrophages) [2]
Inducer for cloning and differentiation secreted by cells [2]
Inducer in conditioned medium from cells (for macrophages and granulocytes) [4,5]
Different inducer for macrophage and granulocyte clones [5]
Cloning and differentiation of macrophages and granulocytes in methylcellulose [5]
Confirmation of cloning and differentiation in agar [3]
Production of inducer for cloning by some leukemic cells [7]
Protein inducer of differentiation that does not induce cloning [8]
Terminology used for proteins that induce cloning and differentiation of normal macrophages and granulocytes
Mashran gm [9]
Colony stimulating factor (CSF) [10]
Colony stimulating activity (CSA) [11]
Macrophage and granulocyte inducer (MGI) [12]
MGI-1 (mashran gm, CSF, CSA) for cloning; MGI-2 for differentiation [13,14]

re-arrangements, and possibly even normal cells under certain conditions, may produce hybrid molecules of different forms of MGI including hybrid molecules with MGI-1 and MGI-2 activity [23].

Induction of Differentiation-Inducing by Growth-Inducing Protein in Normal Cells

Incubation of normal myeloid precursor cells with MGI-1, either MGI-1M or MGI-1G [14] induces the viability and growth of these normal precursors, and results in cell differentiation to macrophages or granulocytes even without adding the differentiation-inducing protein MGI-2. The incubation of normal myeloid precursors with MGI-1 also results in the induction of MGI-2 [13,23]. This induction of MGI-2 by MGI-1 can thus account for the induction of differentiation after adding MGI-1 to the normal cells. The induction of differentiation-inducing protein MGI-2 by growth-inducing protein MGI-1, thus appears to be an effective control mechanism for coupling growth and differentiation in the normal cells.

The multiplication of normal cells is regulated at two control points. The first control is that which requires MGI-1 to produce more cells that can then differentiate by the MGI-2 induced by MGI-1. The second control is the stopping of cell multiplication that occurs as part of the program of terminal differentiation to mature cells induced by MGI-2. There is thus a coupling of growth and differentiation in normal cells at both these points. Mature cells can also produce feedback inhibitors that interfere with the induction of growth of the normal precursors by MGI-1 [9,30,31].

Control of Growth and Normal Differentiation in Leukemia

Normal myeloid precursor cells (myeloblasts), which have been isolated from bone marrow [32], require an external source of MGI-1 for cell viability and growth. There are, however, myeloid leukemic cells that no longer require MGI-1 for viability and growth, so that these leukemic cells can then multiply in the absence of MGI-1 [13,17]. This gives the leukemic cells a growth advantage over the normal cells when there is a limiting amount of MGI-1. Other myeloid leukemic cells constitutively produce their own MGI-1 [7] and these leukemic cells also have a growth advantage compared to normal cells that require an external source of MGI-1. A change in the requirement of MGI-1 for growth, either a partial or complete loss of this requirement, or the constitutive production of MGI-1, can thus both give a growth advantage to leukemic cells. The growth advantage of leukemic cells with a constitutive production of MGI-1 would be even greater, if in these leukemic cells MGI-1 did not induce MGI-2.

The existence of myeloid leukemic cells that either no longer require MGI-1 for viability and growth or constitutively produce their own MGI-1, raises the question of whether these leukemic cells can still be induced to differentiate to mature cells by the normal differentiation-inducing protein MGI-2. This question has been answered by showing that there are clones of myeloid leukemic cells that no longer require MGI-1 for growth, but can still be induced to differentiate normally to mature macrophages or granulocytes by MGI-2 (Fig. 2) via the normal sequence of gene expression [13,17]. These mature cells are then no longer malignant in vivo [29,33]. Injection of MGI-2 into animals, or in vivo induction of MGI-2 by a compound that induces the production of this differentiation-inducing protein, results in an inhibition of leukemia development in animals with such leukemic cells [29]. There are also myeloid leukemic cells that constitutively produce their own MGI-1 and that can be induced to differentiate by MGI-2. Our results indicate, that induction of normal differentiation in myeloid leukemic cells by MGI-2 can be an approach to therapy based on the induction of normal differentiation in malignant cells [6,15,17,29,34].

Leukemic clones that can be induced to differentiate to mature cells by MGI-2 have been found in different strains if mice [24,27,35-38] and in humans [6,18]. They are referred to as MGI^+D^+ (MGI^+ to indicate that they can be induced to differentiate by MGI-2; D^+ for differentiation to mature cells). MGI^+D^+ leukemic cells have specific chromosome changes compared to normal cells [39,40]. These chromosome changes thus seem to involve changes in genes other than those involved in the induction of normal differentiation. There are other clones of myeloid leukemic cells that can also grow without MGI-1 but are either partly (MGI^+D^-) or almost completely (MGI^-D^-) blocked in their ability to be induced to differentiate by MGI-2 (Fig. 3) [36,37,41-46]. These differentiation-defective clones have specific chromosome changes compared to MGI^+D^+ cells [40].

There are a variety of compounds, other than MGI-2, that can induce differentiation in MGI^+D^+ clones. Not all these compounds are active on the same MGI^+D^+ clone, and they do not all induce the same differentiation-associated properties. The inducers include certain steroids, lectins, polycyclic hydrocarbons, tumor promoters, lipopolysaccharide, x-irradiation and compounds used in cancer chemotherapy [17]. The existence of clonal differences in the ability of x-irradiation and cancer chemotherapeutic chemicals to induce differentiation, may help to explain differences in response to therapy in different individuals [17].

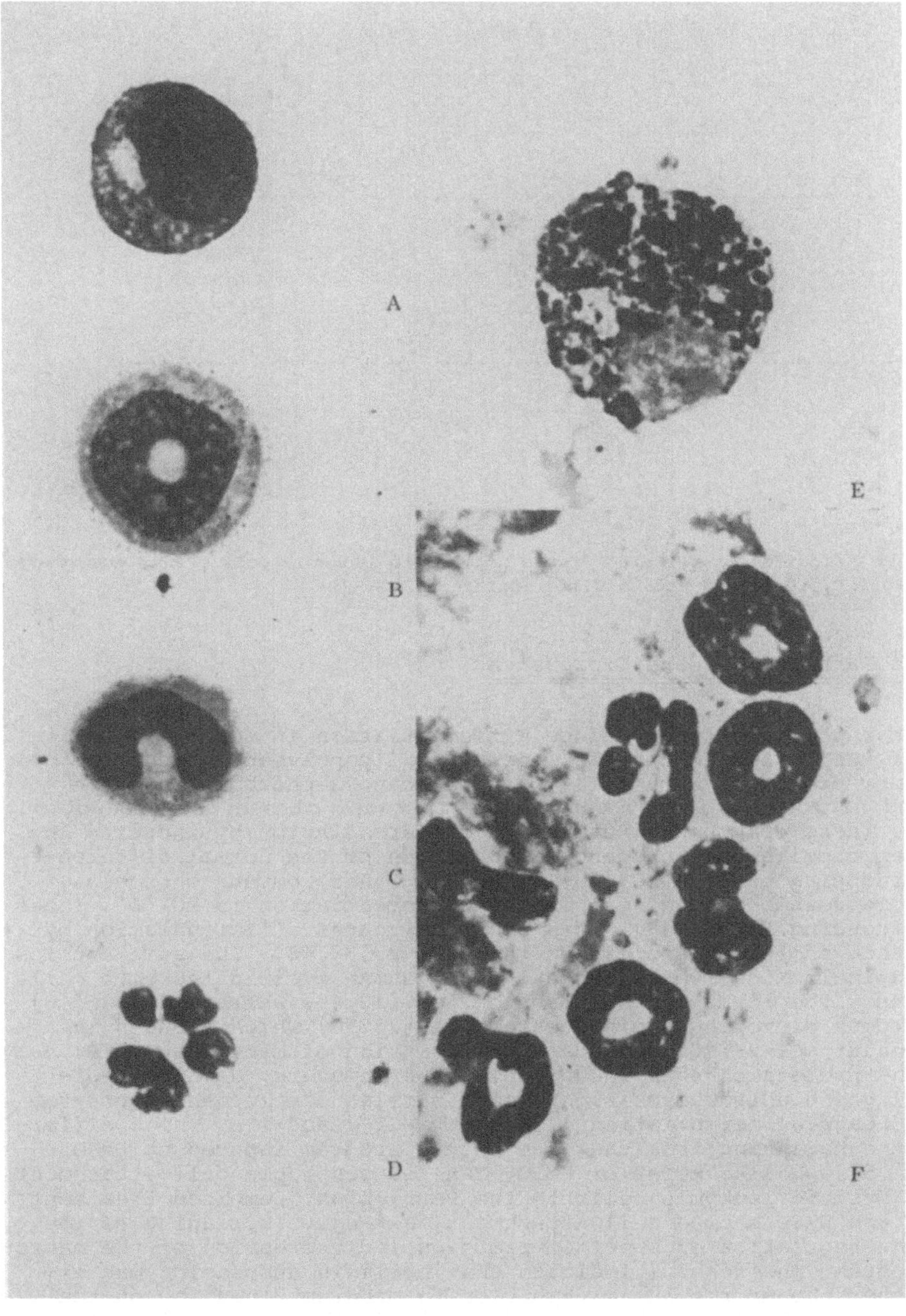

Fig. 2A-F. Differentiation of $MGI^{+}D^{+}$ myeloid leukemic cells to mature macrophages and granulocytes by MGI-2. Leukemic cell (A), stages in differentiation to mature granulocytes (B-D), macrophage (E) and group of granulocytes in different stages of differentiation (F) [17]

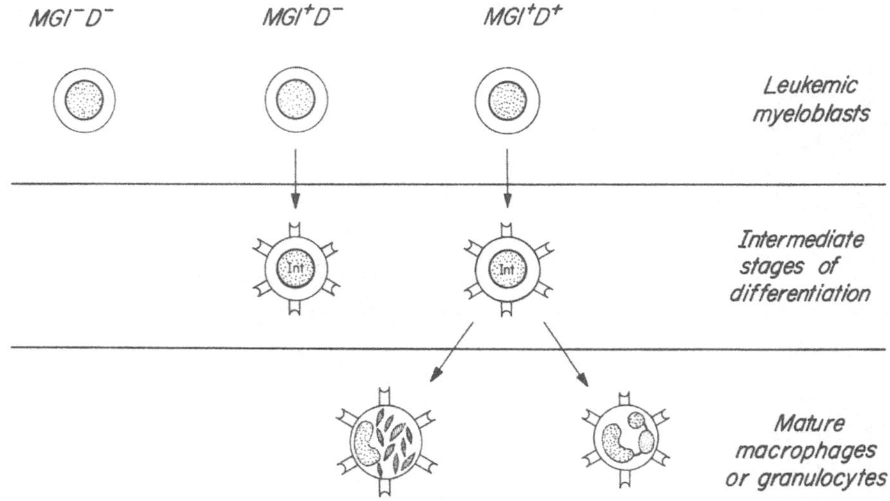

Fig. 3. Classification of different types of myeloid leukemic cell clones according to their ability to be induced to differentiate by MGI-2

Different Pathways of Gene Expression

Some of the compounds that induce differentiation in susceptible clones of $MGI^{+}D^{+}$ leukemic cells, including lipopolysaccharide, phorbol ester tumor promoters such as 12,0-tetradecanoyl-phorbol-13acetate (TPA), and nitrosoguanidine, can induce in these clones the production of MGI-2. These compounds thus induce differentiation by inducing in the leukemic cells the endogenous production of the normal differentiation-inducing protein MGI-2 [18,19,47]. Other compounds such as the steroid dexamethasone, can induce differentiation in $MGI^{+}D^{+}$ clones without inducing MGI-2 [19]. This steroid induces differentiation by other pathways of gene expression than MGI-2 [38,48]. The same applies to dimethylsulfoxide (DMSO). In a line of human myeloid leukemic cells DMSO induces the formation of granulocytes [18,49] whereas MGI-2 [18] and the tumor promoting phorbol ester TPA [18,50] which induces the production of MGI-2 [18], induces the formation of macrophages. Studies on the protein changes induced by these inducers, using two-dimensional gel electrophoresis, showed a similar developmental program for macrophage differentiation induced by MGI-2 and TPA, which differed from the beginning from the granulocyte program induced by DMSO (Fig. 4) [51]. Unlike MGI-2 or DMSO, TPA induces rapid cell attachment of these myeloid leukemic cells to the Petri dish. Combined treatment with TPA and DMSO showed cell attachment, extensive spreading of the cells, the regulation of specific proteins and expression of the macrophage program. The results indicate that cells in suspension can express either the macrophage or granulocyte program depending on the inducer, and that changes in cell shape associated with cell attachment can regulate specific proteins and restrict the developmental program to macrophages (Fig. 5) [51]. The in vivo environment of cells in relation to the possibilities of cell adhesion, may thus play a major role in determining the differentiation program of myeloid and other cell types.

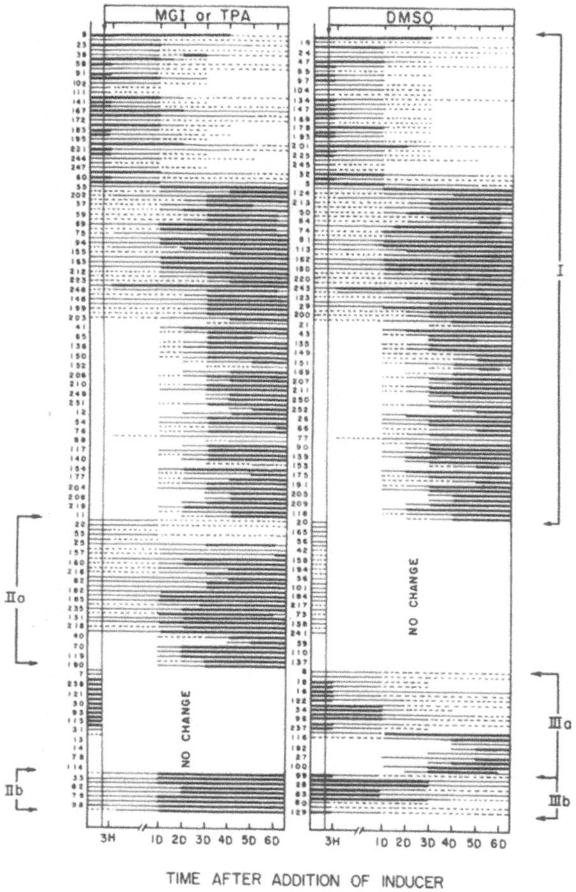

Fig. 4. Schematic summary of cytoplasmic protein changes in human myeloid leukemic cells during macrophage differentiation induced by MGI-2 or TPA and granulocyte differentiation induced by DMSO. The summary is based on analysis of two-dimensional gels of cytoplasmic proteins synthesized in untreated cells and in cells 3 h (*H*) and 1 to 6 days (*D*) after incubation with MGI-2, TPA, or DMSO. High or maximum synthesis ▬ ; intermediate synthesis ── ; low or minimum synthesis ---- . The absence of a line indicates that no synthesis was detected. This summary shows the relative and not the absolute rates of protein synthesis [51]

Complementation of Gene Expression

Induction of differentiation in some myeloid leukemic clones requires combined treatment with different compounds [18,52,53]. In these cases each compound induces changes not induced by others, so that the combined treatment results in new gene expression. This complementation of gene expression can occur both at the level of mRNA production and mRNA translation [54]. With the appropriate combination of compounds, we have been able to induce all our MGI^-D^- leukemic clones for some differentiation-associated properties [53]. It will be interesting to determine whether the same applies to differentiation of erythroleukemic cells [55,56]. It is possible that all myeloid leukemic cells no longer susceptible to the normal differentiation-inducing protein

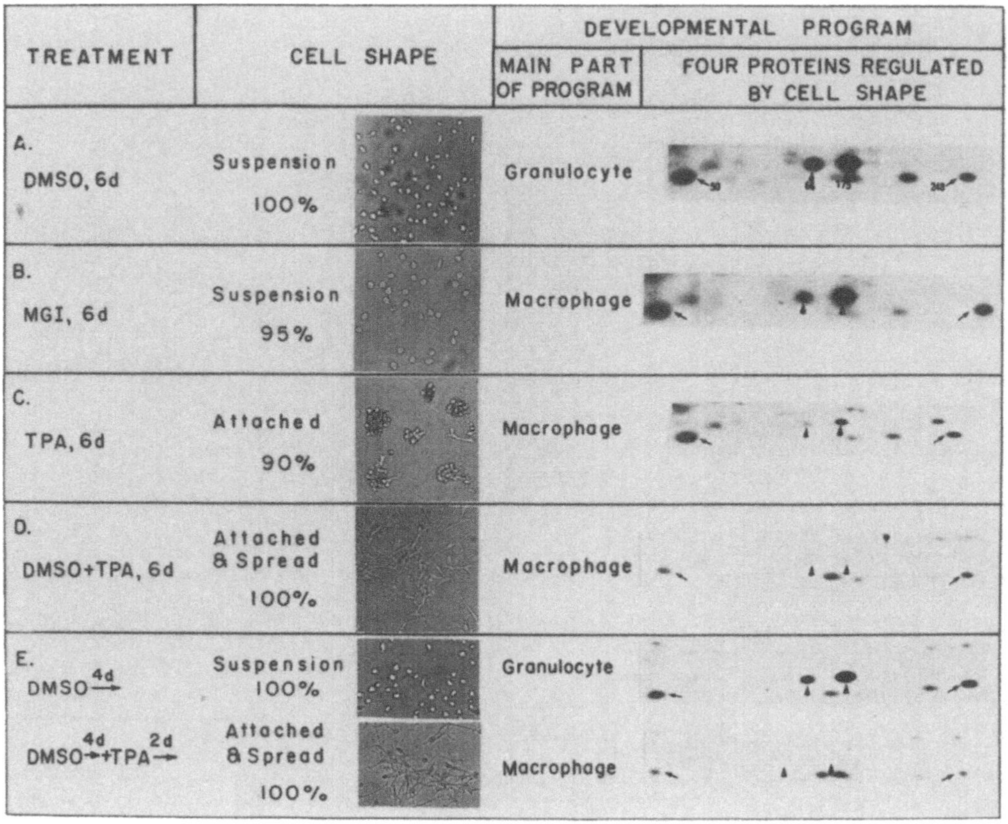

Fig. 5. Cell shape and expression of the developmental programs for macrophages and granulocytes and of four class I proteins in human myeloid leukemic cells. The main part of the program refers to expression of class II or class III proteins (Fig. 4). The four proteins shown are in class I (Fig. 4). Days (*d*) [51]

MGI-2 by itself, can be induced to differentiate by choosing the appropriate combinations of compounds to give the required complementation. This can include the use of hormones such as steroids [41,57] or insulin [53] and different non-physiological compounds [17] together with or without MGI-2.

Uncoupling of Growth and Differentiation in Leukemia

As pointed out above, there are $MGI^{+}D^{+}$ clones of myeloid leukemic cells that no longer require MGI-1 for growth but can still be induced to differentiate normally by MGI-2. These leukemic cells have thus uncoupled the normal requirement for growth from the normal requirement for differentiation. Experiments on the properties of these cells after induction of differentiation by MGI-2 have shown, that the normal requirement of MGI-1 for cell viability and growth is restored in the differentiating leukemic cells [8,58]. MGI-1 added to normal myeloid precursors induces the production of MGI-2, so that the cells can then differentiate by the endogenously produced MGI-2. However in the leukemic cells, MGI-1 did not induce the production of MGI-2 even though, like normal cells, they again required MGI-1 for viability and growth. There was therefore no induction of differentiation after adding MGI-1 [58]. Leukemic cells that constitutively produce their own MGI-1 can also show this lack of induction of MGI-2 by MGI-1, so that the cells do not differentiate [59]. The absence of induction of MGI-2 by MGI-1 therefore uncouples growth and differentiation in these leukemic cells. The lack of requirement of MGI-1 for growth and the absence of induction of the differentiation-inducing protein MGI-2 by the growth-inducing protein MGI-1, are thus mechanisms that uncouple growth and differentiation in MGI^+D^+ leukemic cells [13].

The coupling of growth and differentiation in normal cells is regulated at two control points. The uncoupling of growth and differentiation in MGI^+D^+ leukemic cells is at the first control point, but the coupling at the second control in normal cells, between the initiation of differentiation by MGI-2 and the stopping of multiplication in the mature cells, is maintained. There are differentiation-defective MGI^+D^- leukemic cells, that like the MGI^+D^+ leukemic cells, no longer require MGI-1 for growth. However, in these cells MGI-2 induces only a partial differentiation, mature cells are not produced and the cells do not stop multiplying. In addition to uncoupling growth and differentiation at the first control point, MGI^+D^- leukemic cells thus show a second uncoupling between the initiation of differentiation by MGI-2 and the stopping of cell multiplication that occurs as part of the normal program of terminal differentiation. It has been suggested that leukemia originates by uncoupling the first control and that uncoupling of the second control then results in a further evolution of leukemia [13,17].

Constitutive Gene Expression in Leukemia

Since there are leukemic cells which, unlike normal myeloblasts, no longer require MGI-1 for cell viability and growth, the molecular changes required for viability and growth that have to be induced in the normal cells are constitutive in these leukemic cells. This suggests that the origin of myeloid leukemia can be due to a change from an induced to a constitutive expression of genes that control cell viability and growth [13,17]. Both a change in gene balance, due to chromosome differences that produce differences in gene dosage [15,40], and the integration of an appropriate provirus near growth regulatory sites [60-63], could change gene expression from inducible to constitutive and thus uncouple controls.

Studies on changes in the synthesis of specific proteins in normal myeloblasts, MGI^+D^+, MGI^+D^- and MGI^-D^- leukemic clones at different times after adding MGI-1 and MGI-2, using two dimensional gel electrophoresis [45,46,48], have directly shown that there have been changes from inducible to constitutive gene expression in the leukemic cells. The results also indicate a relationship between constitutive gene expression and uncoupling of the initiation of differentiation by MGI-2 and the stopping of multiplication in the mature cells. The leukemic cells were found to be constitutive for changes in the synthesis of specific proteins that were only induced in the normal cells after treatment with MGI-1. These protein changes, which included the appearance of some proteins and disappearance of other proteins, were constitutive in all the leukemic clones studied derived from different tumors. These have been called C_{leuk}, for constitutive for leukemia. There were other protein changes that were induced by MGI-2 in normal and MGI^+D^+ leukemic cells, and were constitutive in MGI^+D^- and MGI^-D^- leukemic cells. There were more of these constitutive changes in MGI^-D^- than in MGI^+D^- leukemic cells. These have been called C_{def}, for constitutive for differentiation defective (Fig. 6) [45].

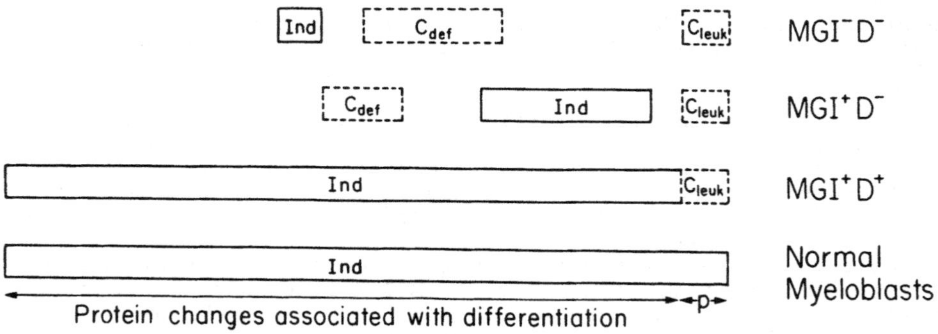

Fig. 6. Schematic summary of protein changes associated with growth (proliferation) and differentiation in normal myeloblasts and different types of myeloid leukemic cells. Constitutive expression of changes in leukemic compared to normal myeloblasts, C_{leuk}. Constitutive expression of changes in the differentiation-defective MGI^+D^- and MGI^-D^- clones compared to MGI^+D^+ clones, C_{def}. MGI induced changes, Ind. Protein changes associated with proliferation to produce more cells before differentiation (p) [45]

These results thus indicate that changes from an inducible to a constitutive expression of certain genes is associated with the uncoupling of growth and differentiation, both at the control which requires MGI-1 to produce more cells and at the control of the stopping of cell multiplication that occurs in the formation of mature cells.

Synchrony of Gene Expression and Regulation of Developmental Programs

The protein changes during the growth and differentiation of normal myeloblasts seem to be induced by MGI-1 and MGI-2 as a series of parallel multiple pathways of gene expression [45]. It can be assumed, that the normal developmental program that couples growth and differentiation in normal cells requires synchronous initiation and progression of these multiple parallel pathways. The presence of constitutive gene expression for some pathways can be expected to produce asynchrony in the co-ordination required for the normal developmental program. Depending on the pathways involved, this asynchrony could then result in an uncoupling of the controls for growth and differentiation and produce different blocks in the ability to be induced for and to terminate the differentiation process (Fig. 7). The suggestion derived from these results [13,17,45] is, therefore, that myeloid leukemia originates by a change that produces certain constitutive pathways of gene expression, so that cells no longer require MGI-1 for growth or constitutively produce MGI-1 without inducing MGI-2. These leukemic cells can, however, still be induced to differentiate normally by MGI-2 added exo-

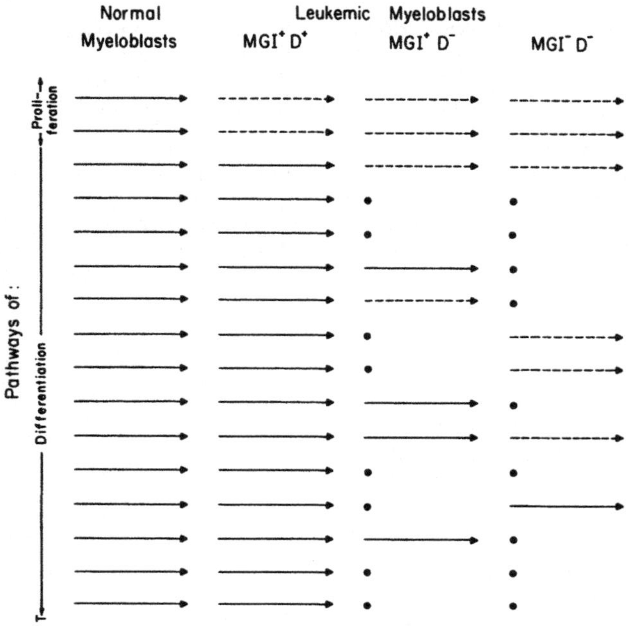

Fig. 7. Schematic relationship between MGI induced and constitutive changes in gene expression and the control of cell growth (proliferation) and differentiation in normal and different types of leukemic myeloblasts. MGI induced changes ⟶; constitutive changes ---->; blocked for induction of changes by MGI-2 •; terminal stage of differentiation (T) [13]

genously or induced in the cells in other ways. The differentiation program induced by MGI-2 can thus proceed normally when it is uncoupled from the growth program induced by MGI-1. This can be followed by constitutive expression of other pathways resulting in the uncoupling of other controls and an asynchrony that interferes with the normal program of terminal differentiation. These second changes then result in the further evolution of leukemia [13].

The MGI-1 independent growth of myeloid leukemic cells seems to proceed through stages [64] probably starting from a small decrease in MGI-1 requirement to ending, finally, in complete independence of MGI-1. There are presumably also stages in the amount of constitutively produced MGI-1 and the degree of lack of inducibility of MGI-2 by MGI-1. These stages would produce different degrees of asynchrony resulting in differences in the degree of hematological abnormality; the final stages of asynchrony then resulting in leukemia.

These conclusions on the origin and evolution of myeloid leukemia may be applicable to malignant tumors derived from other types of cells whose viability, growth and differentiation is induced by other physiological inducers. Identification of the physiological inducers of growth and differentiation for different cell types would be a crucial requirement in extending these conclusions to these other tumors. However, even in the absence of such identification, it appears likely that teratocarcinoma cells [65,66] may be comparable to MGI^+D^+ myeloid leukemic cells. The presence of fetal proteins in certain tumors, may also be due to constitutive gene expression in the tumor of a protein that is induced by the physiological inducer during the developmental program in the normal fetus [13]. There are probably a variety of tumors in which the original malignancy has a normal differentiation program and the cells are malignant because of uncoupling of the requirement for growth from the requirement for differentiation by changing the gene expression required for growth from inducible to constitutive; and where the further evolution of the tumor results from changes from inducible to constitutive of other pathways of gene expression that produce asynchrony in the normal differentiation program, so that mature non-dividing cells are not formed by the physiological inducer of differentiation.

References

1. Ginsburg H, Sachs L (1963) J Natl Cancer Inst 31:1
2. Pluznik DH, Sachs L (1965) J Cell Comp Physiol 66:319
3. Bradley T.R., Metcalf D (1966) Aust J Exp Biol Med Sci 44:287
4. Pluznik DH, Sachs L (1966) Exp Cell Res 43:553
5. Ichikawa Y, Pluznik DH, Sachs L (1966) Proc Natl Acad Sci USA 56:488
6. Paran M, Sachs L, Barak Y, Resnitzky P (1970) Proc Natl Acad Sci USA 67:1542
7. Paran M, Ichikawa Y, Sachs L (1968) J Cell Physiol 72:251
8. Fibach E, Sachs L (1976) J Cell Physiol 89:259
9. Ichikawa Y, Pluznik DH, Sachs L (1967) Proc Natl Acad Sci USA 58:1480
10. Metcalf D (1969) J Cell Physiol 74:323
11. Austin PE, McCulloch EA, Till JE (1971) J Cell Physiol 77:121
12. Landau T, Sachs L (1971) Proc Natl Acad Sci USA 68:2540
13. Sachs L (1980) Proc Natl Acad Sci USA 77:6152
14. Lotem J, Lipton J, Sachs L (1980) Int J Cancer 25:763
15. Sachs L (1974) Harvey lectures 68. Academic Press London, New York, pp 1-35
16. Fibach E, Sachs L (1974) J Cell Physiol 83:177
17. Sachs L (1978) Nature (London) 274:535
18. Lotem J, Sachs L (1979) Proc Natl Acad Sci USA 76:5158
19. Falk A, Sachs L (1980) Int. J. Cancer 26:595
20. Stanley ER, Heard PM (1977) J Biol Chem 252:4305
21. Lipton J, Sachs L (1981) Biochim Biophys Acta 673:552
22. Burgess AW, Camakaris J, Metcalf D (1977) J Biol Chem 252:1998
23. Liebermann D, Hoffman-Liebermann B, Sachs L (1982) Int J Cancer 29:159
24. Fibach E, Landau T, Sachs L (1972) Nature (London) New Biol 237:276
25. Maeda M, Horiuchi M, Numa S, Ichikawa Y (1977) Gann 68:435
26. Hozumi M, Umezawa T, Takenaga K, Ohno T, Shikita M, Yamane I (1979) Cancer Res 39:5127
27. Burgess AW, Metcalf D (1980) Int J Cancer 26:647
28. Nicola NA, Burgess AW, Metcalf D (1979) J Biol Chem 24:5290
29. Lotem J, Sachs L (1981) Int J Cancer 28:375
30. Paran M, Ichikawa Y, Sachs L (1969) Proc Natl Acad Sci USA 62:81
31. Broxmeyer HE, Smithyman A, Eger RR, Myers PA, Sousa de M (1979) J Exp Med 148:1052
32. Lotem J, Sachs L (1977) J Cell Physiol 92:97
33. Fibach E, Sachs L (1975) J Cell Physiol 86:221
34. Lotem J, Sachs L (1978) Proc Natl Acad Sci USA 75:3781
35. Ichikawa Y (1969) J Cell Physiol 74:223
36. Fibach E, Hayashi M, Sachs L (1973) Proc Natl Acad Sci USA 70:343
37. Ichikawa Y, Maeda N, Horiuchi M (1976) Int. J. Cancer 17:789
38. Lotem J, Sachs L (1977) Proc Natl Acad Sci USA 74:5554
39. Hayashi M, Fibach E, Sachs L (1974) Int J Cancer 14:40
40. Azumi J, Sachs L (1977) Proc Natl Acad Sci USA 74:253
41. Lotem J, Sachs L (1974) Proc Natl Acad Sci USA 71:3507
42. Hoffman-Liebermann B, Sachs L (1978) Cell 14:825

43. Simantov R, Sachs L (1978) Proc Natl Acad Sci USA 75:1805
44. Simantov R, Shkolnik T, Sachs L (1980) Proc Natl Acad Sci USA 77:4798
45. Liebermann D, Hoffman-Liebermann B, Sachs L (1980) Dev Biol 79:46
46. Hoffman-Liebermann B, Liebermann D, Sachs L (1981) Dev Biol 81:255
47. Weiss B, Sachs L (1978) Proc Natl Acad Sci USA 75:1374
48. Cohen L, Sachs L (1981) Proc Natl Acad Sci USA 78:353
49. Collins SJ, Ruscetti FW, Gallagher RE, Gallo RC (1978) Proc Natl Acad Sci USA 75:2458
50. Rovera G, Santoli D, Damsky C (1979) Proc Natl Acad Sci USA 76:2779
51. Liebermann D, Hoffman-Liebermann B, Sachs L (1981) Int. J. Cancer 28:285
52. Krystosek A, Sachs L (1976) Cell 9:675
53. Symonds G, Sachs L (1982) J Cell Physiol 111:9
54. Hoffman-Liebermann B, Liebermann D, Sachs L (1981) Int. J. Cancer 28:615
55. Friend C, (1978) Harvey lectures 72. Academic Press, London New York, pp 253-281
56. Marks P, Rifkind RA (1978) Annu Rev. Biochem 47:419
57. Lotem J, Sachs L (1975) Int J Cancer 15:731
58. Lotem J, Sachs L (1982) Proc Natl Acad Sci USA 79:4347
59. Symonds G, Sachs L (in press)
60. Liebermann D, Sachs L (1979) Proc Natl Acad Sci USA 76:3353
61. Liebermann D, Hoffman-Liebermann B, Sachs L (1980) Virology 107:121
62. Hayward WS, Neel BG, Astrin SM (1981) Nature (London) 290:475
63. Payne GS, Bishop JM, Varmus HE (1982) Nature (London) 295:209
64. Collins JS, Gallo RC, Gallagher RE (1977) Nature (London) 270-347
65. Mintz B, Illmensee K, (1975) Proc Natl Acad Sci USA 72:3585
66. Dewey MJ, Martin DW Jr., Martin GR, Mintz B (1977) Proc Natl Acad Sci USA 74-5564

Structure and Tissue-Specific Expression of Members of the Mouse α-Amylase Multigene Family

P. K. Wellauer, U. Schibler, O. Hagenbüchle, R. A. Young, and M. Tosi[1]

One approach to study cell differentiation is to elucidate the molecular mechanisms which govern the expression of tissue-specific genes. Such regulatory mechanisms include rates of transcription, efficiency of RNA processing, mRNA flow rates from nucleus to cytoplasm and mRNA turnover rates. The genes encoding α-amylase isoenzymes of mouse are a useful system to define which of these parameters are responsible for the accumulation of a particular gene product in highly specialized cells. The amylases of mouse have been extensively studied by biochemical and genetic means (for a review see Karn and Malacinski 1978). The α-amylase mRNA's accumulate to very different levels in three different tissues of mouse, the pancreas, the salivary gland and the liver (Schibler et al. 1980). The α-amylase structural genes are located in two closely linked but distinct genetic loci, *Amy*-1 and *Amy*-2, on mouse chromosome 3 (Sick and Nielsen 1964, Eicher 1979). *Amy*-2 is expressed in the pancreas, while *Amy*-1 encodes the α-amylases found in the salivary gland and the liver. We have isolated the two α-amylase genes *Amy*-2^{a} and *Amy*-1^{a} and the various mRNA products using recombinant DNA technology and have studied their sequence organization (Hagenbüchle et al. 1980, 1981, Tosi et al. 1981, Young et al. 1981, Schibler et al. 1982). Recently, we have undertaken a study to elucidate the question of transcriptional versus posttranscriptional regulation of amylase gene expression in various mouse tissues.

The Structure of α-Amylase mRNA's from Pancreas, Salivary Gland and Liver

The various α-amylase mRNA's were cloned in form of their cDNA's in bacterial plasmid vector pBR322 and their complete nucleotide sequences were determined (Hagenbüchle et al. 1980, 1981). The structural relationships of α-amylase mRNA's of pancreas, salivary gland and liver are summarized in Fig. 1. The pancreatic α-amylase mRNA is 1577 nucleotides long and differs by 11% of its sequence from that of its salivary gland (1659 nucleotides) and liver (1773 nucleotides) counterparts. The latter two RNA's are identical except for the 158 5'terminal residues of their 5'nontranslated regions.

In addition to these major α-amylase mRNA's, some minor RNA species were detected in the salivary gland and liver but not the pancreas. One class of minor RNA species, accounting for about 5% of amylase mRNA's in the two tissues, was found to contain an additional 237

1 Swiss Institute for Experimental Cancer Research, CH-1066 Epalinges, Switzerland

33. Colloquium - Mosbach 1982
Biochemistry of Differentiation and Morphogenesis

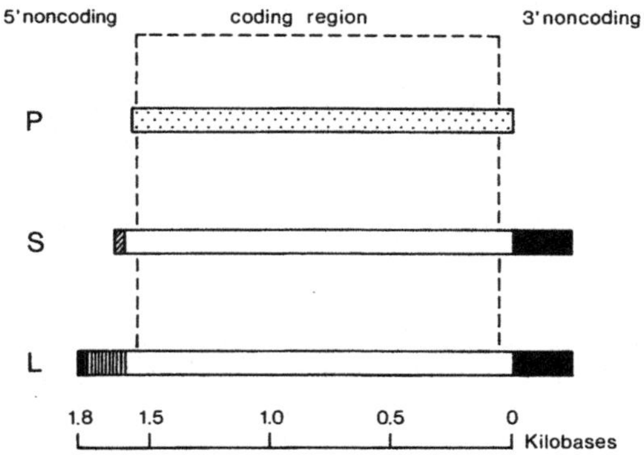

Fig. 1. Schematic sequence comparison of the pancreas, the salivary gland and the liver α-amylase mRNA's. The pancreatic (*P*) α-amylase mRNA which differs in its nucleotide sequence from that of its salivary gland (*S*) and liver (*L*) counterparts is represented by a *dotted box*. The common sequences of salivary gland and liver α-amylase mRNA's are shown as *open boxes* while the *black boxes* designate residues found in the minor mRNA species. Tissue-specific leader sequences are shown as *hatched boxes*

nucleotides at the 3' end. Another type of minor amylase mRNA species occurs in the liver at a frequency of about 25% of the major amylase mRNA's of this tissue. It contains 33 additional residues at its 5' end. The nucleotide sequence of all major and minor RNA species found in the salivary gland and liver is identical within the coding regions. Thus, translation of these mRNA's will result in the production of one single α-amylase protein species. This finding suggested the possibility that all salivary gland and liver amylase mRNA species could be transcribed from a common DNA sequence.

Sequence Organization of Mouse α-Amylase Genes *Amy*-1^a and *Amy*-2^a

To explore the origin of various α-amylase mRNA's of pancreas, salivary gland and liver, genomic amylase DNA sequences were isolated by molecular cloning in phage λ. The sequence arrangement of various amylase clones was studied by electron microscopy, restriction enzymes and DNA sequencing (Young et al. 1981, Schibler et al. 1982). The *Amy*-2^a gene, which specifies pancreatic α-amylase mRNA measures about 10 kb from cap to polyadenylation site and is interrupted by at least 9 intervening sequences (Fig. 2). Two copies of *Amy*-2^a exist per haploid mouse genome. The fact that gene as well as flanking regions of these two gene copies have identical restriction maps suggests that they are the result of a recent gene duplication event. *Amy*-1^a which occurs as a single copy in the mouse genome contains at least 10 intervening sequences (Fig. 2). It specifies both salivary gland and liver amylase mRNA's. The 5' proximal sequences of 5' nontranslated regions of salivary gland and liver mRNA are contained in two different exons. Thus, the two mRNA species are generated by splicing of different tissue-specific leaders onto sequences common to both RNA's. The distance

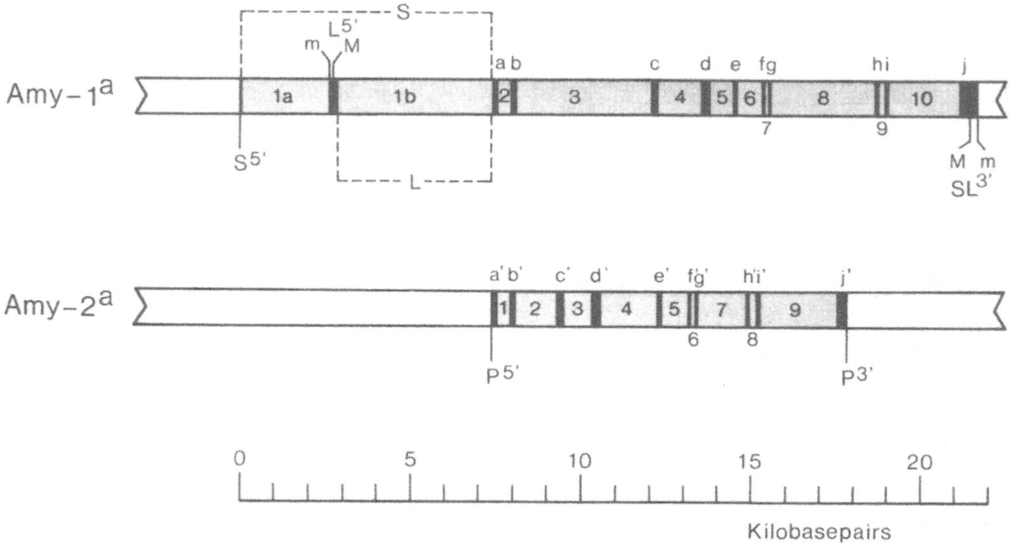

Fig. 2. Sequence organization of mouse α-amylase genes *Amy*-1a and *Amy*-2a. The model is based on restriction enzyme mapping, electron microscopy and sequence analysis. Intervening sequences are shown as *shaded boxes* and are *numbered*. Exons and flanking regions are represented by *black* and *open boxes*, respectively. Comparable exons are designated by lower case letters. The cap (5') and polyadenylation (3') sites are indicated for pancreatic (*P*), for major (*M*) and minor (*m*) salivary gland (*S*) and liver (*L*) α-amylase mRNA's. *Dashed lines* symbolize the tissue-specific RNA splicing events that occur in the salivary gland (*S*) and liver (*L*)

between *Amy*-1a cap and polyadenylation sites used in the salivary gland and liver is 23 and 20 kb, respectively. The DNA sequences specifying the additional 33 nucleotides of the liver 5' minor amylase mRNA species are contiguously linked to the ones that specify the 5' terminal sequences of the 5' major mRNA species in *Amy*-1a. Likewise the sequences that determine the additional 237 nucleotides of 3' minor RNA species found in the salivary gland and liver are contiguously linked to sequences which specify the 3' major RNA species in *Amy*-1a DNA. Therefore, 3' minor RNA species originate by failure to polyadenylate at the major site. These findings demonstrate that at least five, and probably six α-amylase mRNA species are transcribed from the unique *Amy*-1a DNA sequence.

Homologous *Amy*-1a and *Amy*-2a intervening sequences are located at identical positions within the two genes. The first intervening sequence of *Amy*-1a, which separates the salivary gland and liver specific leaders from their first common exonic sequence, has no counterpart in *Amy*-2a. Hybridization studies and nucleotide sequence analysis show that some comparable *Amy*-1a and *Amy*-2a intervening sequences display considerable (50%) sequence homology. This suggests that the two genes have evolved by duplication from a primordial amylase sequence. Some of the *Amy*-1a and *Amy*-2a intervening sequences and flanking regions were found to contain DNA elements that are repeated manyfold elsewhere in the mouse genome. The role of these repetitive elements is at present unknown.

Little sequence homology exists between *Amy*-1^a and *Amy*-2^a 5' and 3' flanking regions. This is not surprising since flanking regions evolve more rapidly than gene sequences. The only common sequence motive in 5' flanking regions of *Amy*-1^a and *Amy*-2^a are TATA-(Goldberg-Hogness) boxes which are presumed to be signals of transcription initiation and are located about 25 nucleotides upstream of the various cap sites. The only cap site which is not preceded by a TATA-box is that for the 5' major liver RNA species in *Amy*-1^a.

Southern blot analysis reveals that, in addition to the one *Amy*-1^a and two *Amy*-2^a copies, several other amylase-like DNA sequences exist in the mouse genome. Some of these may be active in other mouse strains and/or different stages of development, and some may represent pseudogenes.

Tissue-Specific Transcription of *Amy*-1^a in Salivary Gland and Liver

The α-amylase mRNA's accumulate to highly different levels in the cytoplasm of salivary gland (10^4 molecules) and liver (10^2 molecules) cells (Schibler et al. 1980). In order to elucidate the molecular basis for different steady state concentrations of amylase mRNA's in these tissues, we analysed nuclear amylase RNA's by a sensitive S1 nuclease mapping technique (Schibler, Hagenbüchle, Wellauer, Pittet, and Bovey in preparation).

These experiments demonstrate that the various nuclear amylase transcripts map to the cap sites present in *Amy*-1^a DNA (Fig. 3). RNA molecules that are initiated some distance upstream of the *Amy*-1^a mRNA cap sites are detected in salivary gland and liver nuclei in addition to the tissue-specific nuclear amylase RNA's. They are not processed into stable amylase mRNA's and are also present at variable concentrations in nuclear RNA of tissues which do not produce α-amylase

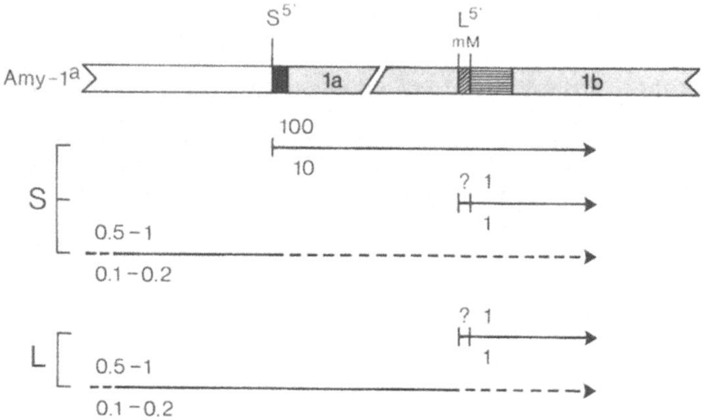

Fig. 3. Summary of nuclear RNA species transcribed from *Amy*-1^a. Part of the *Amy*-1^a DNA sequence containing the cap sites for salivary gland (*S*) and major (*M*) and minor (*m*) liver (*L*) α-amylase mRNA's is shown at the *top*. The tissue-specific leader sequences of salivary gland and liver mRNA's are designated by *black* and *hatched boxes*, respectively. Nuclear RNA transcripts are indicated by *arrows*. *Numbers above the arrows* are relative concentrations of *Amy*-1^a-specific transcripts in steady state nuclear RNA of salivary gland (*S*) and liver (*L*). *Numbers below the arrows* are relative concentrations of nascent RNA chains or RNA polymerase molecules (for methodology see text)

such as brain and kidney. The functional significance of these transcripts is not known at present time.

Amylase mRNA containing the liver type leader is present at a similar concentration in salivary gland and liver. This conclusion is based on RNA gel transfer hybridizations in which equal amounts of cytoplasmic poly A containing RNA from salivary gland and liver were hybridized to a radioactive *Amy*-1^{a} restriction fragment containing exclusively sequences complementary to liver amylase mRNA. However, no salivary gland-specific transcripts are detected in the liver. Other tissues such as lung, heart, spleen, brain and kidney do not contain any of these amylase mRNA species.

We mapped nascent amylase RNA chains to explore whether cap sites for salivary gland and liver amylase mRNA's coincide with sites of transcription initiation in *Amy*-1^{a}. Isolated salivary gland or liver nuclei were incubated in vitro for RNA synthesis in the presence of radioactive uridine triphosphate, heparin and high salt. Under these conditions, only those RNA polymerase II molecules which have initiated transcription in vivo remain bound to the DNA template. Hybridization of labeled, in vitro elongated RNA molecules to an excess of appropriate *Amy*-1^{a} DNA restriction fragments demonstrates that transcription initiation starts indeed at the cap sites for salivary gland and liver amylase mRNA's.

Relative tissue-specific transcription rates, i.e., the efficiency with which *Amy*-1^{a} is transcribed in these two mouse tissues, were measured in similar experiments. Quantitation of the radioactivity associated with in vitro labeled nascent RNA chains reveals a tenfold difference in salivary gland and liver. Since the in vitro elongation rate (15 nucleotides per minute) and the uridine triphosphate pool size were measured to be similar in nuclei of both tissues, about 10 times as many RNA polymerase II molecules are engaged in *Amy*-1^{a} transcription in the salivary gland as compared to the liver. If the in vivo elongation rates are comparable in the two tissues as well this would suggest that the salivary gland promoter is about 10 times more efficient in transcription than the liver promoter (Fig. 3).

Since the observed difference in transcription rate is not sufficient to account for the 100-fold difference of amylase mRNA concentrations in the two tissues, post-transcriptional regulation must be involved in the tissue-specific accumulation of these RNA's as well. The various kinds of post-transcriptional mechanisms which may play a role in regulating amylase gene expression will have to be elucidated in future studies.

Acknowledgements. We thank R. Bovey and A.-C. Pittet for expert technical assistance. This work was supported by grants from the Swiss National Science Foundation (Nos 3.148.77 and 3.148.81).

References

Eicher EM (1979) Mouse News Lett 60:50

Hagenbüchle O, Bovey R, Young RA (1980) Tissue-specific expression of mouse α-amylase genes. Cell 211:179-187

Hagenbüchle O, Tosi M, Schibler U, Bovey R, Wellauer PK, Young RA (1981) Mouse liver and salivary gland α-amylase mRNA's differ only in 5' nontranslated sequences. Nature (London) 289-643-646

Karn RC, Malacinski GM (1978) The comparative biochemistry, physiology, and genetics of animal α-amylases. Adv Comp Physiol Biochem 7:1-103

Schibler U, Tosi M, Pittet A-C, Fabiani L, Wellauer PK (1980) Tissue-specific expression of mouse α-amylase genes. J Mol Biol 142:93-116

Schibler U, Pittet AC, Young RA, Hagenbüchle O, Tosi M, Gellman S, Wellauer PK (1982) The mouse α-amylase multigene family: Sequence organization of members expressed in the pancreas, salivary gland and liver. J Mol Biol 155:247-266

Sick K, Nielsen JT (1964) Genetics of amylase isozymes in the mouse. Hereditas 51:291-296

Tosi M, Young RA, Hagenbüchle O, Schibler U (1981) Multiple polyadenylation sites in mouse α-amylase genes. Nucleic Acids Res 9:2313-2323

Young RA, Hagenbüchle O, Schibler U (1981) A single mouse α-amylase gene specifies two different tissue specific mRNA's. Cell 23:451-458

Actin and Myosin Genes and Their Expression During Skeletal Muscle Myogenesis

M. E. Buckingham[1], A. Minty[1], S. Alonso[1], M. Caravatti[2], B. Robert[1], A. Cohen[1], G. Bugaisky[1], P. Daubas[1], and A. Weydert[1]

The differentiation of skeletal muscle cells is characterized morphologically by the fusion of myoblasts to form multinucleated muscle fibres. This process takes place gradually during skeletal muscle development in vivo. It can also be followed in tissue culture. Mammalian myoblasts will grow in monolayers, either in primary culture or as established cell lines, and will fuse spontaneously when the culture becomes confluent (for review see Yaffé 1968, Buckingham 1977). The formation of muscle fibres is characterized biochemically by the increased synthesis of contractile proteins (e.g. Devlin and Emerson 1978, Garrels 1979) and their organization into sarcomeric structures (Fischman 1970), by the accumulation of enzymes important in muscle metabolism (e.g. Caravatti et al. 1979), and by the appearance of membrane components such as the acetylcholine receptor (e.g. Merlie et al. 1975), essential for nerve-muscle interaction.

It has become evident in the last few years that the contractile proteins present in differentiated muscles are different from those of non-muscle tissues (e.g. Vandekerckhove and Weber 1978) and furthermore that additional isoforms may be expressed during development. These proteins are therefore coded for by multigene families, the different members of which are expressed in a tissue specific manner. The information available at present for the actins and myosins is summarized in Table 1 (for complete reference list see Buckingham and Minty 1982) where a minimum estimate for the numbers of different structural genes is given. For the myosin heavy chain (MHC) embryonic (MHC_{emb}) and newborn (MHC_{NB}) peptides can be distinguished (Whalen et al. 1979, 1981) which are the major forms present at the corresponding developmental stages, in addition to the proteins present in adult muscle tissues. In the case of the myosin light chains, a form of LC_1, LC_{1emb} (Whalen et al. 1978), also found in adult heart atria (Whalen et al. 1982) and embryonic heart (Whalen and Sell 1980), is expressed during skeletal muscle development. From the protein data, there has been no evidence for developmental forms of actin. However, our results on actin coding sequences would suggest that a second actin mRNA accumulates in fetal mouse skeletal muscle and that this is probably the cardiac actin sequence.

The method of regulation of expression of different tissue and developmental programmes is thus of major interest in the muscle system. In addition the co-ordination between the components of any one muscle programme is also an important consideration, since a functional sar-

1 Department of Molecular Biology, Pasteur Institute, 25, rue du Dr. Roux, 75015 Paris, France

2 Friedrich Miescher Institut, Postfach 273, CH-4002 Basel, Switzerland

33. Colloquium - Mosbach 1982
Biochemistry of Differentiation and Morphogenesis

Table 1. Actin and myosin isoforms (15/3/82). This table is compiled from a large number of results in the literature. A complete list of references is given in Buckingham and Minty (1982)

	Actins (α, β, γ)	Myosin heavy chains (MHC)	Myosin light chains: Non phosphorylatable (LC_1, LC_3)	Myosin light chains: Phosphorylatable (LC_2)
Skeletal muscle				
Fast				
Foetal	α_{CM} α_{SM}	MHC_{emb}	LC_{1emb}	LC_{2F}
Newborn	(α_{CM}) α_{SM}	MHC_{NB}	(LC_{1emb}) LC_{1F} (LC_{3F})	LC_{2F}
Adult	α_{SM}	$2MHC_F$ MHC_{SF}	LC_{1F} LC_{3F}	LC_{2F}
Slow				
Foetal		MHC_{emb}	LC_{1emb}	LC_{2F}
Newborn		MHC_{NB}	(LC_{1emb}) LC_{1F} LC_{1S}	LC_{2F}(LC_{2S})
Adult	α_{SM}	MHC_S	LC_{1S} $LC_{1S'}$	LC_{2S}
Cardiac muscle				
Atria				
Adult		MHC_A ? MHC_{V_1}	LC_{1emb}	LC_{2A}
Ventricules				
Foetal	α_{CM}	MHC_{V_3}	LC_{1emb}	LC_{2S}
Newborn	α_{CM}	MHC_{V_1}	LC_{1S}	LC_{2S}
Adult	α_{CM}	MHC_{V_3} (MHC_{V_1})	LC_{1S}	LC_{2S}
Smooth muscle	α_S γ_S	MHC_{SM}	LC_{1NM}	LC_{2NM}
Non muscle	β_{NM} γ_{NM}	$2MHC_{NM}$	LC_{1NM}	LC_{2NM}
Gene number	6	11 - 13	6	4

comere requires a strict stoichiometry of the accumulated contractile proteins. In order to understand how the expression of a muscle phenotype is regulated, information is necessary both on the structure and chromosomal organization of the muscle genes, on their activation and transcription, and on the processing, accumulation, and translational efficiencies of the resultant transcripts. The advent of DNA recombinant technology has made these questions directly accessible to experimental study.

Myosin and Actin Recombinant Plasmids

We have used this approach to obtain purified probes to muscle coding sequences. Recombinant plasmids containing DNA complementary to different sized fractions of poly A^+ RNA from newborn (about 10-day-old) mouse skeletal muscle were constructed by GC tailing and integration at the Pst I size of pBR 322 (for details see Minty et al. 1981). They were characterized by hybridisation of muscle RNA to plasmid DNA immobilized on a filter followed by translation of the bound messenger RNA (mRNA) (Smith et al. 1979). The products were characterized by two-dimensional gel analysis (O'Farrell 1975). Identification was completed by DNA sequencing (Maxam and Gilbert 1980). The plasmids are listed in Table 2. Analysis of the size of the mRNAs and their cross-hybridization with the inserted sequence in the plasmid is based on the results of Northern blots (see Alwine et al. 1979).

In general the coding sequences within any one contractile protein family are sufficiently homologous to cross-hybridize, whereas the non-coding sequences have diverged further and can be used as specific probes for individual members of the multigene family. Thus, for the actins, the non-coding sequence of p91, p91(200) is specific under most conditions for skeletal muscle α-actin. The coding sequence, on the other hand, cross-hybridizes with non-muscle (about 2000 bp) and muscle (about 1600 bp) mRNAs. The actin protein sequences are very highly conserved (e.g. Vandekerckhove and Weber 1979), the maximum variation is 25/375 amino-acid differences between muscle and non

Table 2. Actin and myosin recombinant plasmids isolated from new born mouse skeletal muscle. Construction and characterization of the plasmids was as described in Minty et. al. (1981)

Clone N°	Cross hybridization/ Identification	Size of insertion	Coding/Non Coding		Size of mRNA
p 32	MHC - probably MHC_{NB}	1250	+	+	6900
p 161	LC_{1F}, LC_{3F}(LC_{1emb})	380	-	+	1050, 900 (980)
p 91	Actins - α_{SM}	1360	+	+	2000 (1600)
p 91 (200)	α_{SM}	200	-	+	1600
p 81	Actins - $\alpha_{foetal/cardiac}$	1100	+	(+)	2000 (1600)
p 41	Actin - non muscle (clones from lymphoma)	1150	+	(+)	2000 (1600)

muscle actins. There is significantly more divergence at the nucleic acid level due to differences in codon usage. Cardiac and skeletal muscle actins have 4/375 amino-acid differences, whereas the 3'ends of the coding regions show an overall divergence of about 15% between p91 and p81 (Minty et al. 1982a). In contrast, it is noteworthy that p161, which contains part of the non-coding sequence, cross-hybridizes equally with the mRNAs for LC_{1F} and LC_{3F}. Exceptionally, these mRNAs therefore appear to have very similar or identical 3' non-coding sequences. Since they share a common-COOH terminal amino-acid sequence it is possible that they are transcribed from the same gene in this region. We are currently investigating this situation by sequencing the corresponding genomic fragments which we have isolated.

Contractile Protein Multigene Families: The Actin Genes

We have used the three actin probes, corresponding to skeletal muscle, non muscle, and cardiac/foetal actin sequences to investigate the actin multigene family. Southern blotting (Southern 1975) of restricted mouse genomic DNA and hybridization with actin probes under non-stringent conditions reveals a large number of genomic actin fragments and this result is supported by the number of actin genomic clones detected in recombinant banks, and estimated in the range of 25-100 (see Minty et al. 1982b). This is also seen for actin sequences in the human genome (Humphries et al. 1981). Only six different actin proteins have been described (Table 1), so the question arises as to the nature of the majority of these genomic actin sequences. By washing the blots under more stringent conditions it is clear that many of the bands are not very homologous to our actin coding sequences. Under stringent conditions, or with the specific non-coding sequence probe, only the homologous gene is detected (Fig. 1). In this way using different restriction enzymes it can be shown that there is only one skeletal muscle α-actin gene, and only one cardiac/foetal gene (Minty et al. 1982a). It seems unlikely that the other known actin sequences are present as multiple copies. It also seems unlikely that there are a large number of different actin proteins which have not yet been identified. The most probable explanation is that there are many actin pseudo-genes, which are more or less homologous to the actin coding sequence. Indeed, we have characterized further one group of such putative pseudo-genes which hybridize to actin coding sequences under conditions of low stringency only, as a very strong EcoR1 band in blots of genomic DNA. Recombinant phages (λ), containing mouse DNA that hybridizes with this band on Southern blots, have been isolated on the basis of their weak signal in situ hybridization with the p41 actin probe, and five such recombinants have been shown to contain a related class of genomic actin fragments (see Minty et al. 1982b). Recently, the presence of additional pseudo-genes in the α-globin gene family has been described (Leder et al. 1981). The extent to which this is a general phenomenon is not yet clear. It may be that the proliferation of actin sequences in the genome is a recent event, and consequently they have diverged sufficiently little to make detection of pseudo-genes possible. Preliminary results on the myosin heavy chain genes in our laboratory, suggest that under similar conditions of Southern blotting we see a restricted number of genes which can be accounted for by the known MHC protein sequences.

A second aspect of the actin multigene family which is of particular interest is the question of intron positions in the active genes. As discussed, the coding sequence is very highly conserved between different actin types, and across species. Since actin is a protein pre-

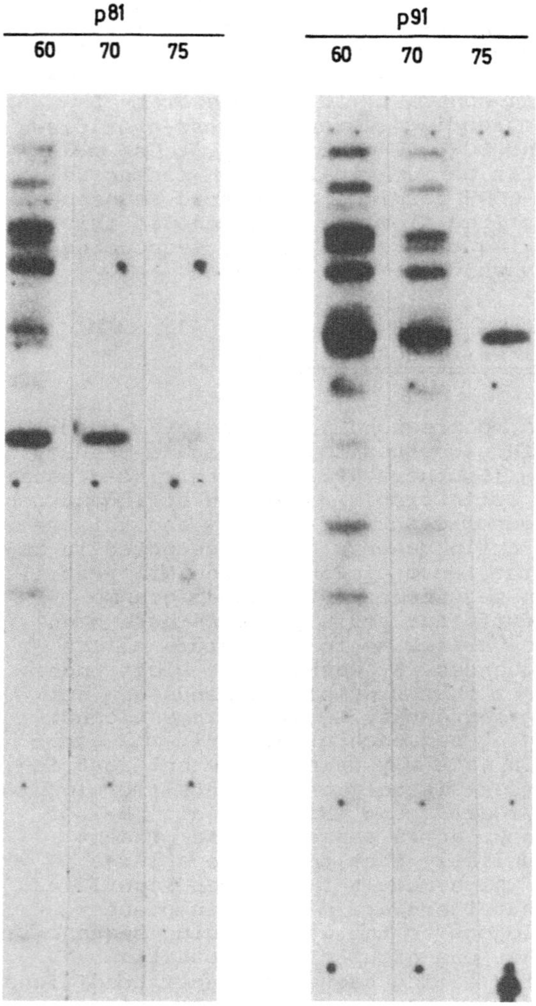

Fig. 1. Southern blot analysis of actin genes. 5 μg of DNA prepared from mouse (129) embryos was digested with EcoRI and separated on 0.7% agarose gels. The DNA was transferred to nitrocellulose (Southern 1975) and hybridized with nick translated p81 or p91 DNA. The blots were washed successively in 0.1 x SSC under increasingly stringent conditions, 60, 70, 75°C, as indicated

sent in eukaryotes as different as yeast and man, it is possible to compare the structure of the actin genes across a wide evolutionary span. Comparison of results obtained in a number of different laboratories (for review, see Buckingham and Minty 1982), shows that between Protista such as yeast (Gallwitz and Sures 1980, Ng and Abelson 1980), Acanthamoeba (Nellen and Gallwitz, personal communication). Dictyostelium (Kindle and Firtel 1978), or between members of the Protostomia grouping of the Animal kingdom, such as Nematoda (Hirsh et al. 1981) or Drosophila (Fyrberg et al. 1981), there is no consistency in intron position for the actin genes. It is only when Deuterostomia are compared, the examples are sea urchins (Durcia et al. 1980, Overbeek et al. 1982, Scheller et al. 1982), chicken (Ordahl et al. 1981) and rat (Nudel et al. 1981), that the intron positions are conserved, although not necessarily all used in any one actin sequence. These examples can therefore be viewed in terms of an ancestral actin gene with introns at all the positions noted, with subsequent elimination events after duplication of the original gene. However, it is more difficult to apply this model to the actin genes which show no consistency in intron position. It seems probable that within this multigene family the introduction as well as the elimination of introns has occurred. How such processes operate, while retaining such a highly conserved coding sequence over more than a thousand million years is unknown. Indeed it is not very clear at the protein level why the actin sequence is so highly conserved, and it is certainly not evident that any equivalence exists between intron positions and functional domains of the protein. Most variation in the amino-acid sequence which distinguishes the different actins is found in the first few residues of the amino-terminal end of the protein (Vandekerckhove and Weber 1979).

Contractile Protein Gene Expression During Skeletal Muscle Myogenesis: Actin and Myosin mRNAs in a Mouse Muscle Cell Line

Using cloned actin and myosin probes we have looked at accumulation of the corresponding mRNAs by the Northern blot technique (Alwine et al. 1979) in muscle tissues at different stages of development and in a mouse muscle cell line. This cell line, T984-Cl10, derived initially from the embryonal body of a mouse teratocarcinoma, forms differentiated muscle fibres in monolayer culture when the cells became confluent (Jakob et al. 1978). This is illustrated in Fig. 2, where

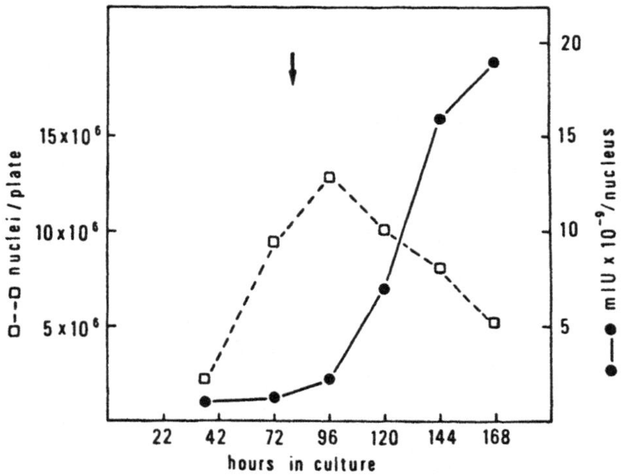

Fig. 2. Proliferation and differentiation in the mouse muscle cell line, T984. Different stages of muscle cell fusion are presented schematically. Myotubes begin to appear after the cultures have become confluent. Large cross-striated fibres are present from about 144 h. The number of nuclei per plate (□ - - - □) was calculated from the DNA content of culture dishes. The T984 cell line is aneuploid (Jakob et al. 1978) and an average value of 8.5 pg DNA per cell/nucleus was used. Total creatine kinase activity per nucleus (●——●) is expressed in international units (IU) where 1 unit of creatine kinase is the amount required to catalyse the formation of 1 μmole ADP per min at 25°C. The arrow indicates when the medium was changed from Dulbecco's modified Eagle's medium containing 15% foetal calf serum to the same medium with 2% foetal calf serum. This stepdown reduces overgrowth of the culture

the accumulation of creatine phosphokinase activity has been examined as a marker of differentiation. The syntheses of most muscle proteins such as myosin heavy chain (MHC_{emb}) and skeletal muscle α-actin, are detectable at 120 h, and increase rapidly thereafter. The major alkali myosin light chain form synthesized initially in the muscle fibres is LC_{1emb}. Synthesis of LC_{3F} is only detectable on two dimensional gels from 144 h, and increases slowly thereafter. The adult form, LC_{1F}, is also synthesized later during muscle fibre formation is these cultures.

RNA extracted from muscle cells on different days in culture has been analyzed on Northern blots with actin and myosin probes as shown in Fig. 3. No mRNA is detectable with p32, p91, (200) or p161 in dividing myoblasts. The messengers for α-actin and the myosin heavy chain are

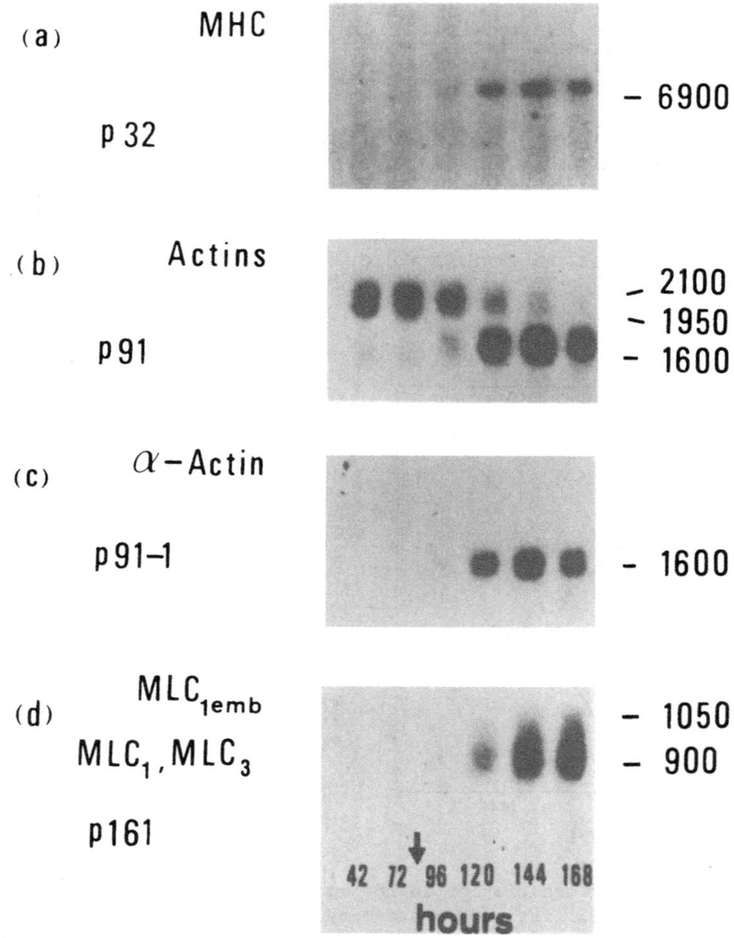

Fig. 3a-d. Northern blot analysis of accumulated actin and myosin mRNAs. Poly $(A)^+$ RNA (2 μg per preparation) from T984 muscle cells at different days after plating was fractionated on 1.25% agarose gels and transferred to DBM paper. (Alwine et al. 1979). The paper was cut into three pieces containing different RNA size classes and hybridized with nick translated plasmid probe as indicated in Fig. 1. The exposure times for the different parts of the blot were as follows: MHC 82 h, α-actin 22 h, the actins 72 h and the MLC 60 h

first detectable at 120 h and continue to accumulate during the next 2 days. The messenger for LC_{1emb}, the alkali light chain initially present, cross-hybridizes weakly with the probe which principally detects LC_{3F} (and LC_{1F}) (Table 2). At 120 h, LC_{1emb} is still the major mRNA. Later accumulation of the mRNA for LC_{3F} (and LC_{1F}) reflects later synthesis of the adult form in vivo. These experiments therefore indicate a close correlation between mRNA accumulation and synthesis of the corresponding muscle protein during myogenesis.

The question of whether a few transcripts of the muscle genes are present in dividing myoblasts (see Buckingham 1977) is not resolved by these experiments. With the p91 (200) probe which is homologous to the mRNA for skeletal muscle actin, assuming that the synthesis of α-actin is 2.5%-5% of total protein synthesis in myotubes, 40-80 molecules per nucleus of the mRNA can be detected by the technique used here (see Caravatti et al. 1982). A minimum estimate for the accumulation of this mRNA is 130-fold during myogenesis. Experiments on DNAse I sensitivity of the muscle genes in myoblasts are in progress in our own and other laboratories and should indicate whether these genes are in a transcriptionally active conformation, by this criterion, immediately prior to terminal differentiation. From the Northern blot results it would appear that non-muscle, β and γ actin mRNAs (2000 bp) cease to accumulate after differentiation. Again it is not yet clear whether these genes are no longer transcriptionally active in muscle fibres. With the p91 probe, but not with the p91 (200) skeletal muscle probe an α-actin type mRNA (about 1500 bp) is detected in myoblasts. It probably corresponds to a smooth muscle actin mRNA (see Minty et al. 1981), synthesized in these cells.

Expression of a Cardiac Isoform of Actin in Skeletal Muscle During Development

Experiments using the actin probe p81 would suggest that skeletal muscle α-actin mRNA is not the only muscle actin sequence to accumulate during myotube formation. The p81 sequence was also cloned from RNA prepared from new born mouse skeletal muscle. Northern blot analysis, however, shows that p81 (Table 2) has more homology with cardiac actin mRNA than with that from adult skeletal muscle. This is in direct contrast to the situation with p91 (Fig. 4). With embryonic skeletal muscle RNA and with RNA from the differentiated cells of the mouse muscle cell line T984 the difference between the hybridizations with the two plasmids is much less striking. We therefore conclude that in RNA from these tissues part of the muscle actin mRNA migrating as a band of 1600 nucleotides is not the α-skeletal muscle form, but a cardiac-like actin mRNA. The degree of hybridization with the p91-200 probe which is specific for α-skeletal muscle actin mRNA would also suggest that in T984 myotubes other forms of actin accumulate. In newborn muscle small amounts of cardiac-like actin mRNA are presumably still present. DNA sequencing of the insertion in p81 demonstrates that it codes for cardiac actin. In Fig. 5 part of the p81 sequence is compared with that of p91. Aminoacid positions 278, 286, and 296 are characteristic of striated muscle actin, while positions 298 and 357 distinguish cardiac from skeletal muscle actin (Minty et al. 1982a).

Although at the positions (89, 298, 357) plasmid p81 has residues characteristic of a cardiac sequence it may nevertheless be the product of a different "embryonic" gene. This has been looked at by Southern blot analysis (Fig. 1). Mouse genomic DNA, digested by EcoR1 was separated

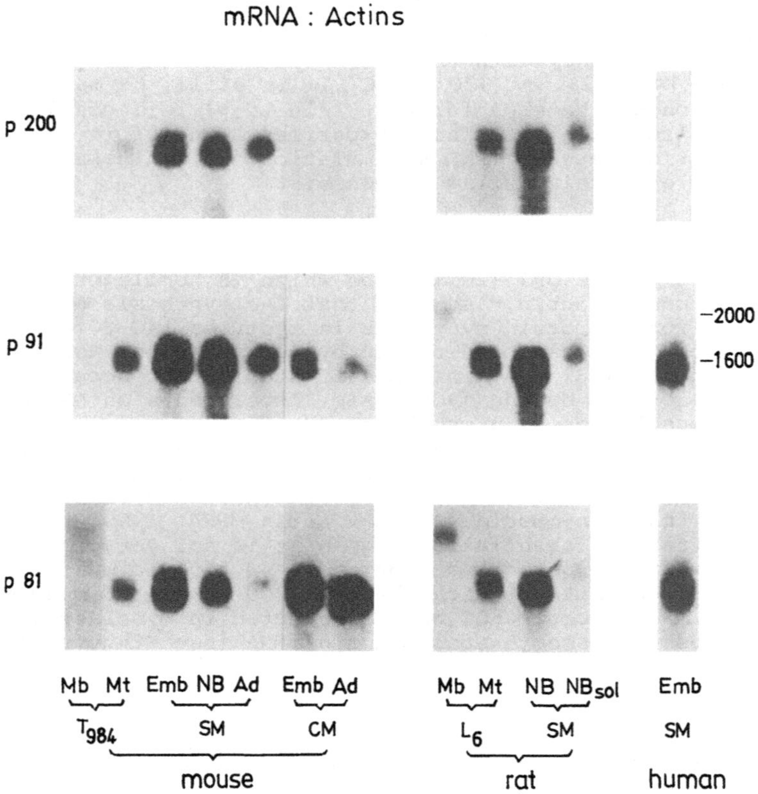

Fig. 4. A Northern blot showing the relative homologies of actin mRNA's from different tissues, with the actin probes p81 and p91. Samples of poly $(A)^+$ RNA were separated on 1.25% agarose gels, transferred to DBM paper and hybridized with the nick-translated probes essentially as described for the actins in Fig. III. The same region of the blot was re-used after decay of the previous probe. The degree of hybridization with the different probes p81, p91 and p91-200 (p200) is therefore directly comparable. The samples shown are from left to right RNA from myoblasts (Mb) and myotubes (Mt) of the mouse muscle cell line T984, from embryonic (Emb), newborn (NB) and adult (Ad) skeletal muscle (SM) and embryonic (Emb) and adult (Ad) cardiac muscle (CM) of the mouse, from myoblasts (Mb) and myotubes (Mt) of the rat muscle cell line L_6, from new-born fast (NB) and new-born slow (soleus) (NB_{sol}) skeletal muscle of the rat and from human foetal (3 months) skeletal muscle (Emb, SM)

on agarose gels, transferred to nitrocellulose filters (Southern 1975) and hybridized with nick-translated plasmids p81 or p91. When the blots are washed at relatively low stringency (60°C, 0.1 x SSC) multiple actin bands are visible corresponding to different actin genes and probably pseudogenes. At higher stringencies the probe hybridizes only to its homologous genomic sequence. In the case of p91 the skeletal muscle actin gene is clearly distinguished as a band at 7.5 Kb after a 75°C wash. Plasmid 81, which appears to be less homologous to the other actin sequences, already gives one major band (5.5 Kb) and two minor bands (7.5 Kb, 1 Kb) after a 70°C wash. The largest band represents slight cross-hybridization with the skeletal muscle gene. The major 5.5 Kb fragment corresponds to the 5' end of the cardiac

MUSCLE ACTINS

	5'									
91	GGG	ATC	CAT	GAG	ACC	ACC	TAC	AAC	AGC	ATC
81	GGC	ATT	CAT	GAA	ACT	ACT	TAC	AAC	AGC	ATC
	gly 272	ile	his	glu	thr	thr	tyr	asn	ser	ile
91	ATG	AAG	TGC	GAC	ATC	GAC	ATC	AGG	AAG	GAC
81	ATG	AAG	TGT	GAC	ATC	GAT	ATC	CGC	AAG	GAC
	met	lys	cys	asp	ile	asp	ile	arg	lys	asp
91	CTG	TAT	GCC	AAC	AAC	GTC	ATG	TCA	GGG	-----
81	CTG	TAT	GCC	AAC	AAT	GTC	CTA	TCT	GGA	
	leu	tyr	ala	asn	asn	val	leu	ser	gly 300	
91	----	ATC	ACA	AAG 3'						
81		ATC	AGC	AAG	----					
		ile	ser 357	lys						

Fig. 5. A comparison between part of the nucleotide sequence of the actin insertions in p81 and p91. DNA sequencing was by the technique of Maxam and Gilbert (1980) as described in Minty et al. (1981). The residues characteristic of striated muscle are indicated. Residues 298 and 357 distinguish cardiac from skeletal muscle

actin gene. The small 1 Kb fragment is the 3' end of the same gene which is cut by EcoR1. We interpret these results to indicate that p81 is homologous with a single gene coding for cardiac actin. Further analyses with other restriction enzymes including PvuII which cuts in the coding sequence of p81 and generates two 4 Kb fragments one of which hybridizes with the 3'end of the p81 insertion and one with the 5'end confirm that there is only one gene (see Minty et al. 1982a).

Thus, in the early stages of muscle fibre formation, a fetal actin mRNA very homologous or identical to cardiac actin mRNA is translated, the situation appears to be somewhat analogous to that observed for the embryonic light chain LC_{1emb} which is also found in heart muscle (Whalen et al 1982). The physiological significance of cardiac like isoforms in embryonic skeletal muscle remains to be demonstrated. From a regulatory point of view the co-expression of parts of two different phenotypic programmes at early stages of development poses some interesting problems. It suggests that the naive view that the programme of any phenotypic group of contractile proteins is regulated as a block is not necessarily correct. The situation for the actins and myosins, as it appears at present mainly from experiments on mouse and rat muscle cell lines and from some in vivo data, can be presented schematically as shown in Fig. 6.

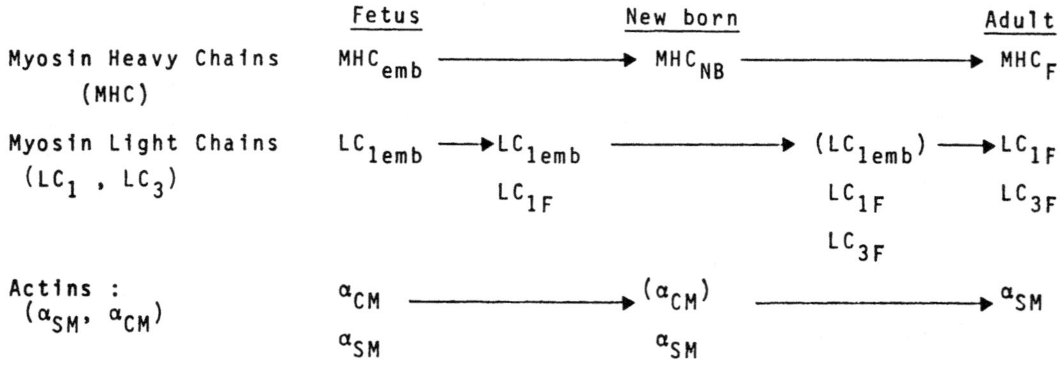

Fig. 6. Changes in actin and myosin gene expression during skeletal muscle development. The results summarized here are derived from data on the myosin heavy chain protein isoforms (Whalen et al. 1979, 1981) and foetal myosin light chain (Whalen et al. 1978) in developing rat skeletal muscle and the rat muscle cell line L_6, together with the results obtained with cloned actin and myosin coding sequences on mouse tissues and the mouse muscle cell line discussed in this paper

Acknowledgements. The laboratory is supported by grants from the Délégation Générale à la Recherche Scientifique et Technique, the Centre National de la Recherche Scientifique, the Institut National de la Santé et de la Recherche Médicale, the Commissariat à l'Energie Atomique, the Ligue Nationale Française contre le Cancer, the Fondation pour La Recherche Médicale Française and the Muscular Dystrophy Association of America. M.C. was the recipient of a fellowship from the Swiss National Research Foundation and from I.N.S.E.R.M., A.M. from the Muscular Dystrophy Associations of America, P.D. from the Ligue Française contre le Cancer and S.A. from the D.G.R.S.T. We are grateful to R. Williamson for providing the sample of human foetal muscle RNA.

References

Alwine JC, Kemp DJ, Parker BA, Keiser J, Renart J, Stark GR, Wahl GM (1979) Methods Enzymol 68:220-242

Buckingham ME (1977) In: Paul J (ed) International review of biochemistry, Biochemistry of cell differentiation II, vol 15. Univ Park Press, Baltimore, 269-332

Buckingham ME, Minty AJ (1982) In: MacLean N, Gregory SP, Flavell RA (eds) Chapter on "Contractile protein genes" in "Eukaryotic genes: their structure, activity and regulation". Butterworths, in preparation

Caravatti M, Perriard JC, Eppenberger HM (1979) J Biol Chem 254:1388-1394

Caravatti M, Minty A, Robert B, Montarras D, Weydert A, Cohen A, Daubas P, Gros F, Buckingham M (1982) J Mol Biol 160:59-76

Devlin RB, Emerson CP (1978) Cell 13:599-611

Durcia DS, Schloss JA, Crain WR (1980) Proc Natl Acad Sci USA 77:5683-5687

Fischman DA (1970) Curr Top Dev Biol 5:235-259

Fyrberg EA, Bond BJ, Hershey ND, Mixter KS, Davidson N (1981) Cell 24:107-116

Gallwitz D, Sures I (1980) Proc Natl Acad Sci USA 77:2546-2550

Garrels JI (1979) Dev Biol 73:134-152

Hirsh D, Files J, Carr S, Krause M (1981) In: Pearson M (ed) Proceedings of the Cold Spring Harbor Meeting on Muscle Development. Cold Spring Harbor, in press

Humphries SE, Whittal R, Minty A, Buckingham M, Williamson R (1981) Nucleic Acids Res 9:4895-4908

Jakob H, Buckingham ME, Cohen A, Dupont L, Fiszman MY, Jacob F (1978) Exp Cell Res 114:403-408
Kindle KL, Firtel RA (1978) Cell 15:763-778
Leder A, Swan D, Ruddle F, D'Eustachio P, Leder P (1981) Nature (London) 293:196-200
Maxam AM, Gilbert W (1980) Methods Enzymol 65:499-560
Merlie JP, Sobel A, Changeux JP, Gros F (1975) Proc Natl Acad Sci USA 72:4028-4032
Minty AJ, Caravatti M, Robert B, Cohen A, Daubas P, Weydert A, Gros F, Buckingham ME (1981) J Biol Chem 256:1008-1014
Minty AJ, Alonso S, Caravatti M, Buckingham ME (1982a) Cell 30:185-192
Minty AJ, Alonso S, Buckingham ME (1982b) submitted for publication
Ng R, Abelson J (1980) Proc Natl Acad Sci USA 77:3912-3916
Nudel U, Shani M, Zahut R, Katcoff D, Calro J, Carmon Y, Finer M, Yaffé D (1981) In: Pearson M (ed) Proceedings of the Cold Spring Harbor Meeting on Muscle Development. Cold Spring Harbor, in press
O'Farrell PH (1975) J Biol Chem 250:4007-4021
Ordahl CP, Fornwald J, Ovitt C, Cooper T, Calmen A (1981) In: Pearson M (ed) Proceedings of the Cold Spring Harbor Meeting on Muscle Development. Cold Spring Harbor, in press
Overbeek PA, Merlino GT, Peters NK, Cohn VH, Moore GP, Kleinsmith LJ (1982) (in press)
Scheller RH, McAllister LB, Crain J, Durcia D, Posakony J, Britten RJ, Davidson EH (1982) (submitted)
Smith DF, Searle PF, Williams JG (1979) Nucleic Acids Res 6:487-506
Southern E.M. (1975) J Mol Biol 98:503-517
Vandekerckhove J, Weber K (1978) Eur J Biochem 90:451-462
Vandekerckhove J, Weber K (1979) Differentiation 14:123-133
Whalen RG, Sell SM (1980) Nature (London) 286:731-733
Whalen RG, Butler-Browne GS, Gros F (1978) J Mol Biol 126:415-431
Whalen RG, Schwartz K, Bouveret P, Sell SM, Gros F (1979) Proc Natl Acad Sci USA 76:5197-5201
Whalen RG, Sell SM, Butler-Browne GS, Schwartz K, Bouveret P, Pinset I (1981) Nature (London) 292:805-809
Whalen RG, Sell SM, Thornell LE, Erickson A (1982) Dev Biol 91:478-484
Yaffé D (1968) In: Moscona AA, Monroy A (eds) Curr Top Dev Biol, vol 4. Academic Press, London New York, pp 37-77

Regulation of Flavonoid and Phytoalexin Accumulation Through mRNA and Enzyme Induction in Cultured Plant Cells

K. Hahlbrock, F. Kreuzaler, H. Ragg, E. Fautz, and D. N. Kuhn[1]

Introduction

Higher plants, in contrast to higher animals, do not perform active movement and do not possess a circulating blood stream for the rapid distribution of antibodies. Therefore, higher plants depend on the existence and function of special defense mechanisms to protect themselves against various kinds of injurious influences from the environment. UV irradiation and invasion by microbial pathogens are two potential dangers to which plants must respond by the rapid induction of appropriate defense reactions.

A large body of evidence indicates that more than one biochemical pathway is activated in the process of active defense, particularly in the case of microbial attack. At present, however, we are far from understanding - even from being able to enumerate - all of the individual reactions which participate in the complex entirety of the various defense mechanisms against phytopathogens. It is possible, though not known with certainty, that the response to UV irradiation involves fewer biochemical reactions than the response to pathogenic organisms. At least in some cases, partially identical reactions take place in the two responses. One such example is the subject of our present studies.

Cell-Culture Systems

We are using cultured plant cells from parsley (*Petroselinum hortense*) and soybean (*Glycine max*) to investigate some of the major defense reactions triggered by UV irradiation or treatment with a phytopathogen. This report will be restricted to our work with cultured parsley cells. For studies of basic induction mechanisms, cell cultures offer very valuable advantages in comparison with intact plants. However, it is important that one is always aware of the limitations of such artificial systems.

For example, it is of great value that large amounts of comparatively homogeneous cell populations can be propagated in culture within a few days. Growth conditions can be precisely controlled and easily manipulated for regulatory studies of biosynthetic pathways. In the ideal

1 Biologisches Institut II der Universität, Schänzlestraße 1, D-7800 Freiburg, FRG

33. Colloquium - Mosbach 1982
Biochemistry of Differentiation and Morphogenesis

case, a cell culture is a system where the high complexity of the biochemical responses triggered by a certain treatment is reduced to a single pathway. One such example, where light induces flavonoid biosynthesis, is discussed below. On the other hand, the second example which will be presented, the induction of phytoalexins with a so-called "elicitor" fraction from a phytopathogen, is a typical demonstration of both the value and the limitations of the use of cell-culture systems. Here, all cells and cell aggregates of the culture are treated with the elicitor and can exhibit a response which, in the intact plant, is limited to a relatively small number of cells surrounding the area of infection. Thus, basic studies of phytoalexin induction are greatly facilitated by the use of cell cultures. However, it is doubtful whether cultured cells show all of the various responses occurring in cells from infected tissue in intact organisms.

Inducible Phenylpropanoid Compounds

Flavonoids are among the most characteristic and widely occurring classes of phenylpropanoid derivatives in higher plants. They serve a variety of functions, including the protection of sensitive tissue against damage by UV irradiation [1,2].

Phytoalexins are antimicrobial substances produced by plants in response to microbial infection or treatment with an elicitor [3,4]. Some phytoalexins, including the furanocoumarins in parsley, are phenylpropanoid derivatives bearing structural similarities with flavonoids, although the occurrence of furanocoumarins in higher plants is much less universal [5]. Figure 1 depicts the common biosynthetic origin of flavonoids and furanocoumarins. 4-Coumaroyl-CoA, a product of general phenylpropanoid metabolism (the "cinnamate pathway"), is the central intermediate in both biosynthetic routes.

Enzyme Induction

Two groups of enzymes (designated as Group I and Group II) catalyze the over-all conversion of phenylalanine and acetyl-CoA, together with a few other substrates derived from intermediary metabolism, to flavone and flavonol glycosides (Fig. 1). The two groups are induced concomitantly in irradiated parsley cells, and the enzymes within each group are coordinately regulated [1]. A summary of our present state of knowledge is given in Fig. 2. The individual enzymes of the two groups and some of their properties are listed in Table 1. With the exception of the first enzyme from each group, they are all similar in size and have similar substrate affinities, as judged from their apparent K_m values [1,6].

When an elicitor, in the form of an autoclaved mycelial cell-wall preparation from the fungal pathogen, *Phytophtora megasperma* f. sp. glycinea, is given to a cell suspension culture of parsley, the enzymes of Group I, but not those of Group II are rapidly induced. The induction of these enzymes is much more rapid with elicitor than with light [7] and is followed by the formation of several furanocoumarins. The chemical structures of the furanocoumarins have recently been elucidated (K. Tietjen and U. Matern, unpublished results). However, none of the enzymes involved in the conversion of 4-coumaroyl-CoA to these products has been identified, and the role of 4-coumaroyl-CoA as an

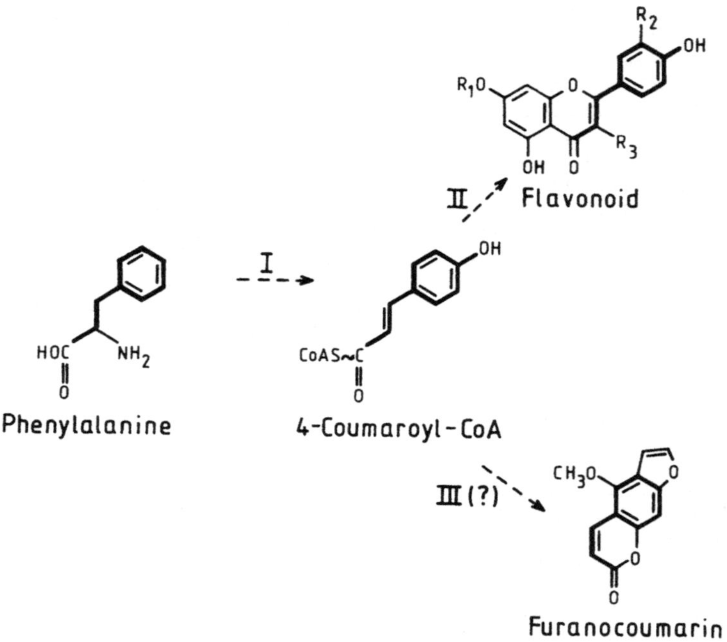

Fig. 1. Common biosynthetic origin of flavonoids and furano-coumarins from phenylalanine. Thick lines indicate the mode of incorporation of the phenylpropanoid unit. The two groups of inducible enzymes (*I* and *II*) leading to the formation of flavonoid glycosides have been studied extensively in cultured parsley cells [1,6]. R_1, R_2 and R_3 indicate the positions where substitution patterns differ among the various flavones and flavonol glycosides isolated from irradiated cells [1]. The third group (*III*) is hypothetical so far and is assumed to catalyze the conversion of 4-coumaroyl-CoA to various furanocoumarins, including the one shown here, isobergaptene (K. Tietjen and U. Matern, unpublished results)

intermediate is only indirectly deduced from the two observations that (a) its biosynthetic enzyme, 4-coumarate:CoA ligase, is induced by the elicitor and (b) the structural element of 4-coumarate is found in all furanocoumarins isolated from elicitor-treated parsley cells. Accordingly, the proposed existence of a third group of enzymes and its possible induction by the elicitor, though very likely in view of the existing data, are so far purely hypothetical. The expected small number of enzymic steps leading to the formation of furanocoumarins from 4-coumaroyl-CoA suggests that the proposed Group III comprises considerably fewer enzymes than Group II.

mRNA Induction

Specific antisera have been prepared for four of the inducible enzymes from parsley cells, two antisera each for enzymes of Group I (phenylalanine ammonia-lyase and 4-coumarate:CoA ligase) and Group II (chalcone synthase and UDP-apiose synthase). These antisera were used to determine the rates of enzyme synthesis, or mRNA activity, by selective immunoprecipitation of the enzyme proteins after synthesis either in vivo or in vitro. The results obtained with both irradiated and elicitor-treated cell cultures [7-10] are summarized in Fig. 3 (A and B,

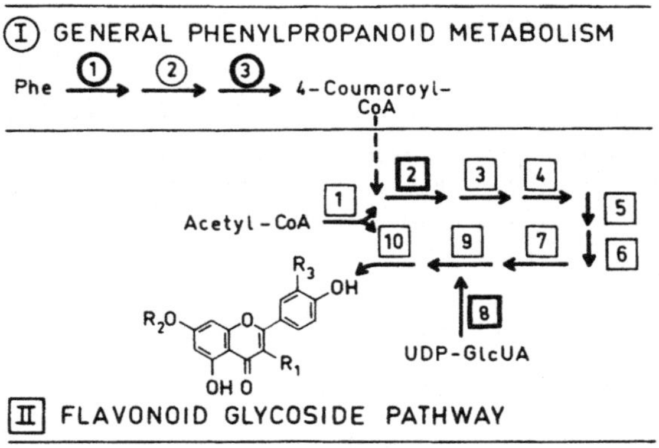

Fig. 2. Schematic representation of the kinetics of induction by light of the enzymes of Group I and Group II in cultured parsley cells. The dotted curve represents flavonoid accumulation. The four enzymes discussed below are marked by broad circles or squares. Group I: *1*, phenylalanine ammonia-lyase; *3*, 4-coumarate:CoA ligase; Group II: *2*, chalcone synthase; *8*, UDP-apiose synthase. For further details, see [1]

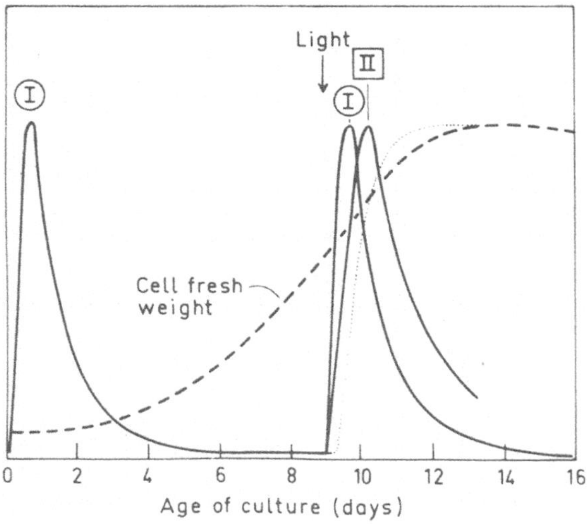

respectively). The high degree of coordination in the induction within each group is even more pronounced at the level of the mRNA activities than at the level of the enzyme activities. Moreover, the extremely rapid induction of the Group I mRNA's and the lack of induction of the Group II mRNA's by the elicitor are clearly demonstrated in Fig.3B. In all cases shown, essentially the same results were obtained regardless of whether the rates of synthesis were measured in vivo or in vitro. This suggests that the rate of synthesis in vivo is regulated mainly, or exclusively, by the availability of translatable mRNA.

We have recently demonstrated that the changes in chalcone synthase mRNA activity are paralleled by changes in mRNA amount. We have cloned cDNA copies of poly(A)$^+$ mRNA's from irradiated parsley cells and isolated several clones with plasmids containing cDNA inserts homologous to chalcone synthase mRNA. The insert from one of these plasmids, pLF56, was characterized by DNA sequencing and is a cDNA copy comprising most of the coding sequence, the 3'-untranslated region and a portion of the poly(A) segment of chalcone synthase mRNA (U. Reimold, M. Kröger and K. Hahlbrock, unpublished results). This cDNA insert was used for

Table 1. Inducible enzymes of phenylpropanoid metabolism in cultured parsley cells. (For references, see Ebel and Hahlbrock [6])

Group	Enzyme	App. K_m (μmol/l)	Approx. M_r	No. of subunits
I	Phenylalanine ammonia-lyase	32	330,000	4
	Cinnamate 4-hydroxlase	70	n.d.	
	4-Coumarate:CoA ligase	14	60,000	1
II	Acetyl-CoA carboxylase	150[a]	800,000	4
	Chalcone synthase	2	80,000	2
	Chalcone isomerase	16[b]	50,000	
	2-Hydroxylase (?)[c]			
	3-Hydroxylase			
	3'-Hydroxylase			
	3'-Methyltransferase	46	50,000	
	7-O-Glucosyltransferase	2	50,000	
	UDP-Apiose synthase	2	86,000	2
	Apiosyltransferase	66	50,000	
	3-O-Glucosyltransferase	n.d.	50,000	
	Malonyltransferase-7	10	50,000	1
	Malonyltransferase-3	4	50,000	1
III	*hypothetic* (enzymes not identified)			

[a] Depending on assay conditions

[b] Determined with enzyme from intact plant

[c] Mechanism of action not fully characterized

the "Northern" blotting experiment shown in Fig. 4A. Hybridization of labeled cDNA with chalcone synthase mRNA was strongest around 10-11 h after induction. Much less hybridization occurred at a later stage, and no hybridization was detected prior to induction. This result is in good agreement with the pattern of changes in mRNA activity, as shown in Fig. 4B.

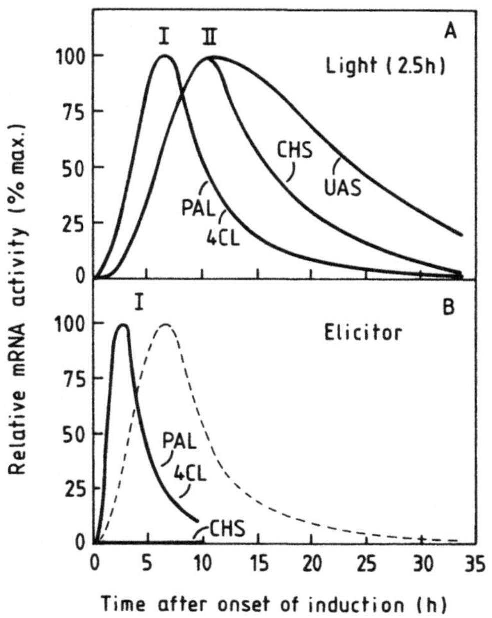

Fig. 3A,B. Kinetics of mRNA induction by 2.5-h irradiation (A) or treatment with elicitor (B) of cultured parsley cells. PAL, phenylalanine ammonia-lyase; *4CL* 4-coumarate:CoA ligase; *CHS* chalcone synthase; *UAS* UDP-apiose synthase. The *broken curve* in panel (B) was taken from panel (A) for comparison of elicitor and light effects on mRNA's of Group I. All curves were normalized, data from [7-10]

Mathematical Calculations

The same clear-cut relationship between mRNA and enzyme activities has been observed in all cases studied so far, independent of whether light or elicitor was the inducing agent [7-10]. This relationship can be expressed by the equation

$$dE(t)/dt = {}^{0}k_s(t) - {}^{1}k_d \cdot E(t) \quad , \tag{1}$$

where changes in enzyme activity (E) with time (t) are determined by the zero-order rate constant of enzyme synthesis (or mRNA activity) (${}^{0}k_s$), which also changes with time, and the first-order rate constant of enzyme degradation (${}^{1}k_d$). The rate constant, ${}^{1}k_d$, which is assumed to remain unaltered upon induction, can easily be calculated from the apparent half-life of the enzyme activity ($T_{1/2}$) with the equation

$${}^{1}k_d = \ln 2/T_{1/2} \quad . \tag{2}$$

The agreement between calculated and experimentally determined data suggests that both light and elicitor act solely, or at least primarily, through mRNA induction.

A much simpler mathematical equation,

$$P(t) = \int_0^t E(\tau)d\tau \quad , \tag{3}$$

can be applied to relate the amount of flavonoid glycosides (P, products) accumulated with time (t,τ) to the induced changes in enzyme

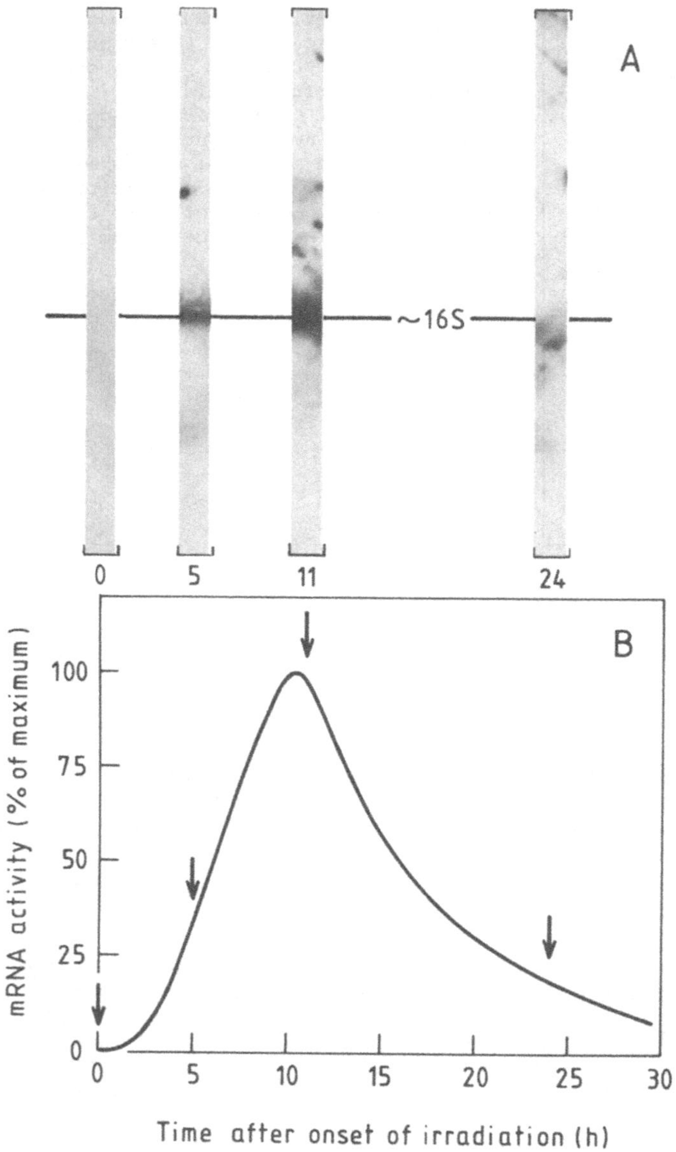

Fig. 4A,B. Comparison of light-induced changes in hybridizable amount (A) and translatable activity (B) of chalcone synthase mRNA. The results shown in panel (A) have not been published previously and will be described in detail elsewhere. The data for panel (B) were taken from Fig. 3A. Irradiation of the cells was for 2.5 h

activity (E). Coincidence of the measured curve for flavonoid accumulation and the curve calculated from Eq. (3) can be expected for the rate-limiting enzyme. Figure 5 summarizes results obtained with irradiated parsley cells under experimental conditions, where, according to this criterion, the rate-limiting enzyme was the first enzyme of Group I, phenylalanine ammonia-lyase [11]. The observed clear-cut

$$\frac{dE(t)}{dt} = mRNA(t) - k_d \cdot E(t) \qquad P(t) = \int_0^t E(\tau)\, d\tau$$

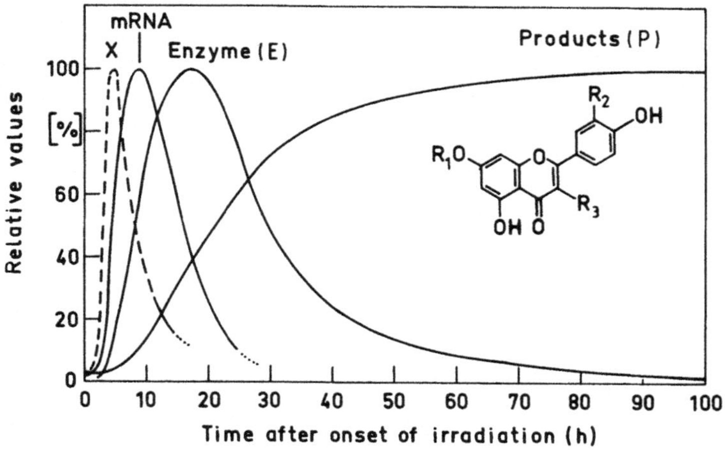

Fig. 5. Time courses of light-induced changes in phenylalanine ammonia-lyase mRNA and enzyme activities and in the amount of flavonid glycosides accumulated in cultured parsley cells during continuous irradiation for up to 100 h. The mathematical equations are those described in the text, with the slight modification that the rate constant, $^{0}k_s$, is replaced by mRNA (activity). For details, see [1]. The *broken curve* (X) represents the expected changes in the rate of phenylalanine ammonia-lyase mRNA synthesis and is purely hypothetical. This curve was calculated from the curve for mRNA activity, using an equation essentially similar to Eq. (1) and an estimated half-life of the mRNA of 3.5 h [7]

relationship between enzyme activity and product accumulation rules out the possibility that end product or other feedback inhibition of the rate-limiting enzyme plays a major role in this system.

No such correlations have been made between the accumulation rates of furanocoumarins and elicitor-induced changes in enzyme activities involved in their synthesis.

Conclusions

Evidence has been obtained for the coordinated induction of metabolically related enzymes and their mRNA's in cultured plant cells. Each of the two inducing agents used in these studies (UV light or elicitor from a phytopathogen) triggers the formation of one class of compounds (UV-protective flavonoids or antimicrobial phytoalexins). The potential function of these compounds is to protect the challenged plant against possible damage by the inducing agent. One sequence of biosynthetic reactions, catalyzed by the enzymes of general phenylpropanoid metabolism (Group I), is the same for both classes of compounds. The subsequent, more specific reactions are carried out by the enzymes of the flavonoid glycoside pathway (Group II) and a hypothetical third group (Group III) comprising the enzymes of furanocoumarin synthesis. A schematic representation of the results obtained so far with cultured parsley cells is given in Fig. 6.

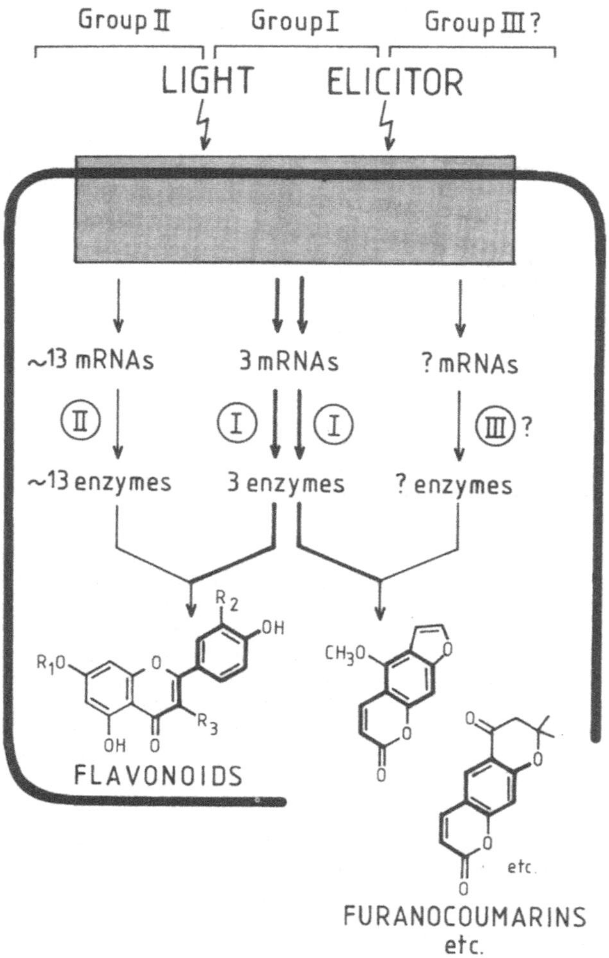

Fig. 6. Scheme summarizing the results presented in this article

Acknowledgements. Work in our own laboratory was supported by Deutsche Forschungsgemeinschaft (SFB 46) and Fonds der Chemischen Industrie. We thank K. Tietjen and U. Matern for making their data on the structural identification of furanocoumarins from parsley cells available to us prior to publication.

References

1. Hahlbrock K (1981) In: Stumpf PK, Conn EE (eds) The biochemistry of plants, vol VII. Academic Press, London New York, pp 425-456
2. Wellmann E (1982) In: Shropshire Jr W, Mohr H (eds) Encyclopedia of plant physiology. New series: photomorphogenesis. Springer, Heidelberg Berlin New York, in press
3. Grisebach H, Ebel J (1978) Angew Chem 90:668-681
4. West CA (1981) Naturwissenschaften 68:447-457
5. Brown SA (1981) In: Stumpf PK, Conn EE (eds) The biochemistry of plants, vol VII. Academic Press, London New York, pp 269-300

6. Ebel J, Hahlbrock K (1982) In: Harborne JB, Mabry J (eds) Recent advances in flavonoid research 1975-1980. Chapman Hall, London, in press
7. Hahlbrock K, Lamb CJ, Purwin C, Ebel J, Fautz E, Schäfer E (1981) Plant Physiol 67:768-773
8. Schröder J, Kreuzaler F, Schäfer E, Hahlbrock K (1979) J Biol Chem 254:57-65
9. Gardiner SE, Schröder J, Matern U, Hammer D, Hahlbrock K (1980) J Biol Chem 225:10752-10757
10. Ragg H, Kuhn DN, Hahlbrock K (1981) J Biol Chem 256:10061-10065
11. Hahlbrock K, Knobloch KH, Kreuzaler F, Potts JRM, Wellmann E (1976) Eur J Biochem 61:199-206

Regulation of Gene Expression During Embryonic Development

W. Knöchel[1]

Introduction

The South African clawed toad, *Xenopus laevis*, is a suitable experimental system to investigate gene expression during the early development of vertebrates. Embryogenesis takes place outside the maternal body and enormous amounts of fertilized eggs can be easily obtained after hormone induced breeding. Embryonic development is characterized by rapid morphogenetic changes which might be induced by different cellular interactions or by the action of morphogenetic factors but is finally based upon differential gene expression. Its regulation can occur at the levels of transcription, processing and translation. In *Xenopus*, translational control mechanisms may play a decisive role during oocyte maturation [1] and also in the first post-fertilization stages. The hormone induced conversion of the stage six oocyte to the egg is accompanied by a 20-fold increase of histone synthesis [2] without detectable changes in the concentration of total translatable histone mRNA [3]. On the other hand it has been shown, that at least in the case of two-cell-stage mouse embryos, the paternal genome is already actively transcribed and expressed into proteins [4]. The present investigation has been started to isolate those genes which are stage-specifically expressed during embryogenesis and to elucidate mechanisms which control their expression.

cDNA Cloning and Analysis of Transcriptional Activity

Initial cDNA/RNA-excess hybridization experiments demonstrated that most, if not all, polysomal poly(A) containing mRNA species present at gastrula stage, are also present at the first tadpole stage [5]. However, part of tadpole mRNA is not present at gastrula stage. To investigate whether such differences also exist at the level of nuclear polyadenylated RNA, we established cDNA clone banks from nuclear poly(A)-containing RNA of various developmental stages. This was achieved by inserting double stranded cDNA into the Pst I cleavage site of plasmid pBR322 [6]. Then the clone banks were screened for the presence and the amount of individual RNA sequences during the course of embryonic development. Cloning of nuclear RNA instead of total RNA, which has been described by [7-9], offers two advantages. The analysis of nuclear RNA more directly reflects stage-specific

1 Institut für Molekularbiologie und Biochemie, Freie Universität Berlin, Arnimallee 22, 1000 Berlin 33

33. Colloquium - Mosbach 1982
Biochemistry of Differentiation and Morphogenesis

transcriptional activities, since maternal transcripts may be stored in the cytoplasm and utilized at least until gastrula stage [10]. Besides that, screening of the clone bank of nuclear origin with cytoplasmic or polysomal RNA or cDNA probes allows the identification of RNA sequences, which are confined to the nucleus or accumulated in the cytoplasm.

Figure 1A shows a clone bank of 768 cDNA clones derived from nuclear polyadenylated RNA of the tadpole stage, that has been hybridized with ^{32}P labelled cDNA from the template RNA (homologous reaction). About 25% of the clones give positive hybridization signals thus representing RNA sequences which were calculated to be present at 40 copies or more per cell. The majority of cloned sequences (75%) is present at copy numbers too low to be detected by a positive hybridization signal. If one compares the screening pattern of the homologous reaction with that of cDNA from the gastrula stage (heterologous reaction; Fig. 1B) it is seen that there are some sequences (indicated by bold squares) which are present only within nuclear RNA of tadpole stage but not in that of gastrula stage. Moreover, a certain number of sequences that give a strong positive signal in the homologous reaction, give only a weak signal in the heterologous reaction. A similar situation was detected when a number of clones selected from a gastrula nuclear polyadenylated RNA clone bank were screened with ^{32}P labelled cDNA, derived from poly(A)-containing RNAs of various developmental stages. According to these experiments the clones that give positive signals at any of the investigated stages can be classified into four different groups. Figure 2 shows a schematic presentation of the relative accumulation of some individual RNA sequences during the course of embryonic development.

Group A. RNA transcripts that are not developmentally regulated. About 50%-60% of the cloned sequences belong to this category (for example clone AB8).

Group B. RNA sequences that accumulate during development. The increase of copy number may originate from an increased rate of transcription or from changes in the stability of RNA transcripts. About 30% of the cloned sequences belong to this group (for example clone CD10).

Group C. Sequences that show a gradual decrease in the copy numbers during development (for example clone L4 G1; 5% of cloned sequences).

Group D. RNA sequences, that show an increase of their copy number during certain stages, but show a decrease at later stages (for example clone EF12; 5% of cloned sequences).

It is interesting to note that the majority of sequences which are developmentally regulated belong to the group B sequences, which show a drastic increase of their copy number during development. This observation also explains the results of earlier cDNA/RNA hybridization kinetics by Perlman and Rosbash [13] indicating a dramatic increase of the copy number of abundant RNA sequences from oocyte until tadpole stage. From the screening experiments of the cDNA clone banks with labelled RNA or cDNA probes of different developmental stages it can be followed, that the amount of many RNA sequences is regulated at the level of transcription. However, cDNA clones are not suitable for more detailed studies on the mechanism of transcriptional control, because they only represent transcribed DNA sequences and, most likely, do not contain regulatory sequences as known to be present on genomic DNA. Therefore we have established genomic DNA libraries of *Xenopus laevis* and, by hybridization with nick translated plasmids containing

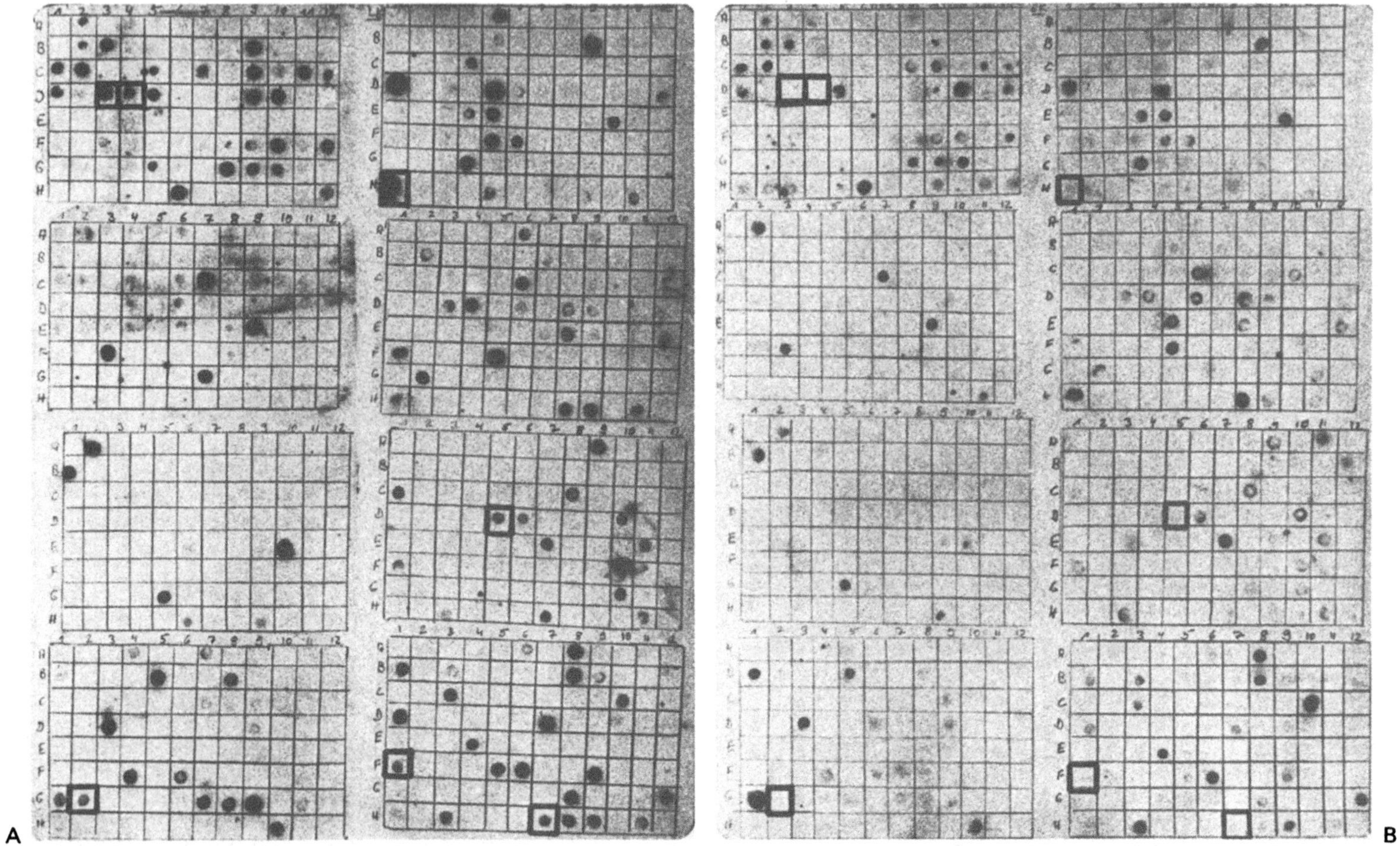

Fig. 1A,B. Autoradiograph of 768 cDNA clones. Nuclear polyadenylated RNA from the early tadpole stage of *Xenopus laevis* was transcribed into cDNA, integrated as double strands by G/C joining into the Pst I cleavage site of plasmid pBR322 and cloned after transformation of E.coli strain HB 101. After transfer to nitrocellulose filters the clones were hybridized with ^{32}P labelled cDNA from (A) tadpole nuclear polyadenylated RNA (homologous reaction), (B) gastrula nuclear polyadenylated RNA (heterologous reaction). After washing the filters they were exposed to X-ray films. Some clones which give a positive signal in the homologous but a negative signal in heterologous reaction are marked by *bold squares*

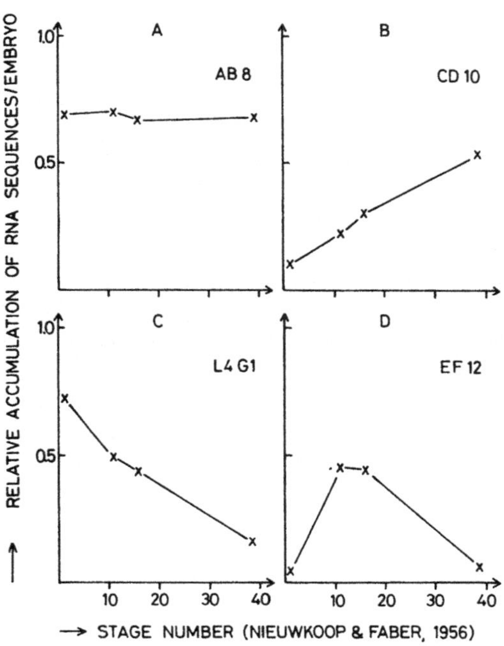

Fig. 2A-D. Relative accumulation of RNA sequences per embryo during different stages of development (stage numbers from normal table of Nieuwkoop and Faber [11]). The relative amount of indicated RNA sequences (A: AB8, B: CD10, C: L4G1, D: EF12) was calculated by the determination of radioactivity of labelled cDNA hybridized to plasmid DNA. For counting in a liquid scintillation counter the clones were cut out from nitrocellulose filters. It was taken into consideration, that the total amount of polyadenylated RNA changes during development [12]

cDNA, we could isolate a series of genomic DNA fragments containing those genes which are developmentally regulated. Their further characterization is under present investigation.

Post-transcriptional Control

Initial evidence, that part of nuclear RNA sequence is retained to the nucleus, was obtained by cDNA/RNA hybridization curves of cDNA derived from nuclear and polysomal polyadenylated RNA of tadpole stage with the homologous and heterologous RNA species [14]. Whereas nuclear RNA completely saturates the polysomal cDNA, the nuclear cDNA has only hybridized to 50%-60% with polysomal RNA. This observation is strongly confirmed by the hybridization behaviour of some of the cloned nuclear sequences. Figure 3 shows the autoradiograph of 6 clones from the tadpole clone bank hybridized with nuclear ^{32}P labelled cDNA and polysomal cDNA from tadpole stage as well with polysomal cDNA from gastrula stage. Clones 4 and 5 are controls, which are present in nuclear and polysomal RNAs of both stages. Clones 1-3 contain sequences present only in nuclear RNA but not in polysomal RNA. However, the sequence of clone 6, which again is not present in polysomal RNA of tadpole stage, is present within the polysomal RNA of gastrula stage. This result indicates that at least part of the sequences, retained to the nucleus at a certain developmental stage, may serve as mRNA molecules which are exported to the cytoplasm and translated into proteins at other developmental stages. It is interesting to note that at least in case of the sequence of clone 6, the cytoplasmic export is accompanied by a significant size reduction of RNA molecules. Figure 4 shows the autoradiograph of nuclear and polysomal RNAs of gastrula and tadpole stages, separated on a denaturing agarose slab gel, transferred to nitrocellulose filter and hybridized with nick translated

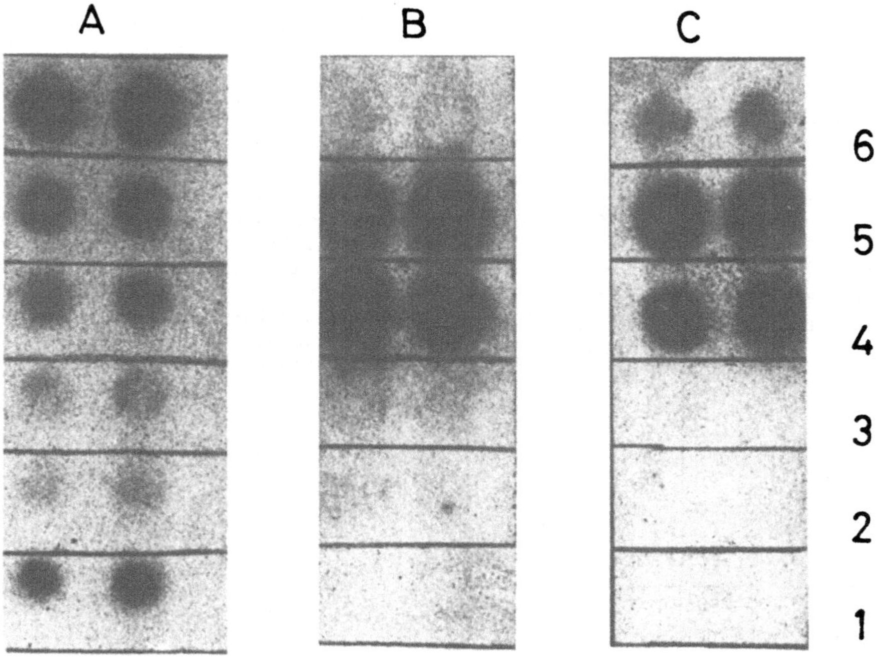

Fig. 3A-C. Autoradiograph of six cDNA clones derived from nuclear polyadenylated RNA of early tadpole stage and hybridized with ^{32}P labelled cDNA from (A) tadpole nuclear poly(A)$^{+}$ RNA, (B) tadpole polysomal poly(A)$^{+}$ RNA and (C) gastrula polysomal poly(A)$^{+}$ RNA

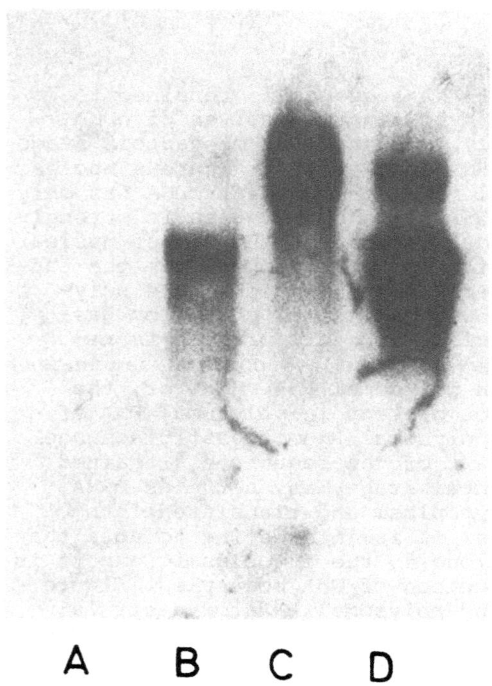

Fig. 4A-D. Size analysis of a cloned RNA sequence. Polysomal [(A) tadpole, (B) gastrula] and nuclear [(C) tadpole, (D) gastrula] RNAs were electrophoretically separated on a denaturing agarose slab gel. After transfer to nitrocellulose paper the RNAs were hybridized with plasmid DNA from clone AB11 (clone 6 in Fig. 3) labelled with ^{32}P by nick translation and subsequently submitted to autoradiography

DNA of the sequence from clone 6. In case of gastrula stage, where the sequence is present within polysomal mRNA, the nuclear RNA migrates as two bands. The smaller one corresponds to the band of polysomal RNA. In the case of tadpole nuclear RNA only the larger band is visible and there is no hybridization signal with tadpole polysomal RNA. From this observation we conclude that the cytoplasmic export of the nuclear polyadenylated RNA sequence is inhibited. This may result from a failure of the processing mechanism. We conclude that in addition to stage specific regulation of gene transcription, post-transcriptional events at the level of processing also regulate the expression of genes by controlling cytoplasmic export.

It is tempting to speculate on the nature of the sequences removed during processing. Obviously, they may simply represent intervening sequences as known from most eucaryotic genes. On the other hand it is possible that repetitive DNA transcripts, that in sea urchin oocytes are known to be co-transcribed with unique mRNA sequences [15] are eliminated. By the formation of intermolecular RNA aggregates after hybridization in formamide we could demonstrate that such complementary repetitive RNA sequences are also present in polyadenylated RNA from different developmental stages of *Xenopus laevis*. Fortunately, some of these repetitive RNA sequences are represented in the cDNA clone banks described. From one sequence, which has been further investigated in our laboratory, we know that it is stage specifically transcribed and present at maximum copy number at neurula stage within many nuclear RNA sequences of different molecular weights but not in polysomal RNAs. It is reasonable to assume, that co-transcription of this type of repetitive sequence, and its elimination during RNA processing, may be involved in some regulatory mechanisms of stage specific gene expression during embryonic development.

Acknowledgements. This investigation was aided by grants from the Deutsche Forschungsgemeinschaft (Sonderforschungsbereich Embryonale Entwicklung und Differenzierung).

References

1. Rosbash M (1981) Dev Biol 87:319-329
2. Adamson ED, Woodland HR (1977) Dev Biol 57:136-149
3. Ruderman JV, Woodland HR, Sturgess EA (1979) Dev Biol 71:71-82
4. Sawicki JA, Magnuson T, Epstein CJ (1981) Nature (London) 294:450-451
5. Knöchel W, Bladauski D (1980) Wilhelm Roux's Arch Entwicklungsmech Org 188:187-193
6. Knöchel W, Bladauski D (1981) Wilhelm Roux's Arch Entwicklungsmech Org 190:97-102
7. Dworking MB, Dawid IB (1980) Dev Biol 76:435-448
8. Golden L, Schafer U, Rosbash M (1980) Cell 22:835-844
9. Jacob E (1980) Nucleic Acids Res 8:1319-1337
10. Davidson EH (1977) Gene activity in early development, 2nd edn. Academic press, London New York
11. Nieuwkoop PD, Faber J (1956) Normal table of Xenopus laevis (Daudin). North Holland Biomedical Press, Amsterdam New York
12. Sagata N, Shiokawa K, Yamana K (1980) Dev. Biol 77:431-448
13. Perlman S, Rosbash M (1978) Dev Biol 63:197-212
14. Knöchel W, John ME (1982) Eur J Biochem 122:11-16
15. Constantini FD, Britten RJ, Davidson EH (1980) Nature (London) 287:111-117

The Yeast Mating Type System – A Model for the Regulation of Gene Expression by the Position of a Certain Gene Within the Genome?

U. Wintersberger and R. Schiestl[1]

Much to the surprise of many molecular biologists it has become evident in recent time that genomes are dynamic structures in which DNA rearrangements can take place leading to sequence variations between cells of one and the same organism. Besides the well known transposable elements in prokaryotes as well as in eukaryotes [1] to which a definite function has not been ascribed until now, there exist mechanisms in which a DNA rearrangement was found involved in changes of gene expression occurring in a regulated manner during development (for review see [2]).

One of the most extensively studied examples is mating-type switching in the yeast, *Saccharomyces cerevisiae* [3]. This unicellular eukaryote can exist in three different cell lineages: there are two cell types in the haploid phase (a and α) which can mate with each other; the resulting third type, diploid a/α cells, cannot mate, but undergo meiosis under unfavourable living conditions and develop into four haploid spores, two of a- and two of α-mating type, which can mate again. The mating-type is determined by the MAT-locus on chromosome III where the information for a or α is expressed. These sex-specific genes alternately occupy the MAT-locus and additionally they are stored as unexpressed "cassettes" at the "library" - loci HML and HMR on chromosome III (Fig. 1). So-called homothallic haploids, under the influence of the gene HO, can change their mating-type frequently so that a few generations after germination of a single homothallic spore the descending haploids, being now of mixed mating-types, can mate and a diploid a/α colony will arise.

The mechanism of the mating-type switching process is well understood at the molecular level: It involves an intrachromosomal recombinational event, in which the sequence at MAT is replaced by a nonreciprocal gene conversion using HML or HMR as donor. The genetic information which inhabited the MAT-locus before the switching event is lost during the gene conversion while the sequences at HML and HMR are conserved (Fig. 1), and switching can be repeated infinitely. Thus mating-type switching is an example for a genome rearrangement involved in a differentiation like process. The status of sex-differentiation in yeast is determined by the position of sex-specific regulatory genes within the genome.

In heterothallic yeast strains, lacking the gene HO-function, mating-type switching is a very rare event, occurring about once in a million cells per generation, so that descendants of single heterothallic spores

1 Institut für Krebsforschung der Universität Wien, A-1090 Wien, Austria

33. Colloquium - Mosbach 1982
Biochemistry of Differentiation and Morphogenesis

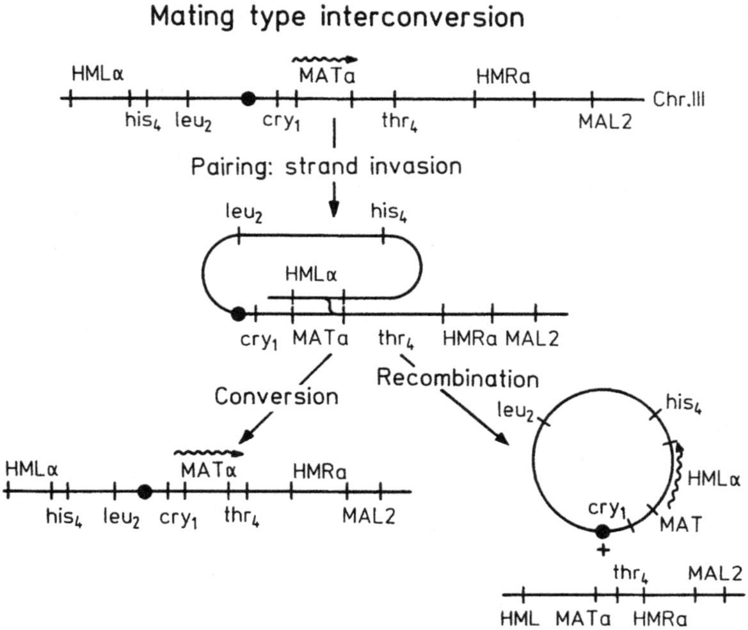

Fig. 1. Events on the chromosomal level which lead to a replacement of a-mating-type specific regulatory genes at the MAT-locus of chromosome III by α-specific genes. After pairing between homologous parts of the sequences at MAT and HML the a-specific information at MAT is replaced by nonreciprocal gene-conversion using the α-specific sequences of HMLα as donor (*left side of the diagram:* conversion!). In rare cases reciprocal recombination takes place leading to a ring-chromosome and to the loss of part of chromosome III (*right side of the diagram:* recombination!)

develop into rather stable haploid populations. We now asked the question whether in such "developmentally fixed" cells the switching frequency can be influenced from outside the cell.

Considering the fact that in the course of the induction of DNA-repair by DNA-damaging agents, mitotic recombination and gene-conversion were found enhanced in yeast [4], we investigated whether turning on DNA-repair might lead to an increase in switching frequency in heterothallic yeast strains. We suspect that induced enzymes involved in gene-conversion and recombination processes might also be able to catalyze DNA-rearrangements, being part of processes called "induced evolution" by Echols [5].

First we had to work out a procedure for precisely monitoring the rare switching events of a heterothallic strain and their possible enhancement by diverse treatments [6]. In short, we trapped switched cells by mating with cells of another strain possessing complementary genetic markers (test cells) and counted diploid colonies which arose on a selective medium. If cells, before mixing with the test cells, were irradiated with X-ray, increasing numbers of diploid colonies were observed in a dose-dependent manner up to 10 Krad which corresponded to a surviving fraction of about 20% (Fig. 2). To convince ourselves that the diploid colonies we counted, really were derived from normally switched haploids we isolated them and studied them genetically. Be-

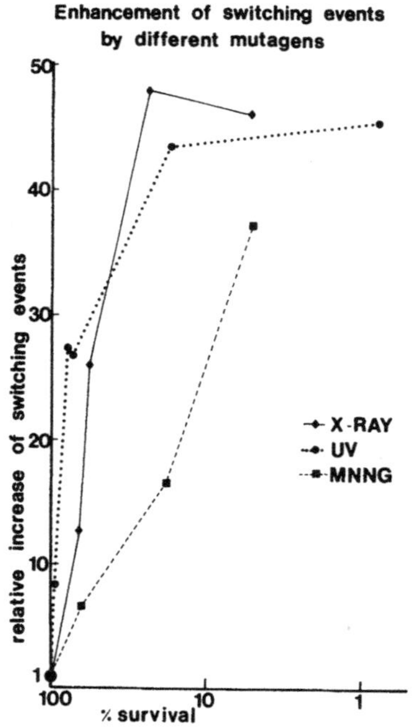

Fig. 2. Summary of data obtained by treating cells with carcinogenic agents before the determination of the number of switching events as described in detail in [6]. Results are depicted as relative increase of switching events in treated cell populations over the control population. Doses of agents applied to cells were as follows: X-ray: 0, 2, 5, 10, and 20 Krad; UV: 0, 50, 100, 150, 300, and 600 erg/mm^2; N-methyl-N'-nitro-nitroso-guanidine (MNNG): 0, 0.1, 0.2, and 0.3 mM

sides bonafide switching by intrachromosomal gene conversion, intrachromosomal recombination leading to a truncated chromosome could be the reason for a change in the expression of mating type specific genes (both possible mechanisms are shown in Fig. 1). However, diploid cells containing a deleted genome would - upon sporulation - develop into two viable and two nonviable spores. Our tetrad analyses of isolated diploid clones did not indicate any increase by X-ray of the percentage of tetrads containing two nonviable spores compared to the spontaneous rate. We also tested the segregation of the mating type, which together with the spore viability data, proved that the rate of normal switching events had been enhanced by the X-ray treatment [6].

In addition we studied the effect of UV irradiation and of the alkylating carcinogen N-methyl-N'-nitro-nitroso-guanidine. Treatment of cells with these agents again led to a dose dependent enhancement of switching events.

From our data we conclude that, at least in yeast, in the course of induction of DNA-repair-systems, DNA-damage indeed can provoke a genome rearrangement which is accompanied by a change in the state of differentiation. These findings could be of relevance for the understanding of the mechanism of chemical and radiation-induced carcinogenesis, as discussed in [2]. Genome rearrangements could also be involved in certain steps during development and differentiation of higher organisms. Induced - illegitimate - switching between gene-location-determined developmental states could lead to cells with a selective advantage over their neighbours. Such cells, in the course of further "induced evolution" [5], could become progenitors of a developing tumour.

Acknowledgements. This project was sponsored in part by a grant from "Fonds Österreichischer Krebsforschungsinstitute" and by an EMBO fellowship to R. Schiestl.

References

1. Calos MP, Miller JH (1980) Cell 20:579-595
2. Wintersberger U (1982) Naturwissenschaften 69:107-113
3. Herskowitz I, Blair L, Forbes D, Hicks J, Kassir Y, Kushner P, Rine J, Sprague G Jr, Strathern J (1980) In: Loomis W, Leighton T (eds) Molecular genetics of development. Academic Press, London New York, pp 79-118
4. Zimmermann FK, Schwaier RV, Laer U (1966) Z Vererbungsl 98:230-246
5. Echols H (1981) Cell 25:1-2
6. Schiestl R, Wintersberger U (1982) Mol Gen Genet 186:512-517

Expression of Gap Junction Protein in Liver and Lens Fiber Tissue

K. Willecke[1], O. Traub[1], U. Janßen-Timmen[1], P. Drüge[1], and R. Dermietzel[2]

Gap junctions have been found in almost all mammalian tissues except hematopoietic cells, some adult nerve cells, and skeletal muscle cells. They are composed of protein subunits which form hemichannels (connexons) in plasma membranes. The hemichannels in the plasma membranes of adjacent cells must recognize each other and undergo a conformational change in order to form a pore through adjacent membranes. Ions and metabolites of a molecular weight < 900 daltons can diffuse through these pores directly from one cytoplasmic compartment to that of the neighboring adherent cell [1,2]. Gap junctions can be recognized on electron micrographs because they form plaques of closely aggregated intermembraneous particles which show a characteristic septilaminar pattern in cross sections of apposing, negatively stained membranes. These lattice-like structures (plaques) can be purified from liver plasma membranes by differential centrifugation, since plaques do not disaggregate in the presence of non-ionic detergents under conditions where most of the other components of plasma membranes are dissolved. Unwin and Zampighi [3] analyzed the structure of isolated gap junction plaques from rat liver by electron microscopy using Fourier and image reconstruction techniques. From these results the authors deduced a model according to which the connexon is a cylinder composed of six subunits which are tilted around its axis. The structure of gap junctions from other tissues has not yet been analyzed to the same extent as that of liver gap junctions. On electron micrographs obtained after freeze fracturing the morphology of gap junction plaques in different tissues appears to be similar. Recently, however, Zampighi and Simon [4] have studied gap junctions from eye lens fiber cells by X-ray diffraction as well as electron microscopy and have reported that the connexons of the eye lens have a tetrameric subunit structure.

Two years ago the isolation of apparently non-degraded subunit protein of liver gap junctions was reported by several laboratories. In earlier investigations the degree of enrichment could only be followed by electron microscopy since isolated gap junction plaques do not show enzymatic activity. Henderson et al. [5] and more recently Finbow at al. [6] concluded from their studies of mouse and rat liver gap junctions that the major subunit protein component has an apparent molecular weight of about 26,000 (26K) daltons after SDS-polyacrylamide gelelectrophoresis. This protein tends to form aggregates and is very sensitive towards proteolytic degradation. Now the identification of the

1 Institut für Zellbiologie (Tumorforschung), Universitätsklinikum Essen, Hufelandstrasse 55, 4300 Essen 1, FRG

2 Institut für Anatomie, Universitätsklinikum Essen, Hufelandstrasse 55, 4300 Essen 1, FRG

33. Colloquium - Mosbach 1982
Biochemistry of Differentiation and Morphogenesis

26K gap junction protein subunit on gels can be used to monitor the degree of purification of gap junction preparations.

We developed and characterized antisera, in rats and rabbits, directed against the purified 26K protein subunit from mouse liver gap junctions. The 26K protein used for immunization had been cut from SDS-polyacrylamide gels. In addition, we developed antisera against gap junction plaques and against the 21K protein which was co-purified with plaques and presumed to be a degradation product of the major 26K protein [5]. We tested the antisera after transfer of the proteins separated by SDS-polyacrylamide gelelectrophoresis onto nitrocellulose paper. The position of the antigen-antibody complex was detected by reaction with ^{125}I-labelled protein A from *Staphylococcus aureus* (Immunoblot) [7].

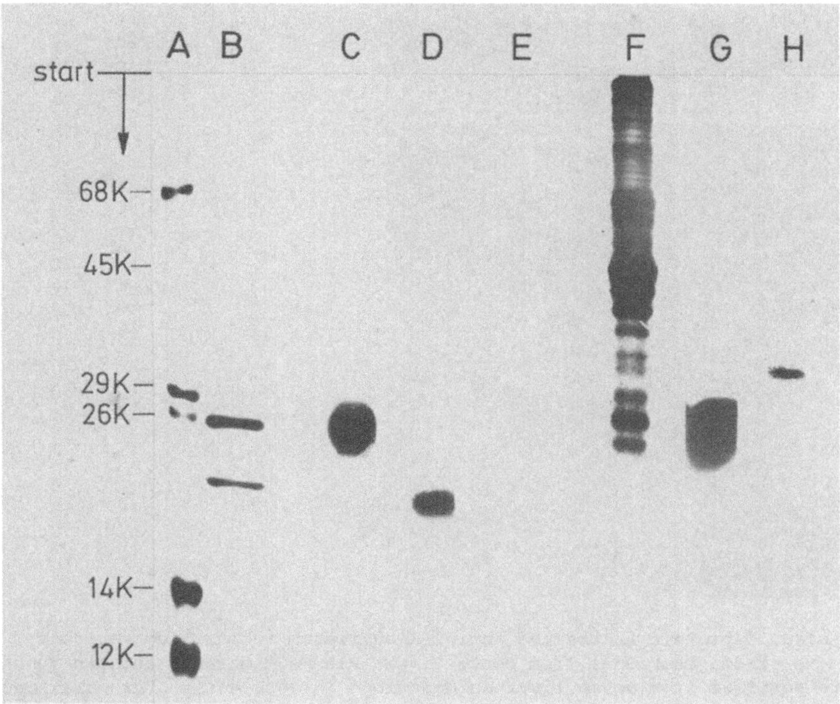

Fig. 1. Immunological identification of gap junction protein from mouse and rat liver with different rabbit antisera. Purified gap junction proteins or total plasma membrane proteins were separated on SDS-polyacrylamide gels and analyzed after transfer onto nitrocellulose paper, incubation with different rabbit antisera and reaction with ^{125}I-protein A according to standard immunoblot conditions (cf. [8]). *Lane A* reference proteins: bovine serum albumin (68K), ovalbumin (45K), carbonic anhydrase (29K), chymotrypsinogen (26K), ribonuclease A (14K), and cytochrome C (12K) - Coomassie brillant blue staining; *lane B* mouse liver plaque proteins stained with Coomassie brillant blue; *lane C* autoradiograph of mouse liver plaque proteins after immunoblot with anti-26K antiserum; *lane D* autoradiograph of mouse liver plaque proteins after immunoblot with anti-21K antiserum; *lane E* autoradiograph of mouse liver plaque proteins after immunoblot with preimmune serum; *lane F* rat liver plasma membranes, stained with Coomassie brillant blue; *lane G* autoradiograph of rat liver plasma membranes after immunoblot with anti-26K antiserum; *lane H* autoradiograph of rat liver plasma membranes after immunoblot with preimmune serum

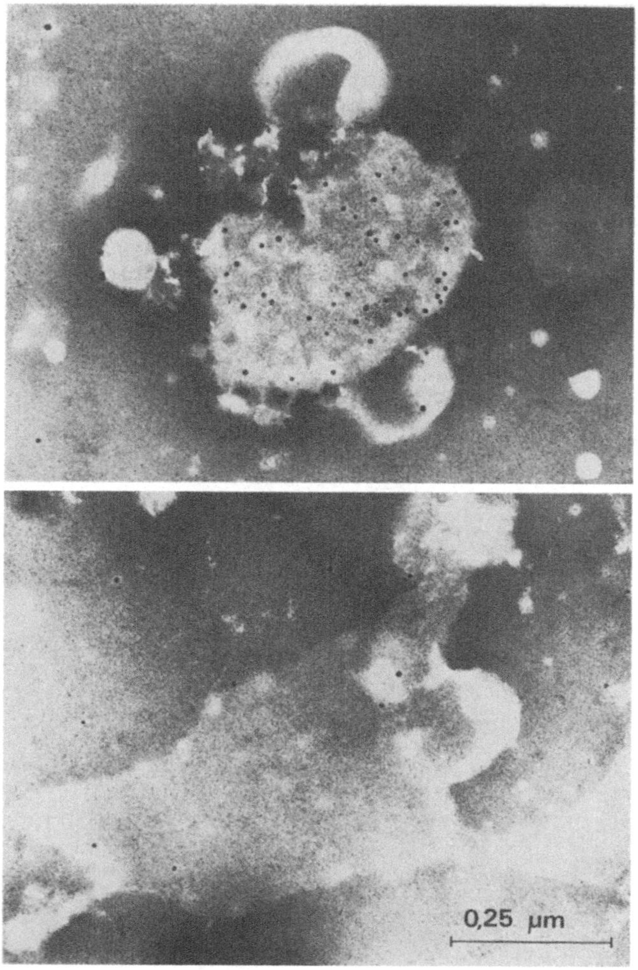

Fig. 2. Binding of gold/protein A-labelled anti-26K antiserum to purified gap junction plaques and to plasma membranes from mouse liver. Plasma membranes and gap junction plaques were purified from mouse liver as described in [8]. Antigen preparations were coated on electron microscopic grids by incubation for 15 min at room temperature and washed in phosphate buffered saline (PBS). Remaining binding sites on the grids were saturated by incubation with bovine serum albumin (3% in PBS). Then the grids were incubated with rabbit anti-26K antiserum (1:5 diluted) for 60 min, followed by washing in PBS. Protein A was labelled with colloidal gold (7-10 nm) prepared with ascorbic acid according to Slot and Geuze [9]. The antigen-antibody complexes on the grids were incubated with gold-labelled protein A for 60 min and washed 4 times in PBS. Finally the coated grids were negatively stained with phosphotungstate. *Panel A* gap junction plaque; the polygonal array of the connexons is visible; *Panel B* plasma membrane

We showed that the anti-26K antiserum did not react with the 21K protein band (Fig. 1, lane C) [cf. 8]. Furthermore, we recently obtained an anti-21K antiserum which did not show, after immunoblot, a reaction with the 26K protein (Fig. 1, lane D). Thus we conclude that the 21K protein cannot be a degradation product of the 26K protein. It must be an independent protein component in preparations of mouse liver gap junction plaques which may have no functional relationship to the 26K subunit polypeptide (see below). Figure 2 illustrates that the anti-liver 26K antiserum reacts with isolated liver gap junction plaques but does not significantly react with sheets of purified liver plasma membranes devoid of plaques. In order to detect the binding of the antiserum by electron microscopy, we used a method by which beads of colloidal gold were electrostatically bound to protein A which reacts with IgG molecules bound to the antigen [9]. When the proteins of total plasma membranes from mouse liver were separated by SDS-polyacrylamide gelelectrophoresis, only the 26K component reacted, after immunoblot, specifically with anti-26K antiserum (Fig. 1, lane G). Under the same experimental conditions no reaction of the pre-immune serum with the 26K protein was seen (Fig. 1, lane H). This result suggested two new experimental approaches:

a) One can quantitate the amount of liver gap junction protein under pathophysiological conditions, for example after partial hepatectomy.
b) One can investigate whether gap junction proteins from different cells or tissues react with the anti-liver 26K antiserum. This should allow one to determine to which extent antigenic similarities exist between gap junction proteins from different cells.

Here we report our results obtained in pursuing these two experimental approaches.

Loss and Reappearance of the Gap Junctions Subunit Polypeptide in Plasma Membranes of Regenerating Rat Liver

Meyer et al. [10] had discovered and analyzed, by morphometric analysis on freeze fracture replicas of regenerating rat liver, changes in the area of hepatocyte membranes occupied by gap junction plaques. At 29-35 h after partial hepatectomy of 3-week-old rats the membrane area of gap junction plaques was found to be reduced about 100-fold. At about 50 h after partial hepatectomy, the initial membrane area of gap junction plaques (about 3% of the hepatocyte membrane area) had been restored [10]. From data of the morphometric analysis, one cannot distinguish between disassembly of gap junction plaques into subunits (which cannot be detected on electron micrographs) and degradation of gap junction protein, followed by resynthesis. We wished to discriminate between these two alternatives by quantitative immunological determination of the gap junction subunit polypeptide. In order to compare the results of our immunological measurements to the data of the morphometric analysis by Meyer et al. [10] we also used plasma membranes from regenerating liver of 3-week-old rats. We found that the anti-mouse liver 26K antiserum reacted with the 26K gap junction protein from rat liver to about the same extent as that from mouse liver. When we analyzed the reaction of increasing amounts of the 26K protein (10-1000 ng) by SDS-polyacrylamide electrophoresis and immunoblot with anti-26K antiserum (densitometric evaluation of the autoradiographs) we obtained a saturation curve which was used in order to quantitate the amount of the 26K protein in plasma membranes of livers from 3-week-old rats after partial hepatectomy [11]. We found, at about

30 h after partial hepatectomy, that the amount of the 26K protein had decreased to about 13% of its level in normal rat liver. At 48 h after hepatectomy we found again about the same amount of 26K protein as in non-hepatectomized rats. For control purposes we also measured the incorporation (1 h pulse at each time point) of ^3H-thymidine into the DNA of the same regenerating rat livers used for immunoblots. We observed two peaks of ^3H-thymidine incorporation (21 h and 32 h after hepatectomy) and concluded that liver regeneration proceeded in our hepatectomized rats as expected [12]. The results of our quantitative measurements of the 26K protein demonstrate that the amount of the gap junction subunit polypeptide first decreased and later increased after hepatectomy. Thus we can exclude that gap junction plaques disassemble and reassemble from pre-existing subunits after hepatectomy. We found that the total amount of 26K protein decreased at a half-life of about 5-7 h before reaching the minimal amount at 30 h after hepatectomy. This half-life is surprisingly close to the half-life of 5 h recently reported by Fallon and Goodenough [13] for the gap junction polypeptide of normal mouse liver. On the other hand, the minimal concentration of the 26K protein which we measured at about 30-35 h after hepatectomy is at least ten times higher than expected from morphometric analysis of gap junction plaques under the same conditions. Probably free subunits of gap junctions or single connexons exist at the time of the minimum in the membranes of regenerating liver which cannot be detected by morphometric analysis. We conclude from our analysis that possibly the synthesis of the gap junctions subunit polypeptide (26K) in regenerating liver is first turned down and after 35 h turned on again. Future investigations are needed to clarify the molecular mechanism which regulates the amount of gap junction protein in regenerating liver.

In a parallel experiment we followed by immunoblot the concentration of the 21K protein originally suggested by Henderson et al. [5] to be a component of mouse liver gap junctions. For this analysis we used the anti-21K antiserum and aliquots of the same regenerating livers in which we had analyzed the concentration of the 26K protein. We found that the amount of the 21K protein did not significantly decrease after hepatectomy. This result confirmed our previous conclusion [8] that the 26K protein and the 21K protein found in preparations of mouse liver gap junction plaques are independent of each other.

Gap Junction Proteins from Liver and Lens Fiber Tissue are Different Proteins with Little or no Antigenic Similarities

The second experimental question we are studying, using the anti-liver 26K antiserum, is immunological cross reactivity with proteins from tissues other than liver. We decided to first investigate the reaction with gap junction protein from eye lens fiber tissue, since this protein is relatively well characterized due to its very high content (> 50%) in plasma membranes of lens fiber cells. The gap junction protein from lens fiber cells has been described unter two names, MIP (i.e., "main intrinsic polypeptide") [13] and MP26 [cf. 14]. Goodenough showed that this protein is the major component of gap junction plaques isolated from lens fiber tissue [16]. Dr. R.M. Broekhuyse (Nijmegen) kindly sent us rabbit anti MIP antiserum directed against MIP purified by chloroform/methanol extraction from bovine lens fiber tissue. We purified gap junction proteins from bovine and mouse lenses according to Goodenough [16] as well as according to Broekhuyse at al. [14]. All lens gap junction preparations were analyzed by SDS gel electrophoresis and subjected to the immunoblot procedure using the

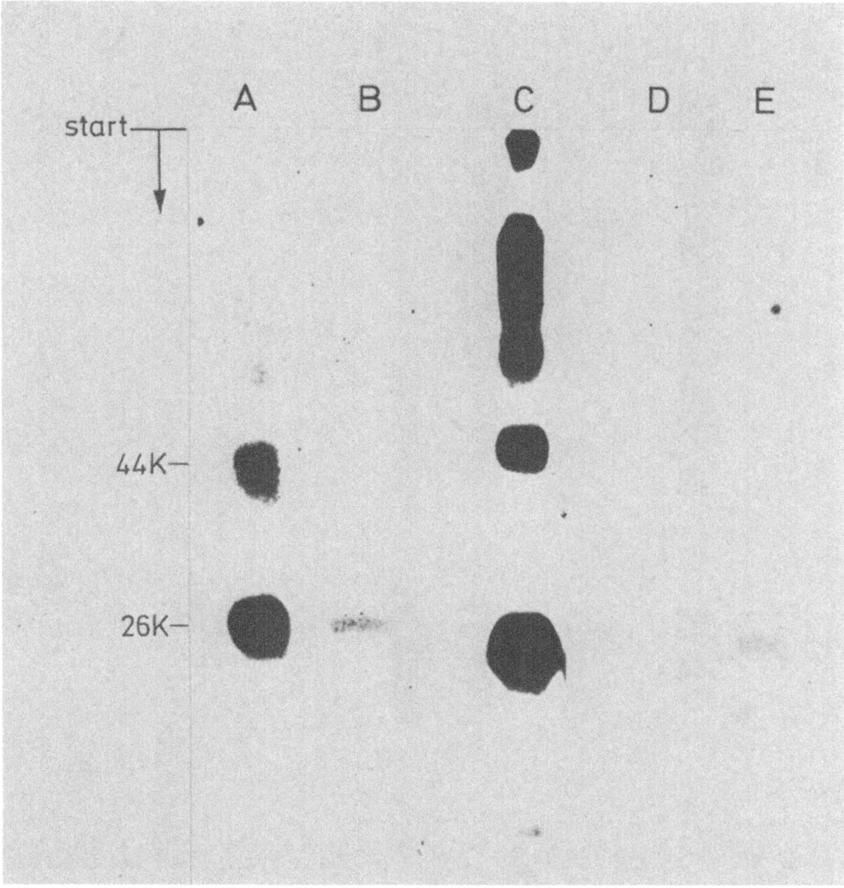

Fig. 3. Immunological reactivity of mouse liver 26K protein and bovine lens MIP protein with different rabbit antisera. SDS polyacrylamide electrophoreses and immunoblots were carried out as described in [8]. All lanes show autoradiographs obtained after incubation of the separated proteins on nitrocellulose paper, with different antisera (1:25 diluted) and ^{125}I-protein A. *Lane A* mouse liver plaque proteins (1 µg) after incubation with anti-26K antiserum; *lane B* bovine MIP (1 µg prepared according to [16]) after incubation with anti-26K antiserum; *lane C* bovine MIP (5 µg obtained from Dr. Broekhuyse [14] after incubation with anti-bovine MIP antiserum [17]; *lane D* mouse liver plaque protein (1 µg) after incubation with preimmune serum; *lane E* bovine MIP (4 µg prepared according to [16]) after incubation with preimmune serum

anti-bovine MIP antiserum obtained from Dr. Broekhuyse. We found that rabbit anti-bovine MIP antiserum reacted with mouse lens gap junction protein prepared according to Goodenough or according to Broekhuyse et al. This result identified the lens gap junction protein in our preparations and extended previous findings [17] that anti-bovine MIP antiserum reacted with lens gap junction proteins of other mammalian species. We found no reaction of mouse liver 26K protein with anti-bovine MIP antiserum. In addition, we analyzed by immunoblot the reaction of anti liver 26K antiserum with lens fiber gap junction protein. Using about the same amount of antigen, we found that the anti-26K antiserum reacted much more weakly with lens gap junction protein than

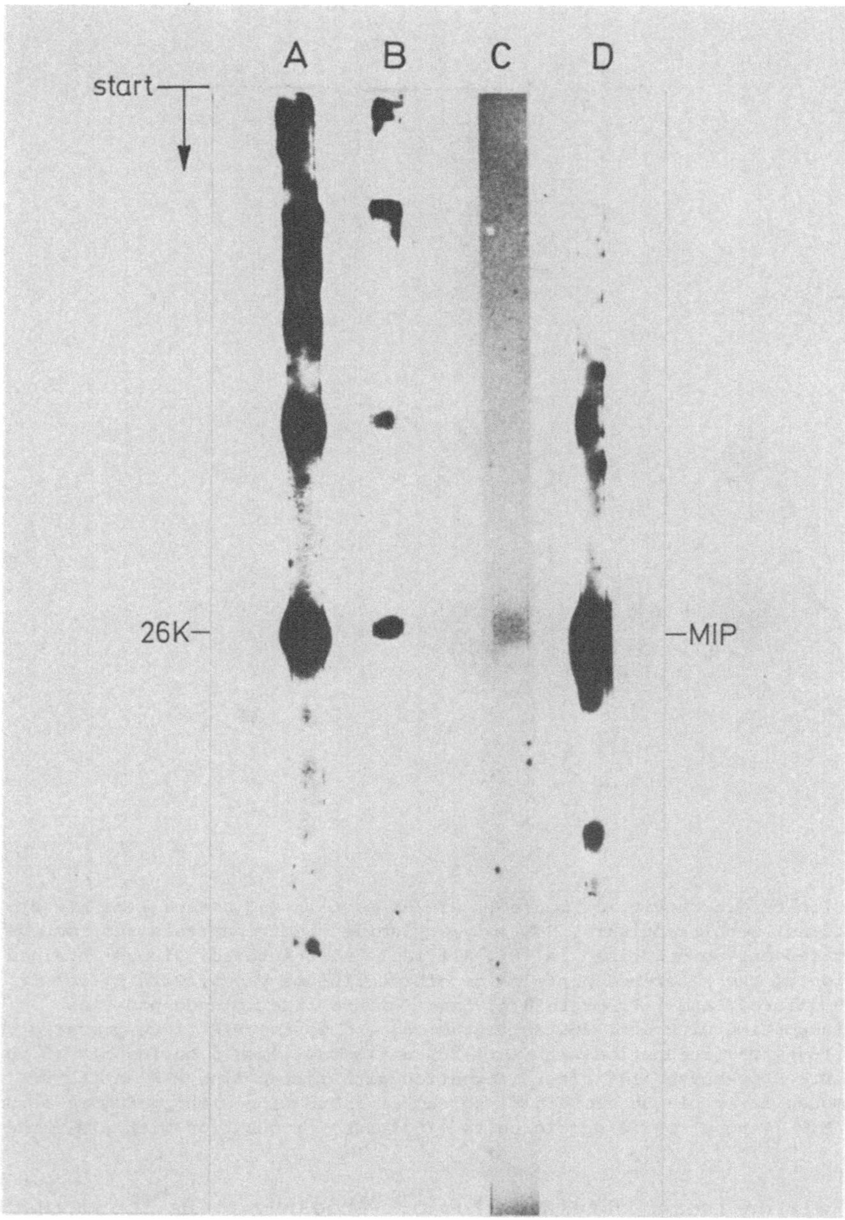

Fig. 4. Immunological reactivity of affinity purified rabbit anti-liver 26K antiserum with bovine MIP. Rabbit anti-liver 26K antiserum was affinity purified with 26K protein on diazobenzoxymethyl paper [8] according to Olmsted [18]. All lanes show autoradiographs obtained after incubation of the separated proteins, transferred onto nitrocellulose paper, with different antisera and ^{125}I-protein A. *Lane A* mouse liver plaque proteins (1 µg) after incubation with crude anti-26K antiserum; *lane B* mouse liver plaque proteins (1 µg) after incubation with affinity purified anti-26K antiserum; *lane C* bovine MIP (6 µg prepared according to [16]) after incubation with affinity purified anti-26K antiserum; *lane D* bovine MIP (6 µg prepared according to [16]) after incubation with crude anti-bovine MIP antiserum [17]

with liver gap junction protein (Fig. 3, lane B). A weak reaction with lens gap junction protein was also found, however, when antisera from different non-immunized rabbits were used for the immunoblot procedure (Fig. 3, lane E). It appears as if control sera from non-immunized rabbits contain a low titer of antibodies reacting weakly with lens gap junction protein in the immunoblot test. In order to decide whether minor antigenic similarities exist between liver and lens fiber gap junction proteins, the anti-26K antiserum was affinity purified according to Olmsted [18] by binding and eluting from liver 26K protein covalently linked to diazobenzoxymethyl paper [8]. Figure 4 illustrates the immunological reactivity of the affinity purified anti-26K antiserum. As expected it reacted with the monomeric 26K protein, with the "dimer" proteins [8] and with higher aggregate forms in a preparation of mouse liver gap junction plaques (lane B). Lane C of Fig. 4 shows the weak reaction of affinity purified anti-26K antiserum with bovine lens MIP. It remains to be checked whether this weak reaction is possibly due to low affinity binding of non-specific immunoglobulin molecules to lens gap junction protein. The large difference in reactivity of liver versus lens fiber gap junction proteins with the corresponding antisera suggests that both proteins are different. The same conclusion was reported very recently by Hertzberg et al. [19] who found different patterns of proteolytic degradation in vitro of gap junction proteins from liver and lens. Furthermore, these authors could not detect antigenic similarities between both proteins using an antiserum directed against the lens fiber gap junction polypeptide. Nicholson et al. [20] compared the amino acid sequence near the N-terminus of rat liver and lens gap junction protein and found that both sequences were different. Presumably only the comparison of the complete amino acid sequences will show whether the gap junction proteins from liver and lens are distantly related. At this time it is surprising to see that, despite of some structural similarities, gap junction proteins of these two tissues appear to be quite different. It will be necessary to purify gap junction proteins from tissue other than liver or lens in order to investigate how many different gap junction proteins are expressed in different mammalian tissues and whether different gap junction proteins are products of a multigene family.

Acknowledgements. We thank Dr. R.M. Broekhuyse (Nijmegen) for rabbit anti-bovine MIP antiserum and for a sample of purified bovine MIP. The technical assistance of Beate Karow is gratefully acknowledged. This work was supported by grants from the Ministerium für Wissenschaft und Forschung (Düsseldorf) and in part by a grant from the Deutsche Forschungsgemeinschaft (SFB 102) to K. Willecke.

References

1. Loewenstein WR (1979) Biochem Biophys Acta 560:1-65
2. Goodenough DA, Gilula NB (1979) In: Nichols JG (ed) The role of intracellular signals: navigation, encounter, outcome. Dahlem Konferenzen. Verlag Chemie, Weinheim, pp 135-154
3. Unwin PNT, Zampighi G (1980) Nature (London) 283:543-549
4. Zampighi G, Simon SA (1981) J Supramol Struct Cell Biochem 16:313
5. Henderson D, Eibl H, Weber K (1979) J Mol Biol 132:193-218
6. Finbow M, Yancey SB, Johnson R, Revel JP (1980) Proc Natl Acad Sci USA 77:970-974
7. Renart H, Reiser H, Stark GR (1979) Proc Natl Acad Sci USA 76:3116-3120
8. Traub O, Janßen-Timmen U, Drüge PM, Dermietzel R, Willecke K (1982) J Cell Biochem 19:27-44
9. Slot JW, Geuze HJ (1981) J Cell Biol 90:533-536
10. Meyer DJ, Yancey SB, Revel J-P, Peskoff A (1981) J Cell Biol 91:505-523

11. Higgins GM, Anderson RM (1931) Arch Pathol 12:186-202
12. Bucher NLR (1963) Int Rev Cytol 15:245-291
13. Fallon RF, Goodenough DA (1981) J Cell Biol 90:521-526
14. Broekhuyse RM, Kuhlmann EO, Stols ALH (1976) Exp Eye Res 23:365-371
15. Benedetti EL, Dunia I, Ramaekers FCS, Kibbelaar MA (1981) In: Bloemendal H (ed) Molecular and cellular biology of the eye lens. John Wiley, New York, pp 137-184
16. Goodenough DA (1979) Invest Ophthalmol Visual Sci 18:1104-1122
17. Bouman AA, DeLeeuw ALM, Broekhuyse RM (1981) Exp Eye Res 32:491-500
18. Olmsted JB (1981) J Biol Chem 256:11955-11957
19. Hertzberg EL, Anderson DJ, Friedlander M, Gilula NB (1982) J Cell Biol 92:53-59
20. Nicholson BJ, Hunkapiller MW, Hood LE, Revel JP, Takemoto L (1980) J Cell Biol 87:ME 1539

Transfer of Genes

Plant Cells Transformed by Modified Ti Plasmids: A Model System to Study Plant Development

J. Schell[1,2], M. Van Montagu[2,3], M. Holsters[2], J. P. Hernalsteens[3], P. Dhaese[2], H. De Greve[3], J. Leemans[3], H. Joos[2], D. Inze[2], L. Willmitzer[1], L. Otten[1], A. Wöstemeyer[1], G. Schröder[1], and J. Schröder[1]

Introduction

Ti plasmids are harbored by a group of soil bacteria (*Agrobacteria*) and are responsible for the capacity of these bacteria to induce so-called "crown gall" tumors on most dicotyledonous plants. (For recent reviews see Gordon 1979, Schell et al. 1979, Schilperoort et al. 1979, Van Montagu and Schell 1979, Schell and Van Montagu 1980, Schilperoort et al. 1980, Van Montagu et al. 1980). The plant tumor cells, unlike nontransformed plant tissues, can be cultured under axenic conditions on synthetic media in the absence of growth hormones, i.e. cytokinins and auxins (Braun 1956); the tumor cells also produce low molecular weight compounds called opines not found in untransformed plant tissues. The type of opine produced defines crown galls as octopine, nopaline, or agropine type tumors (Guyon et al. 1980). Ti plasmids have been found to be responsible for most of the typical properties of *Agrobacteria*: (1) crown gall tumor induction, (2) specificity of opine synthesis in transformed plant cells, (3) catabolism of specific opines, (4) agrocin sensitivity, (5) conjugative transfer of Ti plasmids, and (6) catabolism of arginine and ornithine (Van Larebeke et al. 1974, 1975, Zaenen et al. 1974, Engler et al. 1975, Schell 1975, Watson et al. 1975, Bomhoff et al. 1976, Genetello et al. 1977, Kerr et al. 1977, Firmin and Fenwick 1978, Klapwijk et al. 1978, Petit et al. 1978a,b, Ellis et al. 1979, Guyon et al. 1980). The crown gall tumors contain a DNA segment (called T-DNA) derived from Ti plasmids which is homologous and colinear with a defined fragment of the corresponding Ti plasmid present in the tumor-inducing bacterium (Chilton et al. 1977, Schell et al. 1979, Lemmers et al. 1980, Thomshow et al. 1980a, De Beuckeleer et al. 1981). The T-DNA is covalently linked to plant DNA (Thomashow et al. 1980b, Yadav et al. 1980, Zambyrski et al. 1980) in the nucleus of the plant cell (Chilton et al. 1980, Willmitzer et al. 1980).

The T-DNA is transcribed in the transformed plant cell and T-DNA encoded proteins have been found in several octopine crown gall lines, one of them being the octopine synthesizing enzyme lysopine dehydrogenase (LpDH) (Schröder et al. 1981a). Therefore Ti plasmids are a natural gene vector for plant cells, evolved by and for the benefit of the bacteria that harbor Ti plasmids (Schell et al. 1979). Crown gall cells, as a direct result of genetic transformation by the Ti plasmid, produce various opines. Free-living *Agrobacteria* utilize these opines as sources of carbon and nitrogen. The genetic information both

1 Max-Planck-Institut für Züchtungsforschung, D-5000 Köln 30, FRG

2 Laboratorium voor Genetika, Rijksuniversiteit Gent, Belgium

3 Laboratorium voor Genetische Virologie, VUB Brussel, Belgium

33. Colloquium - Mosbach 1982
Biochemistry of Differentiation and Morphogenesis

for the synthesis of opines in transformed plant cells and for their catabolism by free-living *Agrobacteria* is carried by Ti plasmids. Bacteria are known to be able to conquer an ecological niche by acquiring the capacity to catabolize certain organic compounds, not readily degradable by most other bacterial species. In several cases, the genes determining this degradative capacity have been found to be part of extrachromosomal plasmids. Several groups of soil bacteria, especially those living in and around the rhizosphere of plants, are able to decompose organic compounds released by plants. Clearly, with the advent of Ti plasmids, *Agrobacteria* have carried this capacity one step further, by genetically forcing plant cells - via a gene transfer mechanism - to produce specific compounds (opines) which they are uniquely equipped to catabolize. This novel type of parasitism has therefore been called "genetic colonization" (Schell et al. 1979).

In this paper we want to concentrate on the mechanism by which the plasmid derived T-DNA segment controls the developmental properties of the transformed plant cells with the result that these cells grow as either undifferentiated tumors or as semi-organized shoots (so-called teratomas), or in the case of the so-called hairy-root disease, as roots.

We shall present evidence indicating that the T-DNA contains a number of active genes, the products of which exert a negative control over plant cell organization and organ development.

The T-DNA Region of Ti Plasmids is Stably Integrated in the Chromosomal DNA of the Plant Host Genomes

The T-DNA region is defined as that segment from the Ti plasmids that is homologous to sequences present in crown gall cells. The sequences which are transferred from the Ti plasmid to the plant and determine tumorous growth have been called T-DNA. The T-region of octopine and nopaline Ti plasmids has been studied in great detail both physically and functionally. The T-regions, roughly 23 kb in size, are only a portion of the entire plasmids (Lemmers et al. 1980, Thomashow et al. 1980a, De Beuckeleer et al. 1981, Engler et al. 1981). Southern blotting and cross-hybridization of restriction endonuclease digests of the two types of plasmids as well as electron microscope heteroduplex analyses have revealed that 8-9 kb of the T-DNA regions are conserved and common to both octopine and nopaline types of plasmids (Chilton et al. 1978, Depicker et al. 1978, Engler et al. 1981). All attempts to reveal homology between the T-DNA region and plant DNA have failed.

Detailed analysis of some nopaline lines (Lemmers et al. 1980, Yadav et al. 1980, Zambryski et al. 1980, and submitted) suggests that the mechanism of T-DNA transfer is rather precise since the same continuous segment of the Ti plasmid is always present. Some lines appear to contain a single T-DNA copy, whereas in others the T-DNA occurs in multiple copies which are organized in a tandem array.

Several octopine tumor lines have been studied and the data suggest that the octopine T-DNA is more variable (Thomashow et al. 1980a,b; De Beuckeleer et al. 1981, Ooms et al., submitted). A left T-DNA region (TL) containing the conserved sequence (which is also part of the nopaline T-DNA region) is always present; this region is usually 12 kb in size but one *Petunia* tumor line is shortened at the right end of TL by about 4 kb (De Beuckeleer et al. 1981). In addition, there is often a right T-DNA region (TR) which contains sequences which are adjacent but not contiguous in the octopine Ti plasmid. In some tumor lines TR is amplified whereas TL is not (Merlo et al. 1980, Thomashow et al. 1980a). Recent observations indicate that TL can also be part of a tandem array (M. Holsters, unpublished results).

T-DNAs of nopaline and octopine tumor lines are not identical. However, there are some similarities: both T-DNAs contain a common or conserved DNA and both T-DNAs can be amplified either as a whole as in the nopaline T-DNA or in parts as in octopine T-DNA. It is not known whether the integration is the result of plant or Ti plasmid specific functions but it is likely that both are involved.

T-DNA Consists of DNA Sequences that are Expressed in Transformed Plant Cells

RNA transcripts homologous to T-DNA have been shown to be present in all crown gall tissues studied thus far (Drummond et al. 1977, Gurley et al. 1979, Gelvin et al. 1981, Willmitzer et al. 1981a).

In order to determine whether these transcripts determine well defined functions, three complementary approaches have been used.

1. The number, sizes and location of the transcribed T-DNA segments were studied in both octopine and nopaline tumors (Willmitzer et al. 1982, Willmitzer et al. in preparation). Tumor-specific RNAs were detected and mapped by hybridization of ^{32}P-labeled Ti plasmid fragments to polyadenylated RNA which had been separated on agarose gels and then transferred to DBM paper. The results show that octopine tumors contain a total of ten distinct transcripts which differ in their relative abundance and in their sizes. They all bind to oligo (dT)-cellulose, indicating that they are polyadenylated. Thus, the T-DNA transferred from a prokaryotic organism provides specific poly(A)-addition sites. The direction of transcription was determined, and the location of the approximate 5'- and 3'-ends were mapped on the TL- and TR-DNA.

All RNAs mapped within the T-DNA sequence. This, and the observation that transcription is inhibited by low concentrations of α-amanitin (Willmitzer et al. 1981b) suggests that each transcript is determined by a specific promoter site on the T-DNA recognized by plant RNA polymerase II. Considering the groupwise orientation of several transcripts, the simplest model would assume one promoter site per group of transcripts. If so, one would expect that inactivation of a 5'-proximal gene of a group would also lead to disappearance of the transcripts from the 5'-distal genes. However, analysis of some cell lines containing the T-DNA of Ti plasmid mutants indicated that groupwise inactivation of genes does not occur (Leemans et al. 1982). The results available so far are consistent with the assumption that each gene on the T-DNA has its own signals for transcription in the eukaryotic plant cells.

2. In order to determine whether these mRNAs are translated into proteins a hybridization selection procedure was developped that was sufficiently sensitive and specific to detect mRNAs which represent about 0.0001% of the total mRNA activity in the plant cells. This procedure was used to enrich for T-DNA-derived mRNAs by hybridization to Ti plasmid fragments covalently bound to microcrystalline cellulose; the hybridized RNAs were eluted and translated in vitro in a cell-free system prepared from wheat germ.

The results obtained with this approach (Schröder and Schröder 1982) showed that tumor cells contain at least three T-DNA-derived mRNAs which are translated in vitro into distinct proteins, and the coding regions correlate with those of three of the previously identified transcripts. The protein encoded at the right end of the TL-DNA (Mr 39,000) was of particular interest since previous genetic analysis indicated that this region is responsible for octopine synthesis (Koekman et al. 1979, De Greve et al. 1981, Garfinkel et al. 1981).

The in vitro synthesized protein was shown to be identical in size to the octopine-synthesizing enzyme in octopine tumors. Immunological studies showed that this protein was recognized by antiserum against the tumor specific synthase (Schröder et al. 1981a). These results demonstrate that the structural gene for the octopine-synthesizing enzyme is on the Ti plasmid. So far, this is the only protein product of the T-DNA with known enzymatic properties; the possible functions of two smaller T-DNA-derived proteins are not known. The region coding for the octopine-synthesizing enzyme has recently been sequenced (De Greve et al., in preparation). The 5' end of the octopine synthase mRNA was accurately mapped by sequencing of a T-region DNA fragment hybridized to this mRNA and thus protected from degradation by the single strand specific S1 nuclease (Dhaese et al., unpublished results). The promoter sequence that has thus been identified is more eukaryotic than prokaryotic in its recognition signals and no introns interrupt the open reading frame which starts at the first AUG codon following the 5' start of the transcript. Similar work leading to essentially the same conclusions has also been performed for the nopaline synthase gene (M. Bevan and M.-D. Chilton, L. Willmitzer, H. Goodman, P. Zambryski, A. Depicker, S. Stachel, J. Gielen, P. Dhaese, personal communications). It is important to note that the promoter sequences for the octopine synthase appear to escape possible control mechanisms since they remain active in all tissues of plants regenerated from tobacco cells transformed with mutant Ti plasmids (De Greve et al., in preparation). The 3' polyadenylated terminus of the transcript was also analysed and a polyadenylation signal 5'-AAUAA-3' was found about 10 nucleotides from the start of the poly(A) sequence. This appears to be a general feature of eukaryotic mRNAs since it has been observed in a number of animal mRNAs (Fitzgerald and Schenk 1981). It would therefore appear that these opine synthesizing genes were designed to function in eukaryotic cells rather than in prokaryotic cells.

However, this is not necessarily true for all genes in the T-region, since transcripts were also detected in *Agrobacteria* (Gelvin at al. 1981). It was therefore of interest to determine whether all the T-DNA-derived mRNAs isolated from plant cells shared properties with typical eukaryotic mRNAs. The fact that translation of all three mRNAs was inhibited by the cap analogue pm^{7}G suggests that they contain a cap structure at the 5'-end. This would be typical for eukaryotic mRNA since caps have not been described in prokaryotic RNA.

As noted above the mRNAs for the three in vitro synthesized proteins each represent about 0.0001% of the total mRNA activity in polyribosomal RNA, and this appears to be the detection limit at present for translatable RNA. Some of the other transcripts detected by hybridization experiments are present at even lower concentrations. Assuming that they possess mRNA activity, this is likely to be the reason, why the corresponding proteins could not so far be identified by in vitro translation.

A different approach was therefore developed to search for coding regions on the T-DNA and their protein products. Fragments from the

T-region were cloned into E.coli plasmids and analyzed for gene expression in E.coli minicells (Schröder et al. 1981b). There are at least four different coding regions within the TL-DNA which can be expressed in minicells into distinct proteins from promoters which are active in the prokaryotic cells. The four regions expressed in E.coli correlated with four regions transcribed into RNA in plant cells. The plant transcripts are larger than the proteins in E.coli, and the regions expressed in minicells appear to lie within the regions transcribed in plant cells. One can therefore speculate that plant cells and E.coli at least partly express the same coding regions.

3. Specific mutations were introduced in the T-DNA regions of octopine and nopaline Ti plasmids to produce transformed plant cells in which one or more T-DNA derived transcripts would not be expressed. By observing the phenotypes of the plant cells harbouring such partially inactivated T-DNA's, it was possible to assign functions to most of the different transcripts (Leemans et al. 1982, Joos et al. in preparation). Thus, it was found that none of the T-DNA transcripts was essential for the transfer and stable maintenance of T-DNA segments in the plant genome.

Essentially two different functions were found to be determined by T-DNA transcripts:

a) Transcripts coding for opine synthases: octopine tumors contain either one or two of such genes. One of them is located on the right end of TL and codes for octopine synthase, the other is located at the right end of TR and codes for agropine and mannopine synthase. Tumors that contain both TL and TR-DNA therefore produce both octopine and agropine, whereas tumors only containing TL-DNA produce octopine but not agropine. Nopaline tumors also contain at least two transcripts coding for different opines. One is located at the right end of the T-DNA and codes for nopaline synthase, whereas the other one is located in the left part of the T-DNA and codes for agrocinopine (Ellis and Murphy 1981). Nopaline tumors therefore usually produce both types of opines.

b) Transcripts (after translation in proteins) directly or indirectly responsible for tumorous growth. These transcripts were found to be derived from the T-DNA sequences that are common to both octopine and nopaline tumors (Fig. 1). In total five different well defined transcripts were found to be derived from this common region of octopine and nopaline T-DNAs. Probably all of these transcripts act by suppressing plant organ development. It was observed that shoot and root formation are suppressed independently and by different transcripts. Thus two transcripts (1 and 2) were identified that specifically prevent shoot formation. The effect of these T-DNA gene products is in many ways analogous to that of auxin-like plant growth hormones. Another transcript (4) was found to prevent root formation specifically and the effect of this T-DNA gene can therefore be compared to the effects observed when normal plant cells are grown in the presence of high concentrations of cytokinins.

That both the shoot inhibition and the root inhibition that results from the activity of these genes may be due to the fact that they, directly or indirectly, determine the formation of auxin- and cytokinin-like growth hormones, is further substantiated by our observation that these genes, respectively, inhibit shoot or root formation both in T-DNA containing and in T-DNA negative (normal) cells provided both types of cells grow as one mixed tissue.

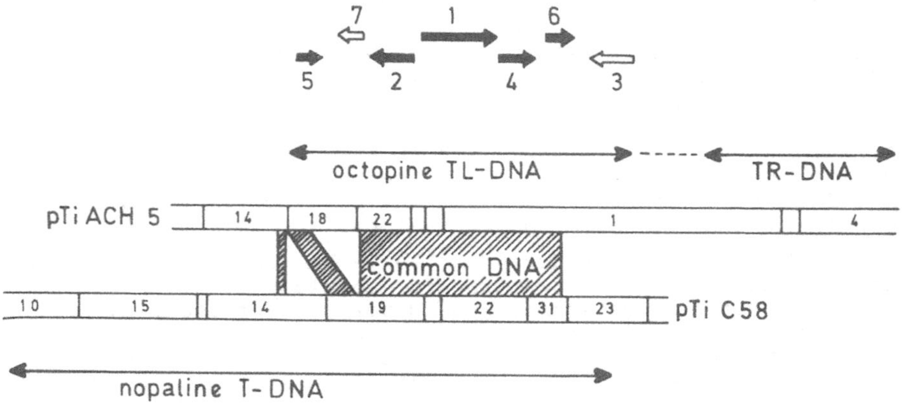

Fig. 1. The T-regions of octopine and nopaline Ti-plasmids have been aligned; *shaded areas* indicate the DNA sequences that are common to both T-regions (Engler et al. 1981). *Numbers* refer to fragments produced following digestion with *Hind* III. *Arrows* above the figure represent different transcripts of T-DNA in octopine crown gall tumors. *Black arrows* represent transcripts derived from common DNA sequences

In addition to this hormone-like activity, the T-DNA codes for at least two other transcripts (5 and 6) that suppress development only in the cells in which they are formed. The combination of one of these transcripts (6) with the cytokinin-like activity of transcript (4) was shown to be sufficient to suppress development of transformed cells and to allow their hormone-independent growth. Another transcript (5) was found to inhibit the organization of transformed cells into teratoma like structures. Elimination of this transcript along with the shoot inhibiting auxin-like genes (1 and 2), resulted in transformed cells organizing themselves as teratomas. Spontaneous deletions were looked for and found in which all five of the tumor controlling genes were simultaneously eliminated (De Greve et al., in preparation). The T-DNA in these plant cells was found to contain and express only the opine synthase genes of the T-DNA. Such tobacco cells were found to regenerate into normal plants. In sexual crosses the T-DNA in such plants behaved as a single dominant Mendelian factor (Otten et al. 1981). Figure 2 summarizes our results and indicates how the different gene products may function.

General Conclusion

Ti plasmids of *Agrobacteria* have provided us with an unexpected and extraordinarily potent system to study the genetic control of plant organ development. Indeed we found that these plasmids contain DNA sequences which are transferred and integrated into chromosomes of plant cells. These so-called T-DNA sequences contain genes that are expressed in plant cells. Some of these genes make products that specifically suppress either shoot or root formation.

These observations indicate that plants have separate genetic programs for shoot development and for root development and that both of these programs must be internally coordinated since they can be controlled by the products of single genes.

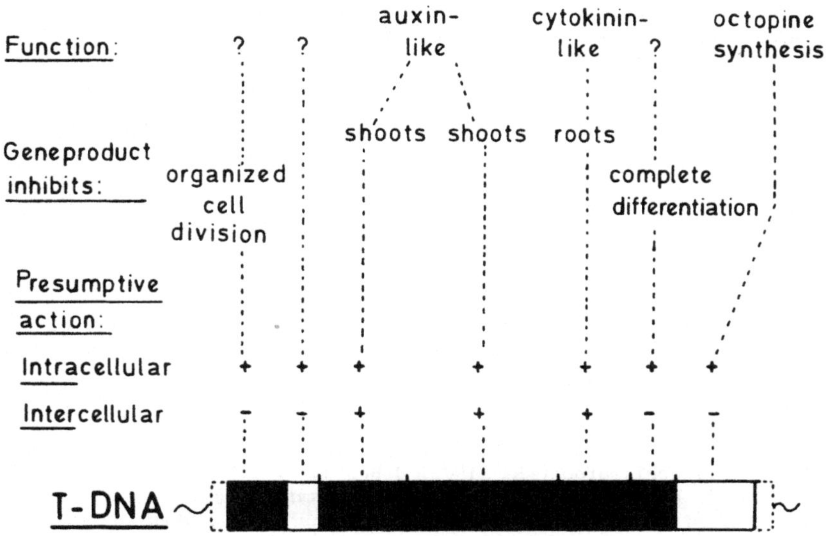

Fig. 2. Function of T-DNA transcripts common to octopine and nopaline tumors

Furthermore the T-DNA remarkably has two independent but complementary mechanisms to suppress plant organ development. Each of these mechanisms alone may be sufficient to suppress the development of transformed cells. One of these mechanisms appears to function via a growth hormone-like mechanism and therefore suppresses both transformed and untransformed cells thus explaining why uncloned primary tumors, that consist of a mixture of T-DNA containing and normal untransformed cells, nevertheless grow as a uniformly undifferentiated callus. This mechanism might well play a role in the production of auxin- and cytokinin-like hormones. These observations are therefore consistent with the observations of Amasimo and Miller (1982) who studied the endogenous levels of auxin and cytokinin in teratoma and unorganized tobacco and found them to be similar to those needed to artificially induce these growth patterns in normal cells. The other mechanism is even more intriguing since it appears to suppress transformed plant-cells only.

References

Amasino RM, Miller CO (1982) Plant Physiol 69:389

Bomhoff G, Klapwijk PM, Kester HCM, Schilperoort RA, Hernalsteens JP, Schell J (1976) Mol Gen Genet 145:177

Braun AC (1956) Cancer Res 16:53

Chilton M-D, Drummond HJ, Merlo DJ, Sciaky D, Montoya AL, Gordon MP, Nester EW (1977) Cell 11:263

Chilton M-D, Drummond HJ, Merlo DJ, Sciaky D (1978) Nature (London) 275:147

Chilton M-D, Saiki RK, Yadav N, Gordon MP, Quetier F (1980) Proc Natl Acad Sci USA 77:4060

De Beuckeleer M, Lemmers M, De Vos G, Willmitzer L, Van Montagu M, Schell J (1981) Mol Gen Genet 183:283

De Greve H, Decraemer H, Seurinck J, Van Montagu M, Schell J (1981) Plasmid 6:235

Depicker A, Van Montagu M, Schell J (1978) Nature (London) 275:150

Drummond MH, Gordon MP, Nester EW, Chilton M-D (1977) Nature (London) 269:535
Ellis J, Kerr A, Tempé J, Petit A (1979) Mol Gen Genet 173:263
Ellis JG, Murphy PJ (1981) Mol Gen Genet 181:36
Engler G, Holsters M, Van Montagu M, Schell J, Hernalsteens JP, Schilperoort RA (1975) Mol Gen Genet 138:345
Engler G, Depicker A, Maenhaut R, Villarroel-Mandiola R, Van Montagu M, Schell J (1981) J Mol Biol 152:183
Firmin JL, Fenwick GR (1978) Nature (London) 276:842
Fitzgerald M, Schenk T (1981) Cell 24:251
Garfinkel DJ, Simpson RB, Ream LW, White FF, Gordon MP, Nester EW (1981) Cell 27:143
Gelvin SB, Gordon MP, Nester EW, Aronson AI (1981) Plasmid 6:17
Genetello Ch, Van Larebeke N, Holsters M, Depicker A, Van Montagu M, Schell J (1977) Nature (London) 265:561
Gordon MP (1979) In: Marcus A (ed) Proteins and nucleic acids. The biochemistry of plants, vol VI. Academic Press, London New York, p 531
Gurley WB, Kemp JD, Albert MJ, Sutton DW, Callis J (1979) Proc Natl Acad Sci USA 76:2828
Guyon P, Chilton M-D, Petit A, Tempé J (1980) Proc Natl Acad Sci USA 77:2693
Kerr A, Manigault P, Tempé J (1977) Nature (London) 265:560
Klapwijk PM, Scheuldermon T, Schilperoort RA (1978) J Bacteriol 136:775
Koekman BT, Ooms G, Klapwijk PM, Schilperoort RA (1979) Plasmid 2:347
Leemans J, Deblaere R, Willmitzer L, De Greve H, Hernalsteens JP, Van Montagu M, Schell J (1982) EMBO Journal 1:147
Lemmers M, De Beuckeleer M, Holsters M, Zambryski P, Depicker A, Hernalsteens JP, Van Montagu M, Schell J (1980) J Mol Biol 144:355
McPherson JC, Nester EW, Gordon MP (1980) Proc Natl Acad Sci USA 77:2666
Merlo DJ, Nutter RC, Montoya AL, Garfinkel DJ, Drummond MH, Chilton M-D, Gordon MP, Nester EW (1980) Mol Gen Genet 177:637
Otten L, De Greve H, Hernalsteens JP, Van Montagu M, Schieder O, Straub J, Schell J (1981) Mol Gen Genet 183:209
Petit A, Dessaux Y, Tempé J (1978a) In: Proceedings IVth international conference plant pathogenic bacteria (station de phytologie végétale et phytobactériologie) Angers, INRA, p 143
Petit A, Tempé J, Kerr A, Holsters M, Van Montagu M, Schell J (1978b) Nature (London) 271:570
Schell J (1975) In: Ledoux L (ed) Genetic manipulations with plant materials. Plenum Press, New York,p 163
Schell J, Van Montagu M (1980) In: Leaver CJ (ed) Genome organization and expression in plants. Plenum Press, New York, p 453
Schell J, Van Montagu M, De Beuckeleer M, De Block M, Depicker A, De Wilde M, Engler G, Genetello C, Hernalsteens JP, Holsters M, Seurinck J, Silva B, Van Vliet F, Villarroel R (1979) Proc R Soc London Ser B 204:251
Schilperoort RA, Hooykaas PJJ, Klapwijk PM, Koekman BP, Nuti MP, Ooms G, Prakash RK (1979) In: Timmis K, Pühler A (eds) Plasmids of medical, environmental and commercial importance. Elsevier, Amsterdam, p 339
Schilperoort RA, Klapwijk PM, Ooms G, Wullems GJ (1980) In: Cleton FJ, Simons JW (eds) Genetic origins of tumor cells . Martinus Nijhoff, The Hague p 87
Schröder J, Schröder G (1982) Mol Gen Genet 185:51
Schröder J, Schröder G, Huisman H, Schilperoort RA, Schell J (1981a) FEBS Lett 129:166
Schröder J, Hillebrand A, Klipp W, Pühler A (1981b) Nucleic Acids Res 9:5187
Thomashow MF, Nutter R, Montoya AL, Gordon MP, Nester EW (1980a) Cell 19:729
Thomashow MF, Nutter R, Postle K, Chilton M-D, Blattner FR, Powell A, Gordon MP, Nester EW (1980b) Proc Natl Acad Sci USA 77:6448
Van Larebeke N, Engler G, Holsters M, Van den Elsacker S, Zaenen I, Schilperoort RA, Schell J (1974) Nature (London) 252:169
Van Larebeke N, Genetello C, Schell J, Schilperoort RA, Hermans AK, Hernalsteens JP, Van Montagu M (1975) Nature (London) 255:742
Van Montagu M, Schell J (1979) In: Timmis K, Pühler A (eds) Plasmids of medical, environmental and commercial importance. Elsevier, Amsterdam, p 71

Van Montagu M, Holsters M, Zambryski P, Hernalsteens JP, Depicker A, De Beuckeleer M, Engler G, Lemmers M, Willmitzer L, Schell J (1980) Proc R Soc London Ser B 210:351
Watson B, Currier TC, Gordon MP, Chilton M-D, Nester EW (1975) J Bacteriol 123:255
Willmitzer L, De Beuckeleer M, Lemmers M, Van Montagu M, Schell J (1980) Nature (London) 187:359
Willmitzer L, Otten L, Simons G, Schmalenbach W, Schröder J, Schröder G, Van Montagu M, De Vos G, Schell J (1981a) Mol Gen Genet 182:255
Willmitzer L, Schmalenbach W, Schell J (1981b) Nucleic Acids Res 9:4801
Willmitzer L, Simons G, Schell J (1982) EMBO Journal 1:139
Yadav NS, Postle K, Saiki RK, Thomshow MF, Chilton M-D (1980) Nature (London) 287:458
Zaenen I, Van Larebeke N, Teuchy H, Van Montagu M, Schell J (1974) J Mol Biol 86:109
Zambryski P, Holsters M, Kruger K, Depicker A, Schell J, Van Montagu M, Goodman HM (1980) Science 209:1385

The Dualism of Expression of onc-Genes: Transformation versus Differentiation

H. Bauer, A. Barnekow, C. B. Boschek, R. R. Friis, and A. Ziemiecki[1]

Introduction

Tumor cells differ from their normal counterparts in two important functional properties: they are capable of indefinite proliferation and they display a more or less drastically altered metabolism. The fundamental molecular mechanisms responsible for these dramatic phenotypic changes are not yet understood, but can be directly investigated by the use of tumor viruses. Many tumor viruses are known which contain a so-called *onc* or tumor gene that is solely responsible for tumor formation in vivo and cell transformation in vitro. Some of these genes have been named according to the histopathological type of tumor they induce, for instance *myb* for a gene which induces myeloblastosis and *src* for one which results in a sarcoma (Coffin et al. 1981). Of special interest are the RNA tumor viruses, because homologous counterparts of the *onc* gene of these viruses are found in the genome of normal cells (Bishop 1981). These cellular *onc* genes (c-*onc*) are not only detected in the natural host of the virus but also in other animal species, having been highly conserved during evolution.

A breakthrough in this field was the first identification of the *onc*-gene product of an RNA tumor virus. Erikson's group detected the protein, encoded for by the *src*-gene of Rous sarcoma virus (Brugge et al. 1978, which turned out to be associated with an enzyme activity (Collett and Erikson 1978) that appears to be of major importance in cell transformation.

The role of this protein in cell transformation will be discussed and evidence will be presented for the possible function of its cellular counterpart under physiological conditions.

The Transforming Protein of Rous Sarcoma Virus

The genome of Rous sarcoma virus, a single-stranded RNA molecule, contains four identified genes (Bishop 1978) (Fig. 1). Two of these (*gag* and *env*) code for structural proteins of the virion, the third (*pol*) codes for the reverse transcriptase that after infection transcribes the viral RNA into the DNA provirus, and the fourth gene (*src*), which is not required for virus replication, alone mediates cell transformation. This *src*-gene codes for a phosphoprotein of about 60,000 *d*

1 Institut für Virologie, Fachbereich Humanmedizin, Justus-Liebig-Universität Giessen, Frankfurter Str. 107, D-6300 Giessen, FRG

33. Colloquium - Mosbach 1982
Biochemistry of Differentiation and Morphogenesis

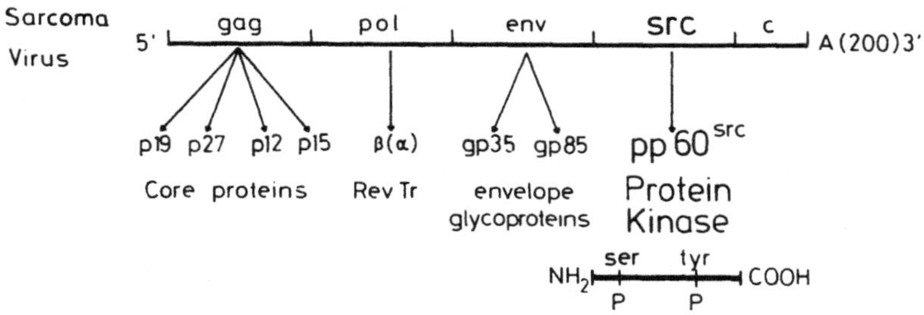

Fig. 1. Genome and gene products of avian sarcoma virus

called pp60^{src} which was first identified by Erikson and coworkers by immunoprecipitation with antisera from Rous sarcoma tumor-bearing rabbits (TBR-serum) (Brugge et al. 1978). Pp60^{src}contains a phosphoserine in the amino terminal and a phosphotyrosine in the carboxy terminal part, and was the first protein detected to be associated with a tyrosine-specific phosphokinase activity (Brugge et al. 1978, Collett and Erikson 1978, Hunter and Sefton 1980). It is further distinguished from most of the other known phosphoprotein kinases by its cAMP-independent activity. There is ample evidence that the kinase activity is an intrinsic property of pp60^{src}, and it appears that the phosphotyrosine on this molecule results from autophosphorylation (see Purchio 1982). Phosphorylation of tyrosine is a novel event and is responsible for less than 0.1% of the phosphorylated amino acid residues in normal chick embryo cells (Sefton et al. 1980). Upon RSV-transformation, however, this value increases up to tenfold, evidence that tyrosine phosphorylation may have an important function in the transformation mechanism. Indeed, several proteins have been identified which, upon transformation are phosphorylated in tyrosine, as will be discussed below. In addition, the transforming proteins of three further, unrelated RNA-tumor viruses have also been found to be associated with tyrosine-specific phosphokinase activities (Patschinsky and Sefton 1981).

Furthermore, a novel phosphotyrosine phosphatase activity has been detected having properties significantly different from those of previously described phosphatases in that its activity is Zn^{2+} sensitive but fluoride resistant (Table 1) (Brautigan et al. 1981, Foulkes et al. 1981). This phosphotyrosine phosphatase activity may function as an antagonist in the regulation of tyrosine phosphorylation in RSV-transformed cells.

A convenient technique to assay the activity of pp60^{src} kinase is based on its property to phosphorylate the heavy chain of the immunoglobulin in an immunoprecipitate of pp60^{src}. This can be demonstrated by the addition of gamma-(^{32}P)ATP to the immune precipitate, followed by analysis on a polyacrylamide gel (Collett and Erikson 1978) (see Fig. 3). It should be mentioned in this context that the in vitro phosphorylation of any protein by the pp60^{src} kinase is not sufficient evidence for a natural target protein for this kinase. A variety of proteins such as casein or actin are phosphorylated in vitro by pp60^{src} kinase but not in vivo, i.e. intracellularly.

One of the important advantages of the Rous sarcoma virus system is the existence of temperature-sensitive transformation-defective (td-ts) mutant viruses, which have a temperature sensitive defect in the *src*-gene, i.e. these viruses transform cells at the low temperature of

Table 1. Characteristics of a novel phosphotyrosine activity

Additions	Phosphorylase phosphatase activity	Phosphotyrosine phosphatase activity
None	100	100
1 mM $MnCl_2$	90	78
1 mM Mg acetate	104	85
100 μM $CaCl_2$	103	97
100 μM Zn acetate	107	15
50 mM NaF + 1 mM EDTA	0.2	95
1 mM EDTA	93	92

Rat muscle extracts were incubated with either ^{32}P-labelled IgG heavy chain (phosphotyrosine phosphatase activity) or ^{32}P-labelled phosphorylase a (phosphorylase phosphatase activity) for 20 min at 30°C. Enzyme activity in the presence of the various additions is given as a percentage of the activity in the absence of any addition. (For full details see Foulkes et al. 1981)

35°C, but not at elevated temperature of 42°C (Fig. 2). By temperature shift experiments from the permissive to the non-permissive temperature, or vice versa, not only the transformation specificity of any given phenotypic alteration can be studied, but also the rate at which the various transformation-associated changes occur.

This allows the study of the significance of the kinase activity for the transformed phenotype. Thus, it has been shown that at high temperature, i.e. when the cells are phenotypically normal, the kinase activity is much lower than in phenotypically transformed cells, and that within minutes after downshift to the permissive temperature of 35°C, the kinase activity reappears (Fig. 3) (Friis et al. 1980). In another experiment in which new protein synthesis was inhibited by the addition of cycloheximide before downshift, the same result was obtained, indicating that the temperature sensitive defect of $pp60^{src}$ and of the kinase activity, is reversible and independent of de novo protein synthesis (Ziemiecki and Friis 1980).

In recent experiments, in which the quantity of $pp60^{src}$ was estimated by pulse-chase labelling, the half-life of $pp60^{src}$ was determined to be less than 60 min (Ziemiecki et al. 1982) (Fig. 4).

The Phenotype of RSV-transformed Fibroblasts

The most obvious and easily detectable characteristics of RSV-transformed cells are their altered morphology and proliferation capacity. The morphological changes appear to partly result from the disarrangement of the cytoskeleton. Noteworthy are also the reduced substrate dependency and the independance from growth factors for proliferation

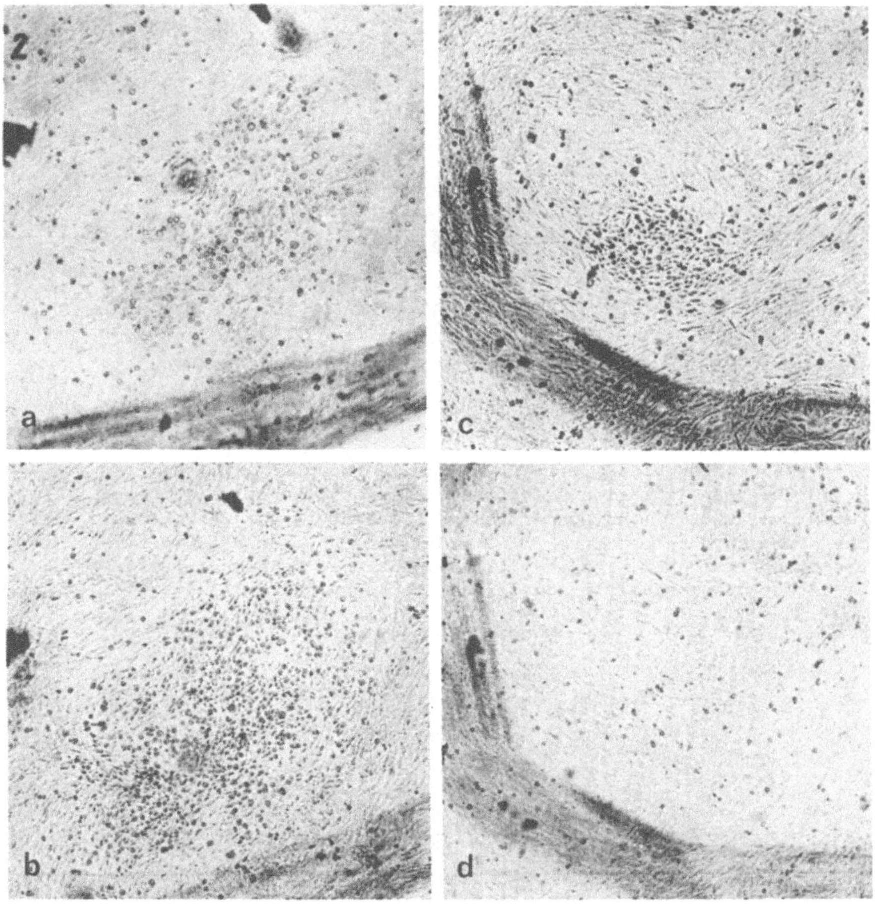

Fig. 2a-d. Sister cultures of chick embryo cells infected with either wild type RSV (a, b) and td-ts RSV mutant (c, d). An area of the tissue culture has been photographed at permissive temperature (a, c) and 24 h after shift to the non-permissive temperature (b, d). In the latter case, the mutant virus-infected cells revert to the normal phenotype while the wild-type infected cells retain the transformed morphology

in vitro. Many of the biochemical and metabolic properties of transformed cells, as summarized schematically in Table 1, are also characteristic for normally dividing cells, and it is uncertain which of these is the consequence of a primary effect of $pp60^{src}$ during transformation. The multitude of alterations in the transformed cell (Table 2), on the other hand, suggests that $pp60^{src}$ is multifunctional or has a pleiotropic activity.

In fact, the detection of RSV-mutants that are temperature-sensitive only for certain of the various phenotypic properties of a transformed cell and which therefore induce different phenotypes specific for a given viral mutant, has led us to postulate a pleiotropic and/or multifunctional effect of $pp60^{src}$ (Calothy and Pessak 1976, Becker et al. 1977, Friis et al. 1977, Weber and Friis 1979). Further evidence

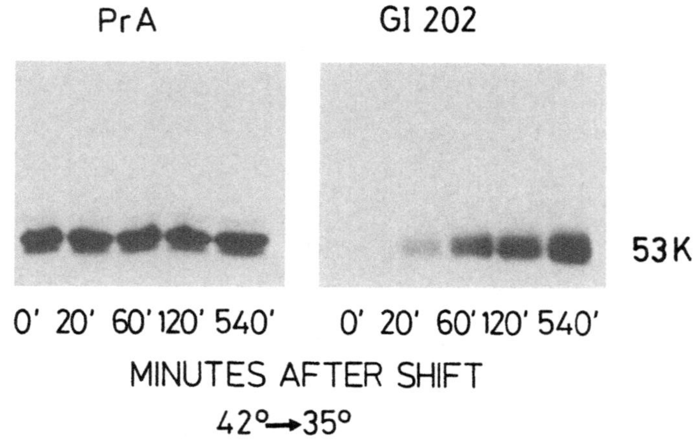

Fig. 3. Pp60^{v-src}-associated kinase activity in cells infected with the Prague A strain of RSV and the transformation-defective, temperature-sensitive mutant GI 202. Sister cultures grown at 42°C (the nonpermissive temperature for GI 202) were shifted to 35°C (the permissive temperature for GI 202) and the kinase activity was determined at the times indicated. The in vitro demonstrable kinase activity is reflected by the extent of phosphorylation of the IgG heavy chain (53K)

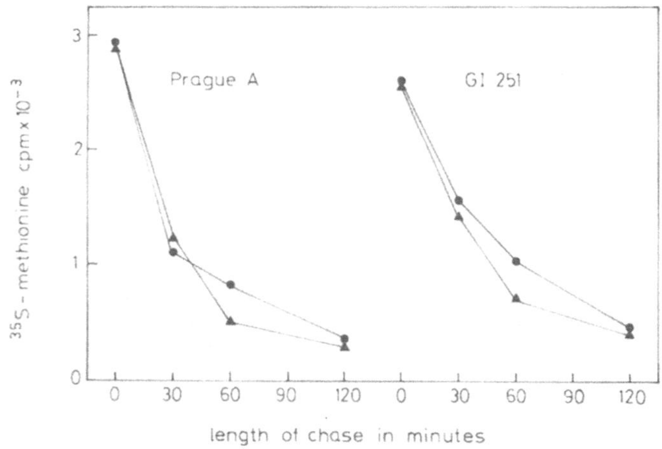

Fig. 4. The half-life of ^{35}S-methionine labeled pp60^{v-src} of the Prague A strain of RSV and of the transformation-defective, temperature-sensitive mutant GI 251 at the permissive (35°C) and nonpermissive (42°C) temperatures. Sister cultures were pulselabeled and chased for the times indicated. Cells were lysed in RIPA-buffer, the lysates pre-absorbed with antiserum against viral structural proteins and the pp60^{v-src} immunoprecipitated with TBR-serum. Pp60^{v-src} was localized after resolution on a SDS-polyacrylamide gel, excited and quantitated by scintillation counting. *Closed triangles:* 35°C, *closed circles:* 42°C

Table 2. Structural, biological and biochemical alterations in RSV-transformed cells

Morphology of the cells
1. Rounded
2. Development of many short surface projections
3. Reduced number of adhesion plaques
4. Disorganization of actin-containing microfilament bundles
5. Changes in the interaction between microtubules and intermediate filaments
Biological cell behaviour
1. Contact insensitive cell movement
2. Growth to high saturation density
3. Anchorage independent proliferation, i.e. formation of colonies in suspension
4. Decreased requirement for growth factors for proliferation
Biochemical and metabolic properties
1. Increased uptake of metabolites
2. Increased lactate production
3. Altered cation metabolism
4. Increase in cell surface proteases
5. Decrease in fibronectin
6. Decreased collagen synthesis
7. Altered glyosylation of proteins and lipids
8. New antigenic cell surface structures
9. Increased level of tyrosine phosphorylation in many cell proteins

for this has been provided since that time by several reports on different candidate target molecules for $pp60^{src}$ as will be discussed below. Whether a pleiotropism of $pp60^{src}$ is only due to the interaction of the tyrosine kinase, with more than one protein involved in different metabolic processes, or whether $pp60^{src}$ may exert further functions, either enzymatically or by steric interaction, is not yet known.

Possible Targets for pp60^{src}

The Cytoskeleton

It has previously been reported that several hours after transformation the actin containing microfilament network becomes severely disorganized (see Boschek 1982). More recently we have shown, by using td-ts mutants in temperature shift experiments, that the distribution of microfilament proteins is altered very early after the initiation of transformation. By downshift of RSV-infected cells from the restrictive to the permissive temperature it is possible to rapidly initiate transformation thus allowing the study of possible early alterations.

In such experiments, within 15 min after downshift, i.e. after the initiation of transformation, large whorllike structures could be observed at the cell surface (Fig. 5). These structures, which are called dorsal-surface ruffles or flowers, were shown to contain a variety of proteins which are otherwise constituents of the cytoskeleton, namely actin, alpha-actinin, myosin, tropomyosin and vinculin, but not tubulin or vimentin (Boschek et al. 1981). Since this is the earliest observable alteration after the initiation of transformation and since

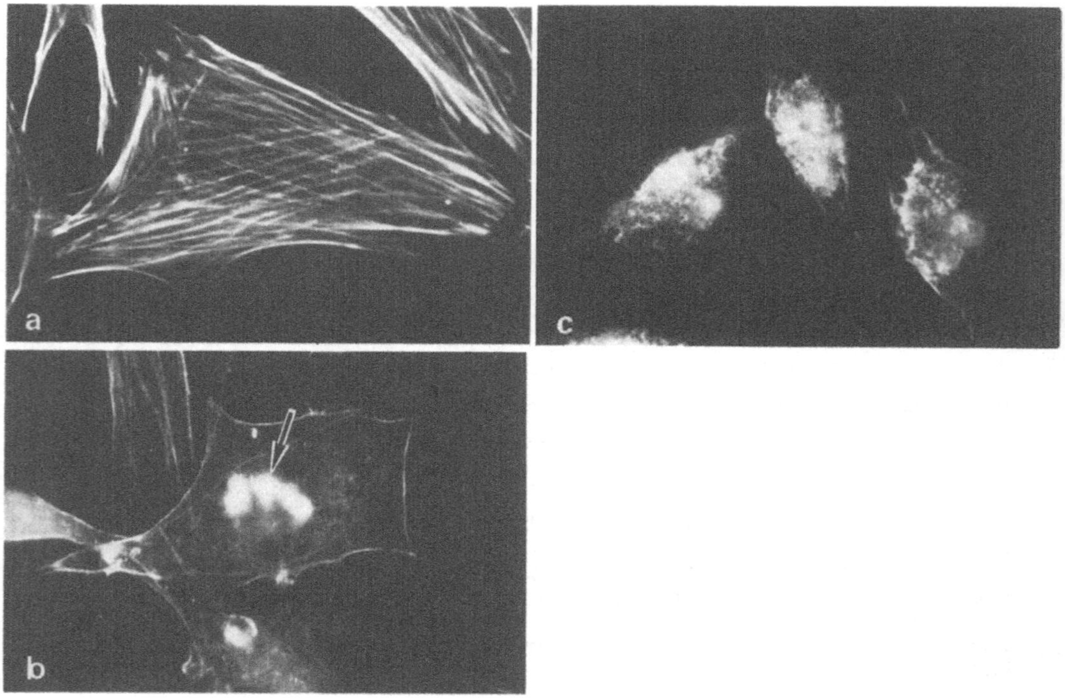

Fig. 5a-c. Td-ts RSV-infected chick embryo cells labeled for immune fluorescence with affinity purified antibodies to smooth-muscle actin. (a) Nonpermissive temperature (42°C). Cells display normal phenotype. (b) 30 min after downshift to the permissive temperature (35°C). Note large actin-containing *"flower" (arrow)* on dorsal cell surface. (c) 12 h after downshift. Cells display transformed phenotype

re-phosphorylation and kinase activation upon downshift with td-ts mutants also occurs within 15 min, it is likely that the microfilament system is one of the cell structures that is earliest affected by $pp60^{src}$.

This conclusion appears to be in agreement with recent findings by Sefton et al. (1981). They demonstrated that the 130,000 *d* vinculin shows a tenfold higher tyrosine-phosphorylation in RSV-transformed cells as compared to normal cells. Vinculin has been proposed to act as a linker between actin microfilaments and other yet unidentified anchorage proteins in the plasma membrane (Geiger et al. 1980). An interaction between vinculin and actin is suggested by the work of Jockusch and Isenberg (1981) showing that purified vinculin causes a bundling type of microfilament aggregation. Thus, the phenomenon we have earlier observed in transformation may be the consequence of an altered function of vinculin due to the effect of $pp60^{src}$. In this connection it is interesting that the bulk of $pp60^{src}$ has been localized within close proximity to the inner surface of the cell membrane (Willingham et al. 1979). In additon by use of several techniques, such as $pp60^{src}$ kinase assay at the surface of intact cells, immunofluorescence with $pp60^{src}$-specific antisera on intact cells (Barnekow et al. 1980, 1981), and staining of cryosections with gold-labelled antibodies (unpublished results), we have found that $pp60^{src}$ can also be detected at the outer cell surface and is released into the culture medium by the cell.

In a recent paper by Ball and Singer (1981) evidence has been presented for a molecular linkage between microtubules and vimentin-intermediate filaments which is modified in RSV-transformed cells, and it was suggested that the modification of these constitutents of the cytoskeleton is also due to one of the functions carried out by $pp60^{src}$.

Carbohydrate Metabolism

There is convincing evidence that changes in the activity of specific glycolytic enzymes are affected after transformation of chick embryo cells by RSV (Singh et al. 1974, Carroll et al. 1978). Chick embryo cells contain the M_2 type isoenzyme of pyruvate kinase (M_2-PK), one of the key enzymes in glycolysis. The kinetic properties of M_2-PK are characteristically changed when chick embryo cells are transformed by RSV; the enzyme has a lower affinity for phosphoenol pyruvate, is highly stimulated by fructose, 1,6-diphosphate, and is rapidly inactivated by magnesium-ATP (Presek et al. 1980). These changes of M_2-PK in RSV-transformed CEC are assumed to be caused by phosphorylation of the M_2-PK by a cyclic adenosine 3':5'-monophosphate independent protein kinase, which is known to phosphorylate and thereby inactivate M_2-PK. Evidence has been presented to suggest functional similarities between the $pp60^{src}$ kinase and the protein kinase which participates in the M_2-PK regulation. Neither of these kinases can be stimulated by cyclic AMP; they use CTP and ATP as phosphate donors; both are inhibited by P^1, P^5-di (adenosine 5') pentaphosphate, and M_2-PK inhibitor protein, and inhibition by fructose 1,6-diphosphate of both protein kinases is not competitive with respect to ATP. Thus it has been assumed, that transformation by RSV may affect the regulatory system of M_2-PK (Presek et al. 1980, Glossmann et al. 1981). These results are difficult to interpret in terms of metabolic consequences. One would expect that inactivation of M_2-PK would lead to decreased lactate production, whereas lactate production is in fact increased in transformed cells. However, increased lactate production may be the consequence of glutamine oxidation which in HeLa cells has been shown to provide more than

98% of the cell energy (Reitzer et al. 1979). Human diploid cells have also been shown to use glutamine as their major energy source (Zielke et al. 1976). The increase of FDP, as a consequence of M_2-PK inactivation, may therefore activate protein biosynthesis (Lenz et al. 1978) and enhance nucleic acid synthesis through the formation of increased levels of phosphoribose (Reitzer et al. 1980).

Another candidate cellular protein with an apparent molecular weight of 34,000-36,000 *d* has been identified as a possible substrate for the $pp60^{src}$-kinase. This protein is abundant, highly conserved during evolution, cytosolic and unglycosylated. It is phosphorylated at tyrosine within 1 h after shift from the non-permissive to the permissive temperature in chick embryo fibroblasts (CEC) infected with a ts-td RSV (Radke et al. 1980). In an attempt to relate this protein to an enzymatic function, it has been found to be inseparable from cytosolic malic dehydrogenase activity when purified by extensive chromatography (Rübsamen et al. 1982). In a separate study by Cassmann and Vetterlin (1974), cytosolic malic dehydrogenase has been found to be more active in the phosphorylated form; phosphorylation of this enzyme after transformation could therefore be one important step in the alteration of the cell metabolism as required for cell proliferation.

Calcium Metabolism

Together with cyclic nucleotides, calcium is considered as a second messenger with regulating functions for various metabolic processes as well as for cell proliferation. Balk et al. (1979) have shown that normal chick embryo fibroblasts are unable to divide when the extracellular calcium concentration is reduced to 1/5 of the concentration normally used in cell culture. RSV-transformed cells, on the other hand, retain the capacity for cell division under those circumstances, suggesting that the calcium metabolism is altered in virus-transformed cells. Recent studies in our laboratory have demonstrated significant differences in calcium transport and calcium content between normal and RSV-transformed chick embryo cells. The uptake rate of $^{45}Ca^{2+}$ has been found to be decreased by 20% in RSV-transformed cells when compared to untransformed cells (Fig. 6), and the determination of the intracellular calcium content by atomic absorption spectroscopy showed twice the calcium content in transformed cells as compared to CEC (unpublished results). As in the calcium uptake measurements, in

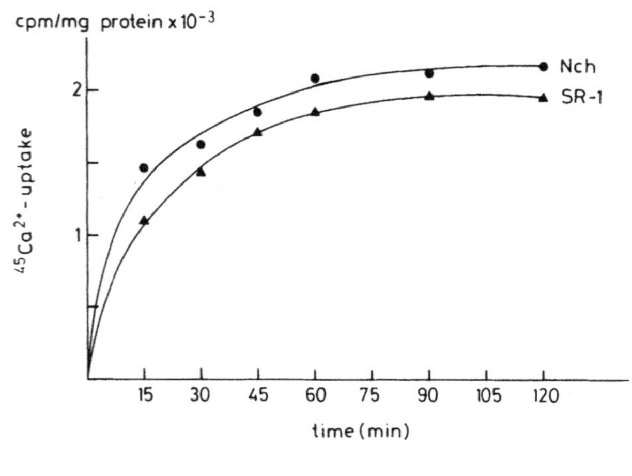

Fig. 6. $^{45}Ca^{2+}$ uptake by RSV-transformed (▲ SR-1) and normal sister cultures (● Nch) of chick embryo cells. Sister cultures were incubated with 1.5 µCi $^{45}CaCl_2$/ml for 30 min before the individual cultures were harvested at the indicated time intervals and the $^{45}Ca^{2+}$ content of the cells was determined

control experiments it was shown that this difference was neither temperature-dependent nor due to virus replicaton or proliferative status of the cells (A. Barnekow, G. Rose, and H. Bauer, unpublished).

At the moment it is difficult to explain the mechanism underlying this alteration of calcium metabolism in RSV-transformed cells. Experiments with active transport inhibitors suggest that the calcium transport mechanism in normal and RSV-transformed CEC is not basically different. Thus, the decreased calcium uptake could be the result of a reduced gradient between the extracellular and intracellular calcium concentration and thus be secondary to an event which causes a higher retention of intracellular calcium. This, on the other hand, may be due to reduced activity of a calcium pump responsible for calcium extrusion and thus calcium homeostasis. Whether this is the consequence of other transformation-associated metabolic processes, or a direct effect of pp60^{src}, remains to be determined.

These results are probably significant because of the important regulatory functions performed either by soluble calcium or through the activation of calmodulin and other calcium-binding and calcium-activated proteins. These have been recognized to regulate a broad spectrum of metabolic processes (see Marx 1980). For instance, part of the increased lactate production in tumor cells appears to be induced by calcium (Bossi et al. 1979). Calcium regulates the synthesis and degradation of the alternative type of second messenger, cyclic nucleotides, which are also altered in transformed cells. Furthermore, constituents of the cytoskeleton have been described either to bind calmodulin or to interact in a calciumdependent manner (Bretscher and Weber 1980, Glenney and Weber 1980).

An Hypothesis on the Pleiotropic Effect of pp60^{src}

Somatic cell division requires the replication of the cell genome as well as the increased biosynthesis of cellular organelles and structural constituents, i.e. cell growth. Little is known about the biochemistry of the initiation of these events and of their possible interrelationships. Cell growth appears to precede DNA synthesis. However, stimulation of DNA synthesis may independently require an unknown cascade of events which is initiated at the same time as cell growth. Cell growth may be induced by a mechanism which is not automatically linked to DNA synthesis as is shown in studies with growth factors (Gospodarowicz and Moran 1976).

The experimental evidence for a pleiotropic effect of pp60^{src} suggests that this protein may trigger cellular DNA synthesis and cell growth by independent effects (Fig. 7). Thus, the effect of pp60^{src} on the cytoskeleton, for instance through phosphorylation of microfilament proteins and/or possible Ca^{2+} effects on other constituents of this system, may finally lead to DNA synthesis and chromosomal reduplication. This may actually involve the direct function of specific structures of the cytoskeleton, the microtubules. The regulation of microtubule assembly by Ca^{2+} dependent calmodulin has been described in normal cells, and evidence has been presented that microtubule depolymerization is involved in signal transduction for DNA synthesis (for ref. see Chafouleas et al. 1982). This, taken together with the calcium requirement for the progression of the cell cycle into the DNA synthesizing phase (Chafouleas et al. 1982), justifies speculation that similar alterations in the cytoskeleton, part of which may be

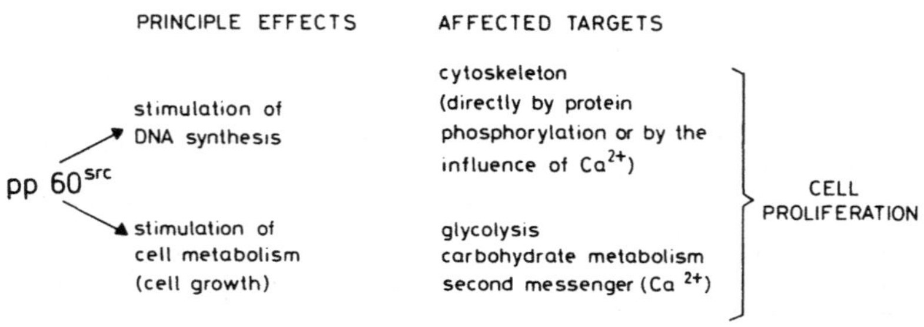

Fig. 7. The mechanism of transformation by $pp60^{src}$

calcium-dependent, and others caused by protein phosphorylation, represent one cluster of important effects exerted by $pp60^{src}$.

On the other hand, the suggested effects of $pp60^{src}$ on the function of pyruvate kinase and malic dehydrogenase as well as on calcium metabolism may stimulate metabolic steps required to supply enough energy, nucleotides, amino acids, and other prerequisites for cell division.

The Possible Function of Endogenous *src*-Gene

As mentioned above, normal cells contain a gene which is homologous to the viral *src*-gene, called c-*src*. In chick embryo cells in vitro, c-*src* is expressed at only between 1/50 and 1/100 of the amount of v-*src*. Since there is no functional difference known between $pp60^{c-src}$ and $pp60^{v-src}$, the transformation effect by RSV appears to be a dosage effect. It can also be assumed, that the metabolic alterations induced by this phosphoprotein are not unique to the tumor cell but also play a role under certain physiological conditions, i.e. at a given state of development or differentiation. Such a situation exists during embryonic development when various undifferentiated cells show a fundamental property of tumor cells, namely the ability to proliferate. In contrast to the controlled and limited proliferation of embryonic cells, however, the proliferation of tumor cells is unconstricted. Whether neoplastic cells may be related to undifferentiated cells can be tested with the help of tumor-virus systems which allow a thorough comparison of transformed cells with normal sister cells.

In many systems it has been shown that cell differentiation is paralleled by cell division and that a dividing, differentiating cell does not yet express the specific function or functions of its differentiated cell counterpart. On the other hand, it is noteworthy that many cells lose their specific function after viral transformation. This holds true for instance for myoblasts, melanoblasts, and chondroblasts, when transformed by Rous sarcoma virus (for ref. see Bauer 1982). A specific function of fibroblasts, the synthesis of collagen, appears to be fully expressed only in differentiated cells (Diegelman and Peterkofsky 1972) and is distinctly reduced upon transformation by RSV (Levison et al. 1975). A metabolic property of embryonic cells, increased lactate production, is also characteristic for RSV-transformed fibroblasts (Carroll et al. 1978). Furthermore, RSV-transformed fibroblasts express cell-surface antigens which are characteristic for muscle cells at an early embryonic development and have been named *onco-fetal antigens* (OFA) (Yoshikawa et al. 1979).

These common properties of RSV-transformed fibroblasts and early embryonic cells allow the speculation that fibroblasts, upon RSV-tranformation, acquire metabolic and growth properties similar to those of embryonic cells (Fig. 8). Therefore, cell transformation by RSV appears to involve cellular development towards the stage of an incomplete form of differentiation, which has been referred to in this and other systems as *retrodifferentiation* (Uriel 1976, Bauer and Yoshikawa 1980).

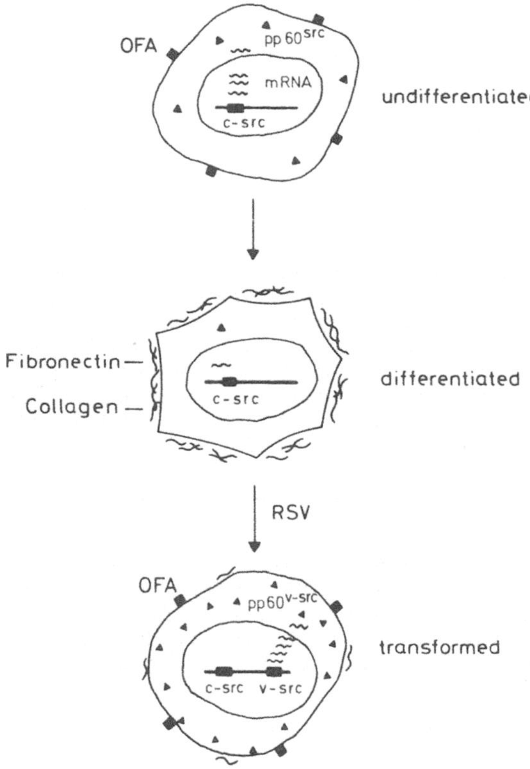

Fig. 8. Schematic comparison of the phenotype of undifferentiated (embryonic), differentiated, and RSV-transformed chick fibroblasts with respect to the expression of endogenous and exogenous pp60^{src}, fibronectin and collagen synthesis, and the expression of oncofoetal antigens

Based on these experimental data we questioned whether this correlation between embryonic and RSV-transformed cells is paralleled by an expression of c-*src* during embryogenesis. We investigated this by a quantitative analysis of pp60^{c-src} and its associated kinase activity. In fact, we found quantitative differences, first by comparing early and late passages of chick embryo fibroblasts in tissue culture, showing that primary cultures prepared from chicken embryos contain more pp60^{c-src} and associated kinase activity than cells after several subcultivations. A comparison of the amount of muscle tissue from pp60^{c-src} kinase activity in embryos with high-passage fibroblasts in vitro is shown in Fig. 9.

These results suggest that at an early stage of differentiation of mesenchymal cells, the relatively high expression of the endogenous *src*-gene may be responsible, at least in part, for the control of cell

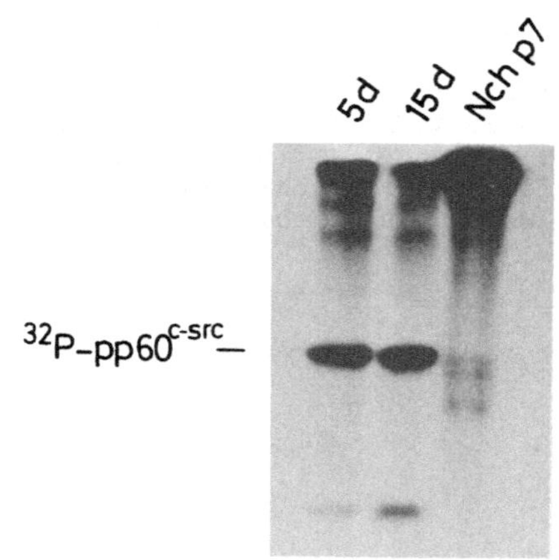

Fig. 9. Quantitative comparison of endogenous $pp60^{src}$ in chick muscle tissue from 5-day and 15-day-old embryos with $pp60^{c-src}$ in chick embryo cells after 7 subcultivations (about 6 weeks) in vitro

metabolism and proliferation. An essential difference between RSV-transformed cells and their embryonic cell counterparts could then be that the endogenous *src*-gene is under the physiological control of the organism and the cellular microenvironment, allowing only limited proliferation capacity of the embryonic cell. The viral *src*-gene, on the other hand, would not be restricted by any cellular control mechanism due to its association with another genetic structure (the viral genome) allowing its unlimited expression (see Fig. 8).

Interestingly, increased expression of the endogenous *src*-gene is not necessarily correlated with cell proliferation. This is seen from experiments with brain tissue from embryos as well as from adult birds, which in contrast to other tissues contain relatively large amounts of $pp60^{c-src}$. Whereas in embryonic cells the $pp60^{c-src}$ kinase activity is very similar in brain and muscle tissue (Bauer et al. 1982), adult animals show a reduced activity in muscle but a relatively high activity in brain extracts and also in other organs (Table 3). Although the

Table 3. The organ-specific distribution of the $pp60^{c-src}$ kinase activity (cpm/mg protein)

	Mouse	Chicken	Fish
Muscle	<100	<100	<50
Heart	170	120	n.t.
Skin	n.t.	n.t.	150
Liver	320	330	200
Kidney	400	330	n.t.
Lungs	600	220	n.t.
Spleen	740	370	300
Brain	720	800	600

n.t. means not tested

significance of these data is not yet understood, they appear to support the theory of multifunctionality of $pp60^{c-src}$ of which one or the other function may be expressed and required even in normal tissue such as brain. Further analysis of the function of a cellular $pp60^{src}$ may help in elucidating the overall function of the $pp60^{v-src}$ in cell transformation.

Finally, it will be interesting to study the regulation of the expression of the c-*src* gene. One system for this purpose appears to be the *Xiphophorus* fish system which has been genetically investigated by Anders and coworkers. Hybrids of these fish spontaneously or upon induction develop melanomas which are mediated by a certain gene, called *Tu*. *Tu* is part of the normal genome which may be present in several copies per cell and is controlled by genetically well-defined regulator genes. By selective crossbreeding, genotypes have been obtained which display quantitatively different expressions of *Tu* as evidenced by different quantities and proliferating capacities of pigmented cells (Anders and Anders 1978, Schartl et al. 1982). Various fish species have been investigated for their expression of an endogenous $pp60^{c-src}$ antigenically related to $pp60^{v-src}$ of RSV, and varying amounts of tyrosine-specific kinase activity have been found in all species of fish tested (Barnekow et al. 1982). In an investigation of a possible relationship between the *Tu*-gene and the *src*-gene it was found that inheritance and gene dosage effects of *Tu* coincide to some degree with the measured $pp60^{c-src}$ kinase activity, suggesting some linkage or epistasis between these two genes (Schartl et al. 1982).

In summary, it appears that both $pp60^{v-src}$ and $pp60^{c-src}$ are powerful tools in correlating the molecular mechanism of malignant cell transformation with cellular events which are physiological under certain conditions, especially during differentiation.

Acknowledgements. Our work reported here was supported by the Deutsche Forschungsgemeinschaft (SFB 47, Virologie).

References

Anders A, Anders F (1978) Etiology of cancer as studied in the platyfish-swordtail-system. Biochim Biophys Acta 516:61-95

Balk SD, Polimeni PI, Hoon BS, LeStourgeon DN, Mitchell RS (1979) Proliferation of Rous sarcoma virus-infected, but not of normal, chicken fibroblasts in a medium of reduced calcium and magnesium concentrations. Proc Natl Acad Sci USA 76:3913-3916

Ball EH, Singer SJ (1981) Association of microtubules and intermediate filaments in normal fibroblasts and its disruption upon transformation by a temperature-sensitive mutant of Rous sarcoma virus. Proc Natl Acad Sci USA 78:6986-6990

Barnekow A, Boschek CB, Ziemiecki A, Bauer H (1980) Detection of the *src*-gene product $pp60^{src}$ and its associated protein kinase on the surface of Rous sarcoma virus transformed cells. Biochem Soc Trans 8:735-736

Barnekow A, Bauer H, Boscheck CB, Friis RR, Ziemiecki A (1981) Rous sarcoma virus transformation: action of the *src*-gene product. In: Schweiger HG (ed) International cell biology 1980-1981. Springer, Berlin Heidelbert New York, pp 457-466

Barnekow A, Schartl M, Anders F, Bauer H (1982) Identification of a fish protein associated with a kinase activity and related to the Rous sarcoma virus transforming protein. Cancer Res 42(6):2429-2433

Bauer H (1982) Cell transformation by RNA sarcoma virus. In: Nicolini C (ed) Cell growth. Plenum Press, New York London, pp 653-671

Bauer H, Yoshikawa Y (1980) Oncofetal antigens as markers for retrodifferentiation in malignant transformation. In: Cold Spring Harbor Conferences on Cell Proliferation, vol VII, Cold Spring Harbor, pp 1231-1238

Bauer H, Barnekow A, Rose G (1982) The transforming protein of Rous sarcoma virus, $pp60^{src}$: growth and cell proliferation inducing properties. Hormones and Cell Regulation, vol 6, INSERM European Symposium. Dumont JE, Nunez J, Schultz G (eds). Elsevier Biomedical Press, pp 187-205
Becker D, Kurth R, Critchley D, Friis RR, Bauer H (1977) Distinguishable transformation defective phenotypes among temperature sensitive mutants of Rous sarcoma virus. J Virol 21:1042-1055
Bishop JM (1978) Retroviruses. Annu Rev Biochem 47:35-88
Bishop JM (1981) Enemies Within: the genesis of retrovirus oncogenes. Cell 23:5-6
Boschek CB (1982) Organization changes of cytoskeletal proteins during cell transformation. In: Klein G (ed) Advances in Viral Oncology. Raven Press, New York, in press
Boschek CB, Jockusch BM, Friis RR, Back R, Grundmann E, Bauer H (1981) Early changes in the distribution and organization of microfilament proteins during cell transformation. Cell 24:175-184
Bossi D, Cittadini A, Wolf F, Milani A, Magalini S, Terranova T (1979) Intracellular calcium and magnesium content and aerobic lactate production in intact ehrlich ascites tumour cells. FEBS Lett 104:6-11
Brautigan DL, Bornstein P, Gallis B (1981) Phosphotyrosyl protein phosphatase. J Biol Chem 256:6519-6522
Bretscher A, Weber K (1980) Villin is a major protein of the microvillus cytoskeleton which binds both G and F actin in a calcium-dependent manner. Cell 20:839-847
Brugge JS, Erikson RL (1977) Identification of a transformation-specific antigen induced by an avian sarcoma virus. Nature 269:346-348
Brugge JS, Erikson E, Collett MS (1978) Peptide analysis of the transformation-specific antigen from avian sarcoma virus-transformed cells. J Virol 26:773-782
Calothy G, Pessac B (1976) Growth stimulation of chick embryo neuro-retinal cells infected with RSV: relationship to viral replication and morphological transformation. Virol 71:336-345
Carroll RC, Ash JF, Vogt PK, Singer SJ (1978) Reversion of transformed glycolysis to normal by inhibition of protein synthesis in rat kidney cells infected with temperature-sensitive mutant of Rous sarcoma virus. Proc Natl Acad Sci USA 75:5015-5019
Cassmann M, Vetterlin D (1974) Allosteric and non-allosteric interaction with reduced nicotinamide adenine dinucleotide in two forms of cytoplasmic malic dehydrogenase. Biochemistry 13:684-689
Chafouleas JG, Bolton WE, Hidaka H, Boyd III AE, Means AR (1982) Calmodulin and the cell cycle: involvement in regulation of cell-cycle progression. Cell 28:41-50
Coffin JM, Varmus HE, Bishop JM, Essex M, Hardy Jr WD, Martin GS, Rosenberg NE, Scolnick EM, Weinberg RA, Vogt PK (1981) Proposal for naming host cell-derived inserts in retrovirus genomes. J Virol 40:953-957
Collett MS, Erikson RL (1978) Protein kinase activity associated with the avian sarcoma virus *src* gene product. Proc Natl Acad Sci USA 77:2021-2024
Diegelmann RF, Peterkofsky B (1972) Collagen biosynthesis during connective tissue-development in chick embryo. Biol 28:443-453
Foulkes JG, Howard RF, Ziemiecki A (1981) Detection of a novel mammalian protein phosphatase with activity for phosphotyrosine. FEBS Lett 130:197-200
Friis RR, Schwarz RT, Schmidt MFG (1977) Phenotype of Rous sarcoma virus-transformed fibroblasts. An argument for a multifunctional *src* gene product. Med Microbiol Immunol 164-155-164
Friis RR, Jockusch BM, Boschek CB, Ziemiecki H, Rübsamen H, Bauer H (1980) Transformation-defective temperature-sensitive mutants of Rous sarcoma virus have a reversibly defective *src*-gene product. Cold Spring Harbor Symp Quant Biol 44:1007-1012
Geiger B, Tokuyasu KT, Dutton AH, Singer SJ (1980) Vinculin, an intracellular protein localized at specialized sites where microfilament bundles terminate at cell membranes. Proc Natl Acad Sci USA 77:4127-4131
Glenney JR Jr, Weber K (1980) Calmodulin-binding proteins of the microfilaments present in isolated brush borders and microvilli of intestinal epithelial cells. J Biol Chem 255:10551-10554

Glossmann H, Presek P, Eigenbrodt E (1981) Association of the *src*-gene product of Rous sarcoma virus with a pyruvate-kinase inactivating factor. Mol Cell Endocrinol 23:49-64

Gospodarowicz D, Moran JS (1976) Growth factors in mammalian cell culture. Anna Rev Biochem 45:531-558

Hunter T, Sefton BM (1980) Transforming gene product of Rous sarcoma viruses phosphorylates tyrosine. Proc Natl Acad Sci USA 77:1311-1315

Jockusch BM, Isenberg G (1981) Interaction of alpha-actinin and vinculin with actin: opposite effects on filament network information. Proc Natl Acad Sci USA 78:3005-3009

Lenz JR, Chatterjee GE, Maroney A, Baglioni C (1978) Phosphorylated sugars stimulate protein synthesis and Met-tRNA binding activity in extracts of mammalian cells. Biochemistry 17:80-87

Levinson W, Bhatnagar RS, Liu T-Z (1975) Loss of ability to synthesize collagen in fibroblasts transformed by Rous sarcoma virus. J Int. Cancer Inst. 55:807-810

Marx JL (1980) Calmodulin: A protein for all seasons. Science 208:274-276

Patschinsky T, Sefton BM (1981) Evidence that there exist four classes of RNA tumor viruses which encode proteins with associated tyrosine protein kinase activities. J Virol 39:104-114

Presek P, Glossmann H, Eigenbrodt E, Schoner W, Rübsamen H, Friis RR, Bauer H (1980) Similarities between a phosphoprotein (pptOsrc)-associated protein kinase of Rous sarcoma virus and a cyclic adenosine 3':5'-monophosphate-independent protein kinase that phosphorylates pyruvate kinase type M_2. Cancer Res 40:1733-1741

Purchio AF (1982) Evidence that $pp60^{src}$, the product of the Rous sarcoma virus *src* gene, undergoes autophosphorylation. J Virol 41:1-7

Radke K, Gilmore T, Martin GS (1980) Transformation by Rous sarcoma virus: a cellular substrate for transformation-specific protein phosphorylation contains phosphotyrosine. Cell 21:821-828

Reitzer LJ, Wice BM, Kennell D (1979) Evidence that glutamine, not sugar, is the major energy source for cultured Hela cells. J Biol Chem 254:2669-2676

Rübsamen H, Saltenberger K, Friis RR, Eigenbrodt E (1982) Cytosolic malic dehydrogenase activity is associate with a putative substrate for the transforming gene product of Rous sarcoma virus. Proc Natl Acad Sci USA 79:228-232

Schartl A, Schartl M, Anders F (1982a) Promotion and regression of neoplasia by testosterone-oromoted cell differentiation in *Xiphophorus* and *Girardinus*. In: Hecker E (ed) Carcinogenesis and biological effects of tumor promoter. Raven Press, New York, submitted

Schartl M, Barnekow A, Bauer H, Anders F (1982b) Oncogenes in *Xiphophorus*: A relation in inheritance and expression between a tumor gene and the cellular homologue of the Rous sarcoma virus transforming gene. Cancer Res (in press)

Sefton BM, Hunter T, Beemon K, Eckhart W (1980) Evidence that the phosphorylation of tyrosine is essential for cellular transformation by Rous sarcoma virus. Cell 20:807-816

Sefton BM, Hunter T, Ball EH, Singer SJ (1981) Vinculin: a cytoskeletal target of the transforming protein of Rous sarcoma virus. Cell 24:165-174

Singh VM, Singh M, August JT, Horecker BL (1974) Alterations in glucose metabolism in chick embryo cells transformed by Rous sarcoma virus: Intracellular levels of glycolytic intermediates. Proc Natl Acad Sci USA 71:4129-4132

Uriel J (1976) Cancer, retrodifferentiation, and the myth of Faust. Cancer Res 36:4269-4275

Weber M, Friis RR (1979) Dissociation of transformation parameter using temperature-conditional mutants of Rous sarcoma virus. Cell 16:25-32

Willingham MC, Jay G, Pastan I (1979) Localization of the ASV *src* gene product to the plasma membrane of transformed cells by electron microscopic immunocytochemistry. Cell 18:125-134

Yoshikawa Y, Ignajatovic J, Bauer H (1979) Tissue-specific expression of onco-fetal antigens during embryogenesis. Differentiation 15:41-47

Zielke HR, Ozand PT, Tildon JT, Sevdalian DA, Cornblath M (1976) Growth of human diploid fibroblasts in the absence of glucose utilization. Proc Natl Acad Sci USA 73:4110-4114

Ziemiecki A, Friis RR (1980) Phosphorylation of pp60src and the cycloheximide insensitive activation of the pp60src-associated kinase activity of transomation-defective temperature-sensitive mutants of Rous sarcoma virus. Virol 106:391-394
Ziemiecki A, Friis RR, Bauer H (1982) Half-life of the Rous sarcoma virus transforming protein pp60src and its associated kinase activity. Mol Cell Biol 4:355-360

The Biology of an Oncogene, Based upon Studies on Neoplasia in Xiphophorus

F. Anders[1]

Xiphophorus, including swordtails and platyfish, is a genus of small viviparous freshwater fish from Central America that serves increasingly as a laboratory animal [1]. For 25 years we have used *Xiphophorus* for studies on neoplasia, which can easily be induced in hybrids between descendants of different provenance [2]. Although neoplasia of these animals is rather well understood in terms of formal genetics, the molecular basis of this phenomenon was extremely resistant to any elucidation. Recently a promising approach to the study of neoplasia in *Xiphophorus* at the molecular level, has been undertaken in a co-operative work of the laboratories of H. Bauer (Institut für Virologie, Giessen), W. and H. Kersten (Institut für Physiologische Chemie, Erlangen), S. Nishimura (National Cancer Center Research Institute, Tokyo), and our laboratories. The present review will trace some steps of our studies that led to the detection of a cellular oncogene and a prominent regulatory gene.

The Taxonomic Groups of Xiphophorus

The genus *Xiphophorus* lives in genetically isolated populations in brooks, rivers, lakes, ponds, and pools, and has evolved into innumerable genotypically and phenotypically distinguishable groups [1]. Based on certain morphological and ecological characters, 17 of these groups have been listed as species [3].

All individuals of this genus, however, can be hybridized in the laboratory without difficulty, and all hybrids are fertile. This, together with the findings on the conformity of genome organization [4], the low degree of enzyme polymorphism [5,6], and the normal chromosome pairing during meiosis in the hybrids [7], led to the conclusion that the taxonomic differences between these groups of *Xiphophorus* are not at the species level, but at the level of elementary local populations as well as ecological and geographical races.

1 Genetisches Institut der Justus-Liebig-Universität, Heinrich-Buff-Ring 58-62, D-6300 Giessen, FRG

33. Colloquium - Mosbach 1982
Biochemistry of Differentiation and Morphogenesis

Insusceptibility to Neoplasia in Wild Populations and Their Purebred Laboratory Descendants

Tens of thousands of individuals from different wild populations of *Xiphophorus* have been collected by several investigators (see [2]), but no tumors were detected. In the progeny of the wild populations, which in the case of *X. helleri* from Rio Lancetilla and *X. maculatus* from Rio Jamapa have been bred in the laboratory since 1939 (about 80 and 120 generations, respectively), no tumors occurred. About 10 000 specimens of purebred descendants of the wild populations have been treated with powerful carcinogens such as benzo(a)pyrene, N-methyl-N-nitrosourea (MNU) and X-rays, but none developed neoplasia. These animals are highly insusceptible to neoplasia (Table 1, first part).

Table 1. Neoplasia in *Xiphophorus* 1 year after treatment with MNU (10^{-3} M; 4 times for 1 h in 2-week intervals) and X-rays (1000 R; 3 times for 45 min in 6-week intervals)

	No. of survivors		No. of neoplasms	
	MNU	X-rays	MNU	X-rays
Purebred				
X. maculatus	410	3405	0	0
X. variatus	ca. 100	ca. 500	0	0
X. xiphidium	ca. 100	ca. 100	0	0
X. helleri	415	ca. 2000	0	0
X. cortezi	ca. 100	ca. 100	0	0
	ca. 7200		0	
Hybrids				
F_1	470	ca. 1000	18 (4%)	0
F_2-F_{24}; BC_n	8258	3587	826 (10%)	163 (5%)
	ca. 13500		1007 (7,5%)	

Susceptibility to Neoplasia in Laboratory Hybrid Populations

In contrast to the animals from purebred wild populations, animals from laboratory hybrid populations derived from crosses between the purebred descendants of wild populations, may be susceptible to neoplasia. Following treatment with carcinogens, depending on the wild populations used for hybridization, about 1% to 4% of the F_1-hybrids develop neoplasia. Tumor incidence increases in the second hybrid generation (F_2) up to about 10% and remains stable in the succeeding generations, which have been tested up to F_{24} (Table 1, second part).

Classification of Neoplasms

As compiled from the results of several investigators in our laboratories [8] 805 of 10 195 (8%) hybrids which survived treatment with MNU and X-rays developed a large variety of different neoplasms. Most of the neoplasms were classified as neurogenic and mesenchymal, with melanoma, neuroblastoma, and fibrosarcoma being the predominant types. Epithelial neoplasms were less frequent but comprised the largest diversity (Table 2).

Table 2. Neoplasms induced in F_2-F_{24} and BC_n generations (MNU: 6608 survivors; X-rays: 3587 survivors; total: 10 195

Type of neoplasm	No. of neoplasms			Incidence (%) based on total No. of survivors	
	MNU	X-rays		MNU	X-rays
Neurogenic					
Melanoma (benign)	135	93	491	2,12	2,6
Melanoma (malignant)	138	34		2,09	0,95
Neuroblastoma	84	7		1,27	0,2
Epithelial					
Squamous cell carcinoma	6	0	78	0,09	0
Epithelioma	19	6		0,28	0,17
Carcinoma (low-differentiated)	3	4		0,05	0,11
Carcinoma (high-differentiated)	2	5		0,03	0,14
Adenocarcinoma (kidney)	8	2		0,12	0,05
Adenocarcinoma (thyroid)	2	3		0,03	0,08
Papilloma	9	0		0,14	0
Hepatoma	5	1		0,07	0,03
Acanthoma	3	0		0,04	0
Mesenchymal					
Fibrosarcoma	190	6	236	2,87	0,17
Rhabdomyosarcoma	33	2		0,5	0,05
Lymphosarcoma	1	0		0,01	0
Reticulosarcoma	4	0		0,06	0
Total	642	163	805		

805 of 10195 (7,9%) hybrids developed neoplasia; 92% of the hybrids were sufficiently protected

Assignment of Cancer Susceptibility to Chromosomes

To study what may be the crucial differences between the fish that were insusceptible, and those that were susceptible, to cancer we attempted to assign the carcinogen-triggered neoplasms to chromosomes. For this study 65 defined genotypes of *X. maculatus*, *X. xiphidium*, *X. variatus*, *X. cortezi*, *X. helleri*, and their hybrids were employed [9,10]. The genotypes exhibit, or lack, specific color patterns or enzyme markers which are due to the expression of specific genes, of which each is located on a different chromosome. We used mainly backcrosses, which were selectively bred for a specific phenotypic marker, and thereby for a specific chromosome. Such backcrosses segregate into 50% animals carrying the marker chromosome, and 50% lacking this chromosome.

Neoplasia could be assigned specifically to many different chromosomes. In the example shown in the scheme of Fig. 1 almost exclusively the backcross (BC) segregants exhibiting the stripes inherited from *X. maculatus* were susceptible. They developed melanoma, neuroblastoma, epithelioma and fibrosarcoma. Some of these hybrids developed several tumors of different types. This is not to say that all fish exhibiting the stripes develop neoplasia; but almost all neoplasms develop in those animals that belong to the group of BC segregants exhibiting stripes. Thus, the susceptibility to develop different kinds of neoplasms apparently depends on the marker chromosome which is responsible for the stripes. There are other examples in which susceptibility to neurogenic and epithelial neoplasia could be assigned to a certain chromosome while susceptibility to neoplasms of mesenchymal origin

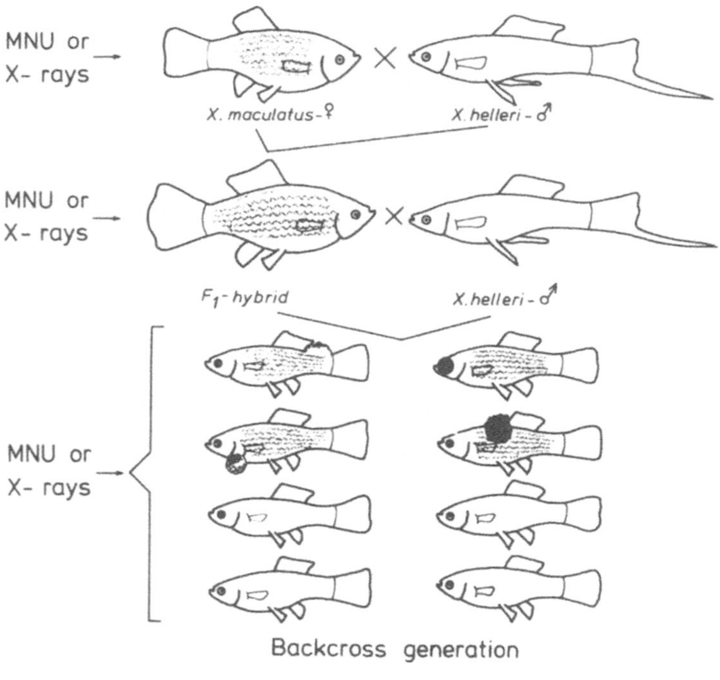

Fig. 1. Crossing scheme showing the assignment of carcinogen-triggered neoplasms to backcross segregants exhibiting stripes that are inherited from *Xiphophorus maculatus*. For details see text

depend on a different chromosome [10]. Other examples have shown that susceptibility to develop melanoma may depend on a Y-chromosome, an X-chromosome, or an autosome. For all backcross hybrids tested so far, susceptibility to develop neoplasia is apparently chromosome-specific.

Assignment of Cancer Susceptibility to Oncogenes and Regulatory Genes

Information about the genes underlying susceptibility to carcinogen-triggered neoplasia, comes from the analysis of the hereditary trait of certain "spontaneously" developing melanomas, pterinophoromas, neuroblastomas, thyroid carcinomas, kidney adenocarcinomas, and reticulosarcomas in hybrids. These neoplasms are very rare compared to the carcinogen-triggered tumors and therefore have been considered as a curiosity by many cancer researchers. They can, however, be produced at the will of the experimenter and have therefore been studied by many investigators [1]. They have contributed many important facts for developing ideas about how other animals, including humans, might inherit cancer [11,12].

In the following chapters we shall mainly deal with the genes involved in melanoma development because their phenotypic effect, due to the enlarged shape and heavy pigmentation of the transformed cells, can easily be observed without killing the fish. Even a single transformed cell can be distinguished from a regular pigment cell.

Crosses of a spotted *X. maculatus* (platyfish) with a nonspotted *X. helleri* (swordtail) result in F_1-hybrids that develop benign melanoma instead of spots (Figs. 2 and 3). Backcrosses of the F_1-hybrids using the swordtail as the recurrent parent result in offspring (BC_1), 50% of which exhibit neither spots nor melanoma while 25% develop benign melanoma (like the F_1), and 25% develop malignant melanoma. Further backcrosses of the fish (not shown in Fig. 3) carrying benign melanoma with the swordtail result in a BC_2 that exhibits the same segregation as the BC_1. The same applies for further backcrosses. Backcrossing of the fish carrying malignant melanoma with the swordtail results in a BC_2 in which 50% of the animals do not develop melanoma, while the remaining 50% develop malignant melanoma. In contrast, backcrossing of the melanoma-bearing hybrids using the platyfish as the recurrent parent results in a gradual suppression and finally disappearance of neoplasia in the succeeding generations.

These results, with the inclusion of cytogenetic findings, were interpreted as follows [2]: The spots and their corresponding genes are a specific accessory of the platyfish. The melanomas of the hybrids are closely related to the spots which actually are extreme benign melanomas. On the other hand the swordtail lacks both the corresponding spots and genes (Fig. 3).

The genetic information for neoplastic transformation of pigment cells is encoded in a "tumor gene" (*Tu*) which is inherited by the platyfish. About 70 cross-overs, deletions, duplications, and translocations show that *Tu* is located at the end of the X-chromosome and is under control of linked and non-linked regulatory genes [13]. In the platyfish used in this experiment a "major" pigment cell-specific regulatory gene (*R*) linked to *Tu* as well as two "minor" regulatory genes, which compartment-specifically suppress melanoma formation in the dorsal fin (R_{Df}) and the posterior part of the body (R_{Pp}), are mutated to R', R_{Df}, and

Fig. 2A-D. Spots and melanomas in *Xiphophorus*. (A) *X. maculatus* from Rio Jamapa (Mexico); "spotted dorsal fin" (mutation R_{Df}') and "spotted posterior part" of the body (mutation R_{Pp}'). (B) F_1-hybrid between *X. maculatus* (A) and *X. helleri* from Rio Lancetilla (Honduras), exhibiting benign melanoma instead of spots. (C) BC-hybrid [derived from backcross of (B) with *X. helleri*] exhibiting malignant melanoma instead of spots. (D) Albino BC-hybrid exhibiting malignant amelanotic melanoma. (A-C) correspond to the respective schematic drawings shown in Fig. 3

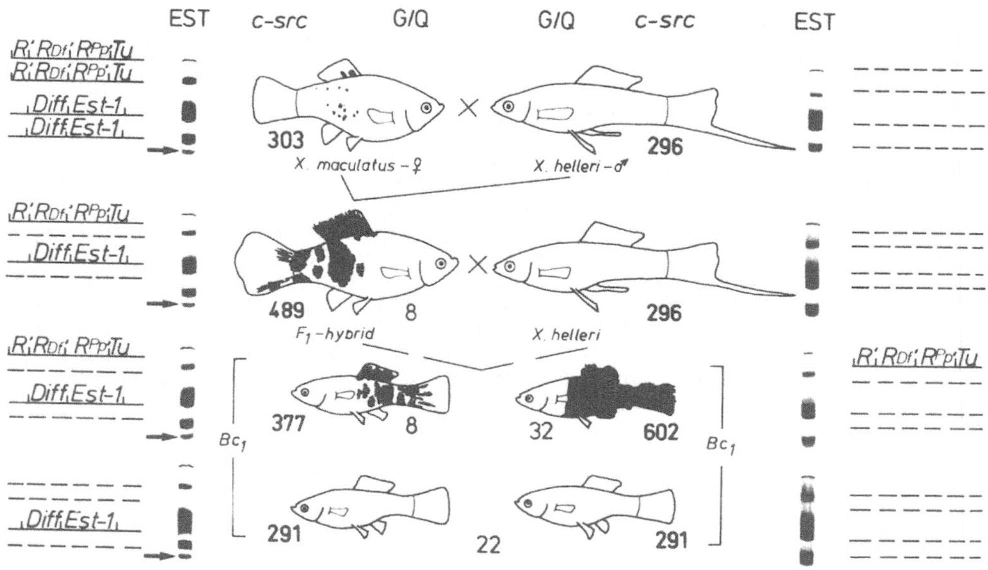

Fig. 3. Crossing scheme which displays the genetic conditions for the "spontaneous" development of spots (Fig. 2A), benign melanoma (Fig. 2B), and malignant melanoma (Fig. 2C). —— chromosomes of *X. maculatus*; --- chromosomes of *X. helleri*; *Tu*, tumor gene; R_{Pp}' and R_{Df}', impaired regulatory genes controlling *Tu* in the compartments of the posterior part of the body (Pp) and the dorsal fin (Df); R', impaired regulatory gene specific to pigment cells but non-specific to the compartments; *Diff*, regulatory gene controlling differentiation of neoplastically transformed cells; *Est-1*, locus for esterase -1 of *X. maculatus*. Est, esterases (polyacrylamid gel electrophoresis from homogenates of the eye); note linkage of *Diff* and *Est-1* (see *arrows*). *c-src*, pp60^{c-src} kinase activity (cpm/mg protein; 53K; see Fig. 11); note basic and excessive activity, and correlation between *c-src* expression and *Tu* expression. G/Q, ^{3}H guanine-incorporation in position 34 (anticodon) of tRNAs for Tyr, Asn, Asp and His (pmol/A_{260}; see Fig. 15); note low incorporation in tumors of *Diff*-animals indicating high Q-content, and high incorporation in the melanoma of *Diff*-lacking animals indicating low Q-content. (Combined from data of [2,6,13, 14,30,31,34,40,46]). For details see text

R_{Pp}', respectively, and can no longer suppress *Tu*. Evidence for this comes from the appearance of some transformed pigment cells in the dorsal fin and in the skin of the posterior part of the body. The regulatory gene that actually suppresses tumor formation in the platyfish used in this experiment is the homozygous non-linked "differentiation gene" (*Diff*) which can easily be detected by the esterase marker (*Est-1*) closely linked to *Diff* [6,14,15]. Further regulatory genes also present in the system are not taken into consideration.

Following crossings and backcrossings using the swordtail as the recurrent parent, the chromosomes of the platyfish are replaced by the homologous chromosomes of the swordtail, resulting in the gradual disintegration of the regulating gene system for *Tu*.

In contrast, following backcrossings of the melanoma-bearing hybrids with the platyfish as the recurrent parent, the chromosomes carrying regulatory genes are re-introduced into the descendants. This results in a reconstruction of the original regulating gene system that suppresses the activity of *Tu*.

"Spontaneous" development of melanoma as well as its suppression following the appropriate crossing precedures were found in several experimental hybrid populations derived from different purebred populations of different geographical or ecological origin.

In order to disclose the genetic basis for the bulk of neoplasms that develop following treatment with a carcinogen, we have modified the experiment shown in Fig. 3 to the experiment shown in Fig. 4: The R' R_{Df}' R_{Pp}' *Tu*-chromosome was replaced by the R R_{Df} R_{Bs}' *Tu*-chromosome, the "major" R of which is nonmutated and active. Since this R is inherited along with *Tu*, neoplasia does not develop spontaneously in the hybrids. Following treatment with carcinogens, those hybrids carrying the *R*-*Tu*-chromosome but lacking *Diff* (determined by the esterase) are highly susceptible to neoplasia. In this case development of neoplasia requires only impairment, or deletion, of a single *R*-gene by the carcinogen in a somatic cell [16].

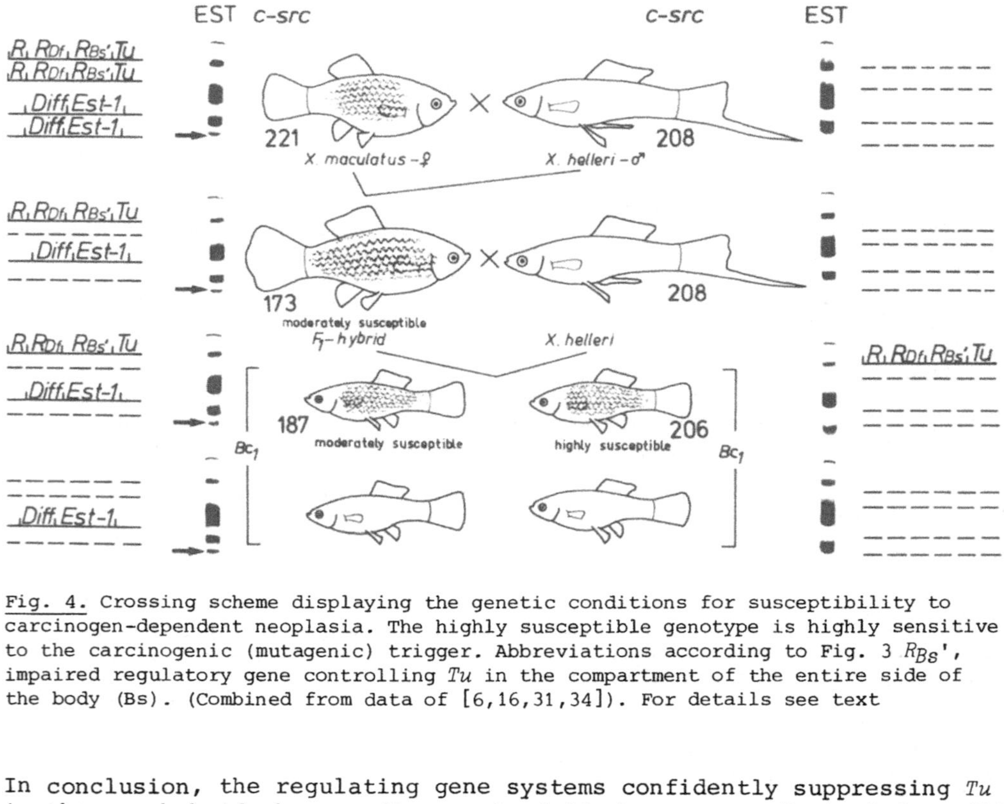

Fig. 4. Crossing scheme displaying the genetic conditions for susceptibility to carcinogen-dependent neoplasia. The highly susceptible genotype is highly sensitive to the carcinogenic (mutagenic) trigger. Abbreviations according to Fig. 3 R_{Bs}', impaired regulatory gene controlling *Tu* in the compartment of the entire side of the body (Bs). (Combined from data of [6,16,31,34]). For details see text

In conclusion, the regulating gene systems confidently suppressing *Tu* in the non-hybrids become disorganized if chromosomes derived from different populations are combined in the hybrids by the experimenter. This implies that, in purebred fish, the *Tu* and its sets of regulatory genes are population-specifically coadapted by natural selection. In any case the genetic information coding for neoplastic transformation can be traced to a *Tu* which is present in the different cell types and is normally under control of population-specific and cell-type-specific polygenic systems of linked and nonlinked regulatory genes, which suppress the development of the various types of potential neoplasms (Fig. 5). According to the formal assignment of different neo-

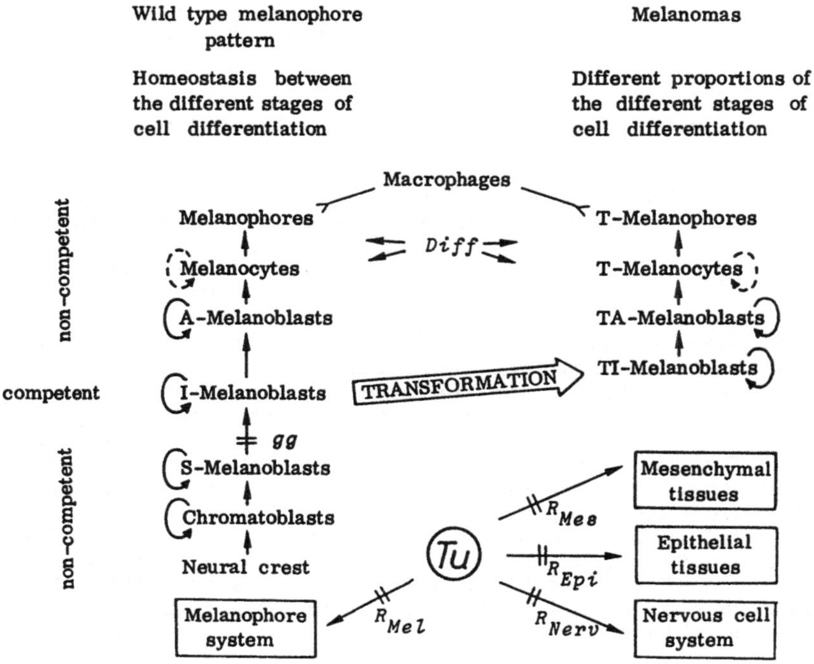

Fig. 5. Schematic presentation of the differentiation of normal and neoplastically transformed pigment cells. Only the intermediate melanoblasts (I-melanoblasts) are competent for transformation. *Tu*, tumor gene; R_{Mel}, pigment cell-specific regulatory genes for control of *Tu*; *gg*, homozygous "golden" mutation that blocks pigment cell differentiation; R_{Nerv}, R_{Epi}, R_{Mes}, regulatory genes that control *Tu* in the nervous cell system, the epithelial tissues and the mesenchymal tissues, respectively; *Diff*, regulatory gene which supports cell differentiation. Macrophages attack only the terminally differentiated pigment cells (both normal and transformed). (According to a scheme in [2], modified). For details see text

plasms to a particular chromosome, the particular *Tu* that codes for neoplastic transformation of pigment cells may also code for transformation of cells of mesenchymal and epithelial origin (see Figs. 1 and 5).

Enhancement of Melanin Pigmentation in Melanoma as an Epiphenomenon of *Tu*-expression

To study the relationship between melanoma formation and melanin synthesis [17,18] we separated both processes by introgression of an albino gene into melanoma bearing hybrids. The result was albinos which developed melanomas that completely lacked the melanin (Fig. 2D). This indicates that enhanced melanin pigmentation of melanomas is an epiphenomenon of neoplastic transformation of pigment cells exerted by *Tu*.

The Competent Cells for the Activity of *Tu* in the Pigment Cell System

The precursors of the melanin-producing pigment cells of *Xiphophorus*, like those of other vertebrates, originate from the neural crest, and migrate to their final destination (see Fig. 5). They divide, and undergo differentiation through the stages of chromatoblasts, stem(S)-melanoblasts, intermediate(I)-melanoblasts, advanced(A)-melanoblasts, melanocytes and, finally, differentiate to melanophores, which are incapable of dividing. At a certain age the melanophores are removed by macrophages. Supply comes from S-melanoblasts [2,19,20].

A-melanoblasts, melanocytes and melanophores have never been observed to undergo neoplastic transformation. Thus, these cells appear to be non-competent for the *Tu*-activity. On the other hand, in genotypes carrying a mutation that arrests differentiation at the stage of S-melanoblasts, melanomas cannot develop until exogenous promoters push differentiation of S-melanoblasts to the stage of I-melanoblasts [21]. These studies show, that also the neural crest cells, chromatoblasts and S-melanoblasts are non-competent. We, therefore, conclude that the only stage of differentiation, in which the pigment cells are competent for the transforming activity of *Tu*, is the stage of I-melanoblasts. These cells become transformed to TI-melanoblasts. We assume that the principle of competence of a cell to the transforming activity of *Tu* applies also for other kinds of neoplasms.

It could not be shown so far whether *Tu* is still active in the transformed cells (T-cells) for the maintenance of the neoplastic state. In any case the TI-melanoblasts continue to differentiate to TA-melanoblasts, T-melanocytes and, finally, to T-melanophores that are incapable of dividing. This process corresponds to differentiation of the non-transformed pigment cells. It is not under the control of *Tu* but under the control of the differentiation gene *Diff*. This, furthermore, implies that *Tu* does not specify the degree of malignancy of the melanoma (see later).

The Genuine Effect of *Tu* in the Pigment Cell System

Information about the genuine effect of *Tu* comes from a balanced laboratory stock carrying a lethal *Tu*-translocation that originated according to Fig. 6. *Tu* from an X-chromosome of *X. maculatus* becomes translocated to an autosome of *X. helleri* and, in its new position, is not under control of its former linked regulatory genes (*R*, R_{Df} etc.). The

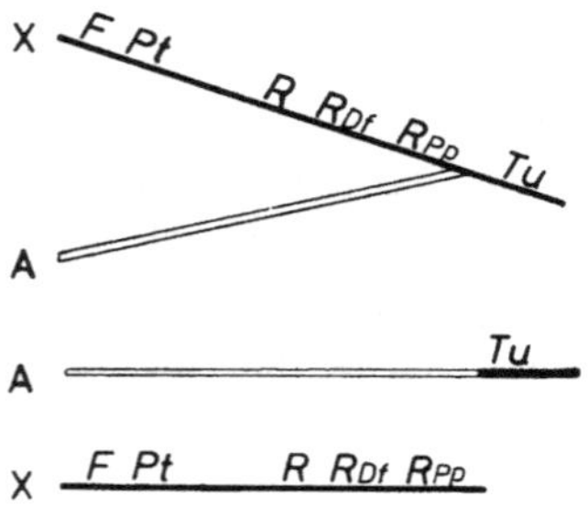

Fig. 6. Translocation of the tumor gene (*Tu*) from the X-chromosome (X) of *X. maculatus* to an autosome (*A*) of *X. helleri*. Note separation of *Tu* from its linked regulatory genes (*R*, R_{Df}, R_{Pp}; see Fig. 3). *F* sex-determining region of the X-chromosome; *Pt* pterinophore locus. For the phenotypic effect see Fig. 7

progeny of this stock segregates into 50% carrying the non-linked *Diff* which survive, while the corresponding 50% lacking *Diff*, is lethal. As a consequence of the Mendelian inheritance of the *Tu*-translocation through germ line, and the lack of *Diff*, *Tu* becomes active in the developing progeny as soon as the pigment cell precursors differentiate to the competent I-melanoblasts.

This process starts in the 5-day-old embryos. Some time later some single dividing T-melanocytes appear at the peduncle of the tail fin of the embryo (Fig. 7). Neoplastic transformation continues in all

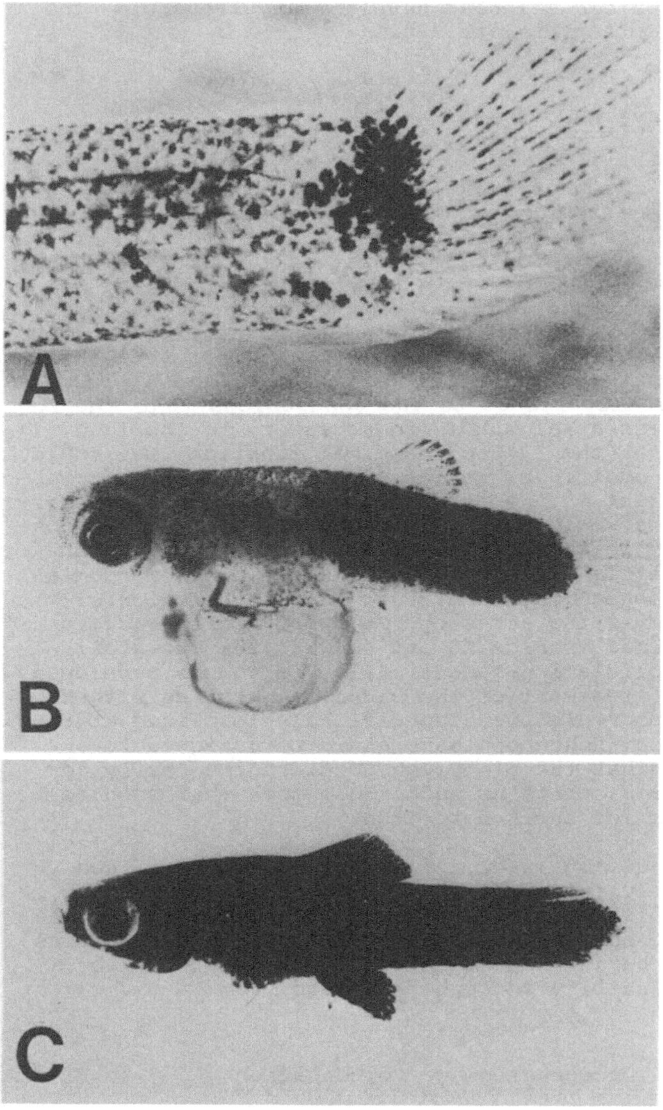

Fig. 7A-C. The genuine effect of the tumor gene *Tu* (corresponding to the scheme of translocation shown in Fig. 6). (A) Tail of a 10-day-old embryo (3 mm in length) exhibiting some T-melanocytes at the peduncle of the tail fin. (B) the same fish, 5 days later (4 mm in length). (C) Neonate of the same genotype (6 mm in length)

areas of the developing embryo, where a pigment cell precursor becomes competent, thus building the lethal "whole body melanoma". This reflects the genuine effect of the completely derepressed *Tu* on the pigment cell system [13].

It is suggestive to assume that *Tu* might exert an essential function in the early embryo which is related to the neural crest and its derivatives. In normal embryogenesis this function becomes switched off or choked by the regulatory genes prior to the fifth day of embryonic life. If, however, the regulatory genes (i.e. the entire switch) are lacking, *Tu* continues to exert its early embryo-specific function, which as a process of misguided cellular development, transforms the competent cells to the neoplastic state.

Indispensable and Accessory *Tu*-Copies

In the sex chromosomes of the platyfish 30 deletions involving *Tu* have been characterized genetically, and some of the major deletions involving both the *Tu* and its linked regulatory genes, additionally to the genetic results, were cytologically observed [13,22]. All deletions are non-lethal. Even the loss of a *Tu*-containing segment of the X-chromosome (one Giemsa-Band) in the homozygous condition in the female, or in the hemizygous state in the male, apparently has no detectable effect on viability. This, together with the fact that the swordtail used in our crossing experiments (see Figs. 3 and 4) lacks the *Tu* population, specifically led us to the conclusions that the *Tu* so far considered is not essential but is an accessory to the fish [2]. This is not to say that the *Tu* has no normal function. One could, for instance, assume that additional copies of *Tu*, that are indispensable to the fish, are present in the autosomes and may compensate for the loss of the sex chromosome-linked *Tu* loci according to a gene dosage compensation mechanism which warrants normal functions. Support for the assumption of multiple copies of *Tu* per haploid genome comes from the following experiment. Platyfish, carrying the deletion of the Giemsa-Band that involves the accessory *Tu*, were crossed according to the procedure outlined in Figs. 3 and 4, with the swordtail lacking the accessory *Tu* population-specifically. No tumors developed in the hybrids. Following treatment of the backcross hybrids with MNU however, melanomas developed which were specifically localized at the upper part of the tail fin. These neoplasms could be assigned to an autosome. Thus it appears that the platyfish, besides the easily detectable accessory *Tu*-copies, contains additional ones that require more intricate experiments for their detection.

Our experience that all individuals of all groups of *Xiphophorus* can contribute to susceptibility to neoplasia in the hybrids, suggests that all individuals contain *Tu*-copies that are indispensable for the fish, and may contain accessory *Tu*-copies. Up to ten accessory copies of the repressed *Tu* could be introduced into a laboratory stock by crossings. No effect on viability could be observed.

Oncogene Dosage Effect and Oncogene Dosage Compensation

Both the X-chromosome of *X. maculatus* containing *Tu* (the X-chromosome according to Fig. 3) and the X-chromosome of *X. maculatus*, having lost the Giemsa-band carrying *Tu*, were introduced into the genome of *X. helleri* lacking the regulatory genes for the accessory *Tu*. X^{Tu} X^{Del} x X^{Tu} Y matings were accomplished. The segregating offspring having none, one, or two, respectively, accessory *Tu*-copies show a clearcut gene dosage effect (Fig. 8). If, however, the experiment was modified by using animals as recipients, having retained the non-linked regulatory genes we observed a clearcut gene dosage compensation (Fig. 9). Dosage effect and dosage compensation of the accessory oncogene *Tu*, therefore, depends on the absence or presence, of the non-linked regulatory genes. Oncogene dosage effect and oncogene dosage compensation has been observed in many experiments of this kind [23].

Fig. 8A-C. *Gene-dosage effect* of the (incompletely) *derepressed* tumor gene (*Tu*). (A) No *Tu* (not a single transformed pigment cell). (B) One dosage of *Tu* (melanoma formation). (C) Double dosage of *Tu* (double effect in melanoma formation). Compare with Fig. 9

Fig. 9A-C. *Gene dosage compensation* of the *repressed* tumor gene (*Tu*). (A) No *Tu* (not a single transformed pigment cell). (B) One dosage of *Tu* (spots consisting of transformed pigment cells; see dorsal fin). (C) Double dosage of *Tu* (effect shows no difference to that of one dosage). Compare with Fig. 8

Transfer of Accessory *Tu*-Copies by Injection of DNA

DNA from laboratory platyfish carrying several accessory copies of *Tu* (derepressed as well as repressed), was injected into the neural crest region of early embryos of the swordtail which lacked both the accessory *Tu*-copies and the regulatory genes (Fig. 10). The injected DNA may maintain its high molecular weight for about 2 h and thereafter becomes degraded to pieces which are too small to contain genetic in-

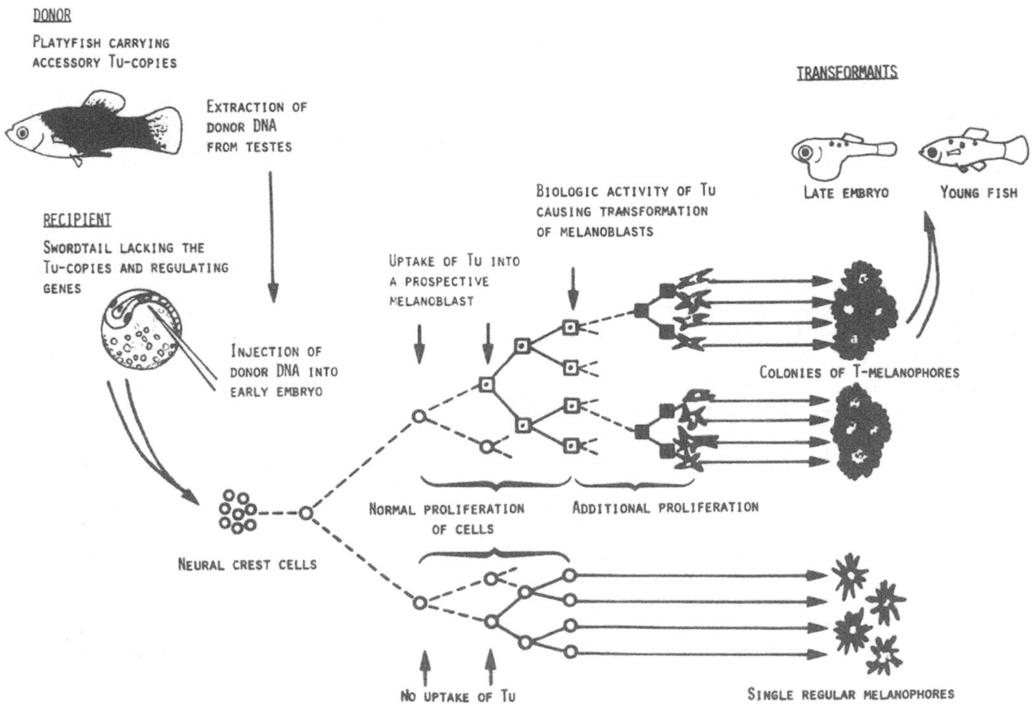

Fig. 10. Schematic presentation of the outcome of *Tu*-transfer by injection of purified DNA. (According to a scheme in [25], modified). For details see text

formation [24]. Following the successful uptake of *Tu* by an S-melanoblast of the embryo, this cell may later become competent to the *Tu*-activity by differentiation to an I-melanoblast, which eventually may become neoplastically transformed to a TI-melanoblast. Additional proliferation of T-cells amplifies the original transforming effect of *Tu*, and the result becomes visible as the transformed cells differentiate to the easily detectable colonies of T-melanocytes and T-melanophores [25].

Depending on the type of the *Tu* donor-DNA (co-transfer of intact or damaged regulatory genes), the percentage of recipients showing T-melanocytes and T-melanophores ranged from 0.4% to almost 8% (total number of survivors tested in these experiments: 1390). Since the number of the target cells (pigment cell precursors) at the time of DNA injection has been estimated to be about 1000, the frequency of the transforming event on the cellular basis is in the range of 10^{-5} [25]. This is the same order of transfection frequencies reported for cell culture systems [25-27].

The *c-src* Oncogene in the *Tu*-Melanoma System

Several virological observations came to our knowledge which could be of interest for a molecular interpretation of the *Tu*-gene: In chicken it was found that the oncogene *v-src* from Rous Sarcoma Virus (RSV) has a counterpart, *c-src*, in the non-infected cells [28]. *c-src* or at least a similar gene was also found in mammals including mouse, calf and humans [29]; and commercial DNA derived from salmon [29] obviously contains the same gene. There is, however, no convincing evidence to relate the cellular *src* or its gene product, a 60 000 Dalton phosphoprotein with kinase activity (pp60$^{c\text{-}src}$), to neoplasia that depends on conditions other than virus infections (see discussions in [30,31]). On this background we started the search for *c-src* in the genome of *Xiphophorus*.

c-src was detected in *Xiphophorus* by molecular hybridization of *src*-specific probe from cloned *v-src* with DNA from fish [32]. To identify pp60$^{c\text{-}src}$, brains of the fish were labelled with ^{32}P-orthophosphate, and brain extracts were immunoprecipitated with antisera from RSV tumorbearing rabbits (TBR-serum) followed by polyacrylamide gel electrophoresis. The 60K protein detected in the gel has a tyrosine-specific kinase activity, and represents the pp60$^{c\text{-}src}$ [31].

The kinase activity was measured according to Fig. 11 [33] and then determined (see legends of Figs. 3 and 11) in several tissues including skin, liver, spleen, testes, brain and melanoma. Brain and melanoma had always the highest kinase activity. Genotype-specific differences in kinase activity showed an identical trend in both brain and melanoma [31]. To compare *c-src* expression in non-tumorous and tumorous fish, kinase activity was mainly determined in the brains of these fish.

To study the possible relation between neoplasia and *c-src* expression we took advantage of the possibility to manipulate neoplasia in *Xiphophorus* according to the three genetic experiments outlined in Figs. 3, 4 and 12 [34].

1. In the experiment recorded in Fig. 3 the purebred *X. maculatus* carrying two repressed copies of the accessory *Tu*, as well as the purebred *X. helleri* and the BC-hybrids lacking the *Tu*, display the same activity of *c-src* kinase. This activity appears to be the basic

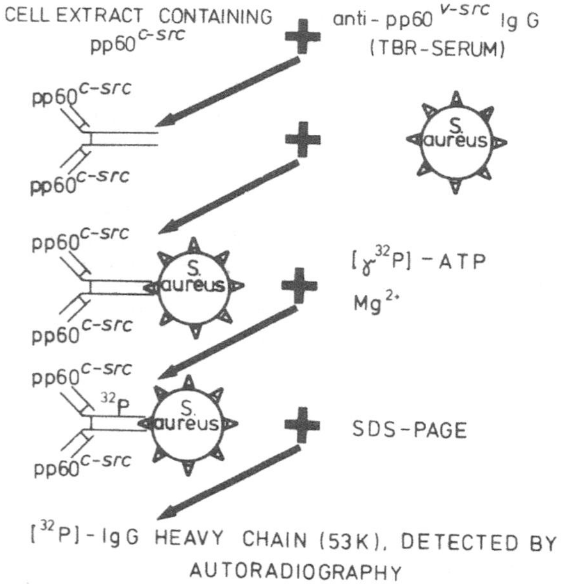

Fig. 11. Assay for pp60$^{c\text{-}src}$ kinase activity according to Collet and Erikson ([33], modified; see also [30,31,34])

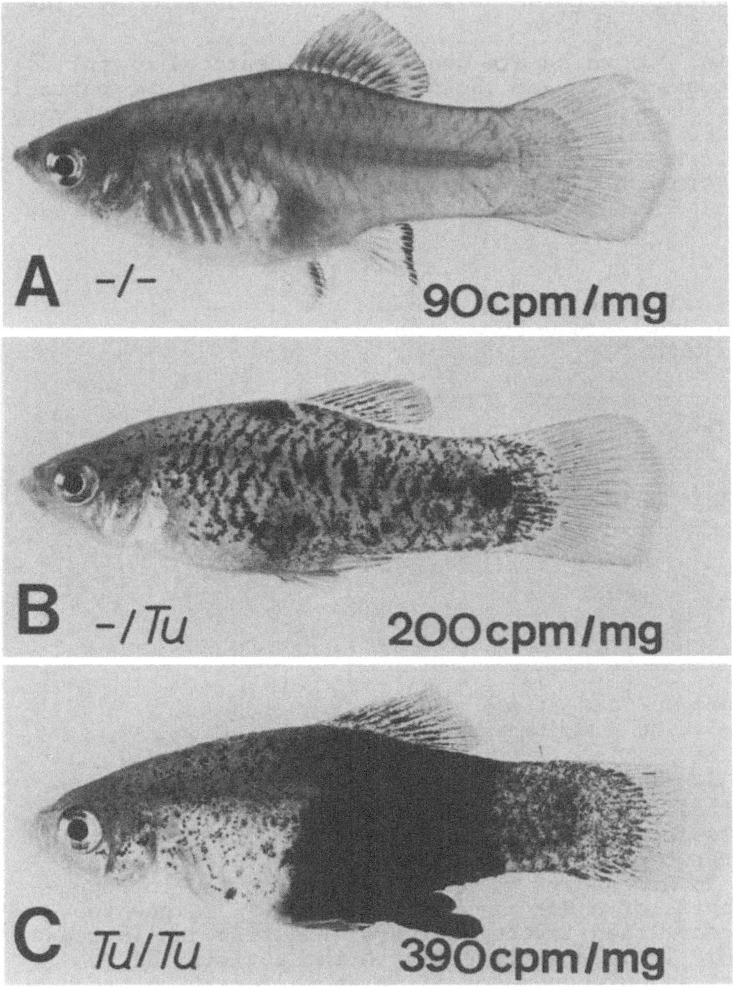

Fig. 12A-C. Correlation between gene dosage effect of *Tu* (phenotype of the tumor) and gene dosage effect of *c-src* (pp60^{c-src} kinase activity) in littermates containing (A) no accessory *Tu*, (B) one dosage of *Tu*, (C) double dosage of *Tu*. The genetic backgrounds of the fish are identical. (A) *Tu* is deleted in the germ line. (B,C) The pigment cell-specific *R* linked to *Tu* is impaired by germ line mutation [13,23]. Kinase data from [34]

expression of *c-src*. In contrast, the melanoma bearing hybrids which contain the derepressed *Tu* show an increase of *c-src* activity, with the malignant melanoma bearing BC-hybrids displaying the highest activities.

2. In the experiment recorded in Fig. 4 all purebred and hybrid animals, irrespective of the lack and the dosage of *Tu* but dependent upon the non-tumorous state exerted either by several regulatory genes or by a linked *R* alone (see the highly susceptible genotype), display a uniform *c-src* activity which seems to represent the basic

c-src expression, as in the purebred animals and *Tu*-lacking hybrids in Fig. 3.

3. In littermates (Fig. 12) which are genetically identical except for the lack of the accessory *Tu* and the presence of one or two partially derepressed accessory *Tu*-copies, *c-src* displays a kinase activity that increases stepwise in parallel to the lack and the dosage of *Tu*, which, in their turn, determine whether the animals will develop no tumors, slowly growing tumors, or fast growing tumors. Table 3 shows additional experiments of the same kind that yielded similar results.

Table 3. $pp60^{c-src}$ associated kinase activity in brain extracts specified by cpm per mg soluble protein[a] in F_2-segregants. (Data from [34])

Tu gene-complex[c]	Dosage of *Tu*[b]		
	-/-	-/*Tu*	*Tu*/*Tu*
Striped'	90	200	390
Dabbed[BR]	170	N.T.	390
Dabbed[RL]	200	260	1240

[a] 3 to 8 brains per measurement. [b] One gel each. [c] Different gels each. BR = Belize River. RL = Rio Lancetilla

The main results of these experiments are that the non-tumorous fish display a basic expression of *c-src* which in the tumorous fish may increase stepwise under two different conditions, namely (a) the stepwise derepression of an accessory *Tu*, and (b) the stepwise introduction of additional copies of a derepressed accessory *Tu*. Since the measurements were accomplished in the brains of the fish the increase of the activity of *c-src* is related directly to the activity of *Tu*, and does not represent an epiphenomenon of melanoma formation, such as the elevated activity of many enzymes, [5,6,14,15] enhanced melanin synthesis [18], certain chromosome aberrations [35], etc.

These findings suggest several possibilities for an interpretation about how *Tu* might be related to *c-src*: (a) *Tu* might be independent from *c-src*, and the correspondence between both *Tu* and *c-src* is due to linkage relationships. (b) *c-src* might represent a regulatory gene for *Tu* or *vice versa*. (c) *Tu* might consist of different oncogenes that are responsible for different kinds of neoplasia and *c-src* is one of these genes. (d) *Tu* might be identical to *c-src*, and this oncogene is capable of coding for a large variety of neoplasms. At present we cannot decide on a special interpretation. Additional data are required.

Distribution of *c-src* in the Animal Kingdom

The presence of *c-src* in the genome of different animals such as chicken, salmon [29] and *Xiphophorus* led to the more systematic search for this oncogene in additional taxomic groups of animals. Firstly, different groups of *Xiphophorus* and different fish genera more or less related to *Xiphophorus* were investigated. All fish tested (Table 4) show a pp60$^{c\text{-}src}$ kinase activity indicating that *c-src* must be present [31]. In addition, *c-src* was evidenced by its kinase activity in a large variety of metazoans other than fish, ranging from mammals to sponges, which, together with the results from other laboratories are listed in Table 5. *c-src* was not found in protozoa, algae, or higher plants [36].

Table 4. Expression of pp60$^{c\text{-}src}$ kinase in brain extracts of different fish species [31]

Xiphophorus helleri	from Belize River
X. helleri	from Rio Lancetilla
X. maculatus	from Belize River
X. maculatus	from Rio Jamapa
X. maculatus	from Rio Usumacinta
X. cortezi	
X. variatus	from Rio Coy
X. variatus	from Rio Panuco
Girardinus falcatus	
Girardinus metallicus	
Poecilia sphenops	
Belonesox belizanus	
Heterandria bimaculata	
Xenotoca eiseni	

Table 5. *c-src* in eukaryotes. (From [36])

Mammals	Cartilaginous fish
Humans [29]	Shark
Calf [29]	
Rat	Jawless fish
Mouse	Lamprey
Birds	Acrania
Chicken	*Amphioxus*
Quail	
Bony fish	Insects
Flat fish	Cockroach
Sea robin	*Drosophila* [37]
Mackerel	Sponges
Roach	Marine sponge
Gudgeon	Freshwater sponge
Xiphophorus	
Salmon [29]	
Codfish	
Cichlid	

The distribution of the cellular counterpart of the viral *v-src* brings about the idea that *c-src* might have evolved together with the multicellular organisation of the animals, and that neoplasia might be a character that is closely related to this evolution.

The Regulatory Gene *Diff* in the *Tu*-Melanoma System

The gene *Diff* is one of the most prominent regulatory genes known in the melanoma system of *Xiphophorus*. As shown in Fig. 3, benignancy and malignancy in the hybridization-conditioned melanomas depend upon the presence or absence of the chromosome carrying *Diff*. Biochemical markers for this chromosome i.e. the esterase *Est-1* and the isozyme A of the glyceraldehyde-3-phosphate dehydrogenase [6] have confirmed that the *Diff*-carrying chromosome is derived from the platyfish, the source of accessory *Tu*.

The Major Characters of *Diff*-Expression

The clearcut Mendelian segregation of benignancy and malignancy has provided the opportunity to study the basic differences between the benign and the malignant state of the melanomas (Fig. 13, Table 6). The cytological, fine structural, biochemical, and biological data suggest that *Diff* promotes the differentiation of T-cells: If *Diff* is

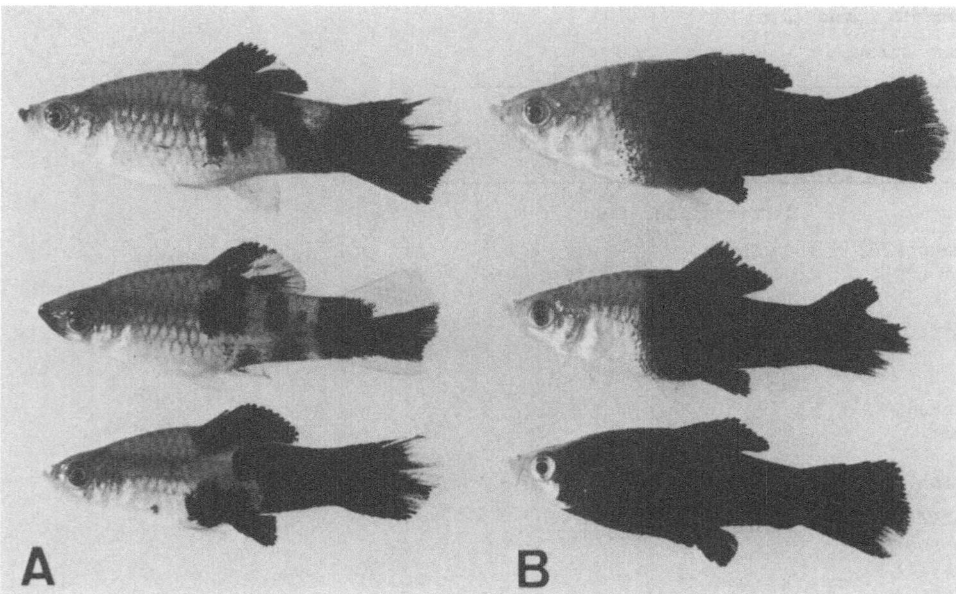

Fig. 13A,B. Segregants of littermates (A) containing one dosage of the differentiation gene (*Diff*) and (B) lacking *Diff*. Segregation of animals carrying benign and malignant melanoma is according to that of the backcross generation shown in the schematic drawings of Fig. 3

Table 6. The Gene *Diff* in *Tu*/- Xiphophorus [2,6,14,18,20,21,38,39,40,46]

Diff/-	-/-
Benign Melanoma	Malignant Melanoma
Differentiated	Poorly differentiated
Slow growing	Fast growing
Non-invasive	Invasive
Non-lethal	Lethal
Difficult to transplant	Easily transplantable
Difficult to promote	Promotion by testosteron, cAMP, corticotropin, BrdUrd, nutrient factors, etc.
Regression following testosteron treatment, etc.	No regression
No vascularization	Vascularization
Weak effect of external factors on growth rate	Drastic effect of external factors on growth rate (temperature; salinity; cyclic Bt_2AMP; corticotropin; BrdUrd; etc)
No effect of nutrient factors	Drastic effect of nutrient factors (amino acids)
Many macrophages	Few macrophages
T-melanophores are prevailing	TA-melanoblasts and T-melanocytes are prevailing
Endopolyploid and multinucleated	Diploid and uninucleated
Mature melanosomes	Immature melanosomes
Lack of ER and Golgi complexes	Well-delveoped ER and Golgi complexes
Low enzyme activities	High enzyme activities (tyrosinase; LDH B4; MDH; etc.)
Low rate of thymidine incorporation	High rate of thymidine incorporation and DNA synthesis
Low pteridine contents	High pteridine contents
First position of the anticodon of $tRNA^{Asp}$, $tRNA^{Asn}$, $tRNA^{Tyr}$, $tRNA^{His}$ contains predominantly Queuosine	First position of the anticodon of $tRNA^{Asp}$, $tRNA^{Asn}$, $tRNA^{Tyr}$, $tRNA^{His}$ contains predominantly Guanosine
Diff-product is diffusible	No product

lacking the majority of the melanoma cells persists in the stage of the poorly differentiated, continuously dividing TA-melanoblasts and T-melanocytes, and only few cells differentiate to the final stage of the T-melanophores. If, however, the *Diff* is present, the majority of the melanoma cells becomes terminally differentiated to T-melanophores, whereas only a few cells remain in the stage of TA-melanoblasts and T-melanocytes. The T-melanophores at a certain age are removed by macrophages([14,18,20,38], see also Fig. 5). This process is antagonistic to the permanent supply of melanoma cells from S-melanoblasts, and thus the melanoma renders benign [20,21].

Diffusiveness of a *Diff*-Dependent Product Involved in Differentiation

Transplantation experiments, including the composition of chimeras by fusion of parts of early embryos, have shown that pigment cell precursors present in the transplants taken from fish carrying *Tu* but lacking *Diff* (see Fig. 3) become incompletely differentiated and give rise to malignant melanoma if transplanted into embryos lacking *Tu* and *Diff*. If, however, the pigment cell precursors of the same genotype were transplanted into *Tu*-lacking embryos that contain the *Diff*, the cells of the developing melanoma become terminally differentiated and regain their distance regulation (Fig.14). Thus the effect of *Diff* on the differentiation of the neoplastically transformed pigment cells can be traced to a diffusible substance [39]. The nature of this substance is unknown.

Fig. 14. Secondary chimera composed by transplantation of tissues containing precursor cells of malignant melanoma that originated from a young BC-hybrid containing *Tu* but lacking *Diff* (see malignant melanoma developing BC-segregant in Fig. 3), to a littermate lacking *Tu* but containing *Diff* (see the non-tumorous BC-segregant at bottom left in Fig. 3). Note terminal dfferentiation and distance (density) regulation of the transformed cells of the transplant according to the *Diff*-genotype of the host. (From [39]). For details see text

Modified tRNAs involved in *Diff*-Dependent Differentiation

There is considerable evidence for the involvement of tRNA containing modifications of the nucleotides in the process of cell differentiation in normal and neoplastic tissues (see the report of H. Kersten, this vol., [40]). Many studies were focussed on a family of tRNAs including $tRNA^{Asn}$, $tRNA^{Asp}$, $tRNA^{His}$ and $tRNA^{Tyr}$ which may contain queuosine (Q) instead of guanosine (G) in the first position of the anticodon (position 34). Q is a hypermodified G. The more the differentiation progresses, the more G is replaced by Q in position 34 [40-46]. The method to estimate the G:Q-ratio in a given population of the tRNA family consists in following the replacement of guanine in position 34 by a labeled guanine exerted by a guanine-transglycosylase of E. coli.

We have measured the incorporation of ^{3}H-guanine in the tRNA for Asn, Asp, His and Tyr in the malignant melanoma, the benign melanoma and the skin of melanoma-free littermates. In addition, F_1-hybrids carrying benign melanomas were studied ([40,46], Figs. 3 and 15):

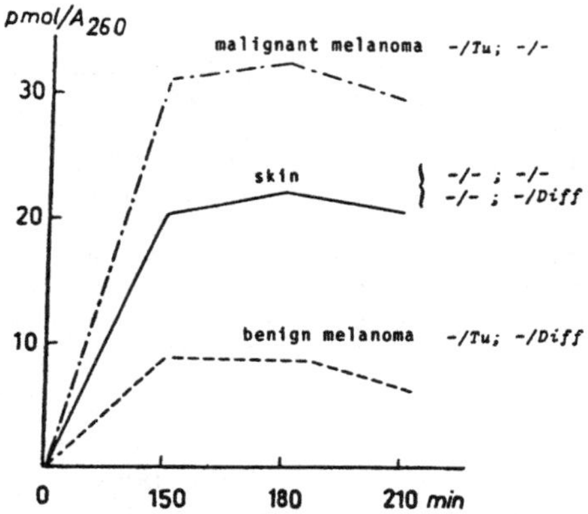

Fig. 15. Incorporation of ^{3}H-guanine in position 34 (anticodon) of tRNAs for Tyr, Asn Asp and His catalyzed by tRNA-guanine-transglycosylase (insertase) of *E. coli*. The melanomas (both malignant and benign) were derived from littermates of BC-hybrids according to Fig. 13 or Fig. 3, respectively. The skin was derived from both *Diff*-containing and *Diff*-lacking non-tumorous BC-segregants according to Fig. 3. Compare with G/Q-data shown in Fig. 3 (From [40,46]). For details see text

^{3}H-guanine incorporation is high if the tRNAs are prepared from malignant melanomas (predominantly poorly differentiated cells). In contrast, the incorporation is low if the tRNAs are derived from benign melannomas (predominantly well differentiated cells). The tRNA family of the malignant melanomas, therefore, is G-rich, whereas the tRNA family of the benign melanomas is Q-rich.

^{3}H-guanine incorporation in the skin of non-tumorous littermates is intermediate between those of the malignant and the benign melanomas. Since the non-tumorous fish (like the tumorous ones) consist of individuals lacking and containing *Diff* in a 1:1 ratio, it is suggestive to assume that the intermediate data represent a mean value of ^{3}H-guanine incorporation in the *Diff*-containing group and in the *Diff*-lacking group. If this is correct the different G:Q-ratios in the different melanomas are no epiphenomena of malignancy and benignancy, but are very closely related to the primary effect of *Diff* that in tumorous fish converts the malignant to the benign state.

Discussion

We have studied neoplasia of *Xiphophorus* at different levels of the biological organization including species, races, populations, generations, littermates, individuals, tissues, cells, genomes, chromosomes, and genes. In doing so we could trace neoplastic transformation to the activity of one or several copies of the oncogene *Tu* which shows a relation to a cellular counterpart of the transforming *src*-oncogene of Avian Sarcoma Virus, the *c-src* [30]. The normal function of *c-src* remained unkown. Since *c-src*, however, was also found in all individuals of all metazoans tested, and was not found in protozoans and plants, it might have some basic functions of life closely related to the multicellular organization of animals including humans [36].

In *Xiphophorus* it further was shown that the cellular oncogene is normally under control of systems of multiple regulatory genes corresponding to regulator genes of bacteria and phages. Some of the regulatory genes are located on other chromosomes than those bearing an oncogene. One of the most prominent regulatory genes appears to be responsible for terminal differentiation of the neoplastically transformed cells exerted via modification of nucleosides in the anticodon of certain tRNAs [40].

Interpopulational or interracial hybridization in preceding generations in *Xiphophorus* is the main event contributing to the disintegration of the regulatory gene-system for the oncogene. Germ line mutations that may also disturb the regulatory gene-systems are probably less important than hybridization because they are always rare, or may become repaired. Somatic mutations and tumor promotion, which are the majority of carcinogenic triggers may complete this disintegration. The majority of the neoplasms of *Xiphophorus* belongs to the types that are triggered by carcinogenes or promoters on a competent genetic background like their counterparts in humans, which represent about 90% of all human neoplasms (see [8]).

The phenomenon of introducing susceptibility to neoplasia by means of hybridization is not limited to *Xiphophorus*. Many examples have been cited from the animal kingdom [8]. It appears that in animals from wild populations neoplasia is difficult to induce and "spontaneously" developing neoplasms are rare, while in animals of hybrid origin (domesticated and laboratory animals; naturally occurring and experimentally produced hybrids) neoplasia is easily inducible and the incidence of "spontaneously" developing neoplasms is high.

While we do not have hybridization in human beings comparable to hybridization of domesticated or laboratory animals such as fish and mice, it is suggestive to speculate how much effect hybridization may have had on the high tumor incidence observable in some of our highly developed nations. Such speculations are probably of little value in the fight against cancer but in our search for the cause of human neoplasia they could help to realize the factors that make an individual susceptible to neoplasia and, therefore, sensitive to the carcinogens of our environment.

Acknowledgements. I am grateful to Dr. Jaenicke to be invited for the review presented at the 33. Mosbacher Kolloquium. Thanks are due to Dr. Heinz Bauer (Giessen) and Dr. Helga Kersten (Erlangen) for their design and realization of the crucial experiments recently performed in their laboratories as well as for the so far unpublished data included in this review. Thanks are also due to Dr. Nishimura (Tokyo) for providing the laboratory of Dr. Kersten with the insertase of E. coli. The critical discussion and assistance of Dr. M. Schartl, Dr. E. Scholl, Dr. A. Barnekow and Dr. A. Anders in preparing the manuscript is gratefully acknowledged. The research was supported by Deutsche Forschungsgemeinschaft through Sonderforschungsbereich 103 "Zellenergetik und Zelldifferenzierung", Marburg, Stiftung Volkswagenwerk, and by Land Hessen through Justus-Liebig-Universität Giessen.

References

1. Kallman K (1975) In: King RC (ed) Handbook of Genetics, vol. 4, p 81, Plenum Press, New York
2. Anders A, Anders F (1978) Biochim Biophys Acta 516:61
3. Rosen D (1979) Bull Amer Nat Hist 162:267
4. Schwab M (1980) Verh Dtsch Zool Ges 73:285
5. Scholl A, Anders F (1973) Archiv für Genetik 46:121

6. Scholl E (1980) Thesis, Giessen
7. Kollinger G, Siegmund E (1981) Verh Dtsch Zool Ges 74:206
8. Anders F, Schartl M, Scholl E (1981) In: Dawe C et al. (eds) Phyletic Approaches to Cancer. Japan Sci Soc Press, Tokyo, p 189
9. Schwab M, Haas J, Abdo S, Ahuja MR, Kollinger G, Anders A, Anders F (1978) Experientia 34:780
10. Schwab M, Abdo S, Ahuja MR, Kollinger G, Anders A, Anders F, Frese K (1978) Z Krebsforsch 91:301
11. Prescott DM, Flexer As (1982) Cancer: The Misguided Cell, Sinauer Associate Inc. Publishers, Sunderland, Mass.
12. Heston WE (1974) Heredity 65:262
13. Anders A, Anders F, Kline K (1973) In: Schröder JH (ed) Genetics and Mutagenesis of Fish, I 33, II 53, Springer, Berlin Heidelberg New York
14. Ahuja MR, Schwab M, Anders F (1980) J of Heredity 71:403
15. Siciliano MJ, Wright, DA (1976) Prog. Exp. Tumor Res. 20:398
16. Anders F, Schwab M, Scholl E (1981) In: Stich HF, San R (eds) Short Term Tests for Chemical Carcinogens, Springer, Berlin Heidelberg New York, p 399
17. Vielkind J, Vielind U, Anders F (1971) Z Krebsforsch 75:243
18. Vielkind U, Schlage W, Anders F (1977) Z Krebsforsch 90:285
19. Anders F, Diehl H, Schwab M, Anders A (1979) In: Klaus SN (ed) Pigmentation, its Genesis and Biological Control, vol 4, p 142
20. Anders F, Diehl H, Scholl E (1980) In: Spearman RIC, Riley PA (eds) The Skin of Vertebrates, Linnean Society Symposium Series Number 9. Academic Press, London p 211
21. Schartl M, Schartl A, Anders A (1981) In: Seiji M (ed) Pigment Cell, University of Tokyo Press, p 507
22. Ahuja MR, Lepper K, Anders F (1979) Experientia 35:28
23. Anders F, Klinke K (1966) Verh Dtsch Zool Ges 30:391
24. Schwab M, Vielind J, Anders F (1976) Mol Gen Genet 144:151
25. Vielind J, Haas-Andela H, Vielkind U, Anders F (1982) Mol Gen Genet 185:379
26. Wigler M, Pellicer A, Silverstein S, Axel R (1978) Cell 14:725
27. Willecke K (1980) In: Celis JE, Graessmann A, Loyter A (eds) Transfer of Cell Constituents into Eukaryotic Cells. Plenum Press, New York London, p 311
28. Stehelin D, Varmus HE, Bishop JM (1976) Nature 260:170
29. Spector DH, Varmus HE, Bishop JM (1978) Proc Natl Acad Sci Wash 75:4102
30. Bauer H, Barnekow A, Boschek CB, Friis RR, Ziemiecki A (This volume)
31. Barnekow A, Schartl M, Anders F, Bauer H (1982) Cancer Res 42:2429
32. Czernilofsky AP, Schartl M (unpublished)
33. Collett MS, Purchio AF, Erikson RL (1980) Nature 285:167
34. Schartl M, Barnekow A, Bauer H, Anders F (in press) Cancer Res
35. Chatterjee K, Kollinger G, Schmidt R-C, Anders A, Anders F (1981) Cancer Genetics and Cytogenetics 3:195
36. Schartl M, Barnekow A (in press)
37. Shilo BZ, Weinberg RA (1981) Proc Natl Acad Sci Wash 78:6789
38. Vielkind U (1976) J Exp Zool 196:197
39. Schartl M (1979) Thesis, Giessen
40. Kersten H 33. Colloquium Mosbach 1982. This volume
41. Okada N, Shindo-Okada N, Sato Sh, Itoh YH, Oda K-I, Nishimura S (1978) Proc Natl Acad Sci Wash 75:4247
42. Kersten H (1982) In: Usdin E, Borchardt R, Greveling B (eds) Elsevier North-Holland Inc
43. Kersten H (in press) J of Cancer Res and Clinical Oncology
44. Shindo-Ikada N, Terada M, Nishimura S (1981) Eur J Biochem 115:423
45. Kasai H, Kuchino Y, Nhei K, Nishimura S (1975) Nucl Acids Res 2:1931
46. Dess G (1982) Thesis, Giessen

Involvement of tRNA and the Modified Nucleoside Queuosine in Cell Development and Differentiation

H. Kersten[1]

Transfer ribonucleic acids, tRNAs, not only function in the fundamental steps of protein synthesis, they are also involved in many other cellular processes, e.g. protein modification, cell wall biosynthesis, viral infection, RNA metabolism, and control mechanisms of gene-expression [1,2]. The modified nucleosides found in tRNA are suggested to be concerned with the metabolic roles of tRNA. Recent experimental evidence supports the view that alterations of nucleoside modifications in specific tRNAs cause structural changes that modulate their biological activities [3].

More than 50 different modified nucleosides have been identified in tRNAs of eubacteria, of archaebacteria, and of eukaryotes [4]. Among these, the deazaguanine-derivative queuosine (7-(((cis-4,5-dihydroxy-2-cyclopenten-1-yl)-amino)-methyl)-7-deazaguanosine) is unique in that its purine skeleton is modified to a 7-deaza structure. Queuine (Q-base) was found to be present in a tRNA family, recognizing the synonym codons NAC and NAU, where N is one of the canonical RNA bases. These tRNAs are: $tRNA^{Asn}$, $tRNA^{Asp}$, $tRNA^{Tyr}$ and $tRNA^{His}$ [5] (Fig. 1).

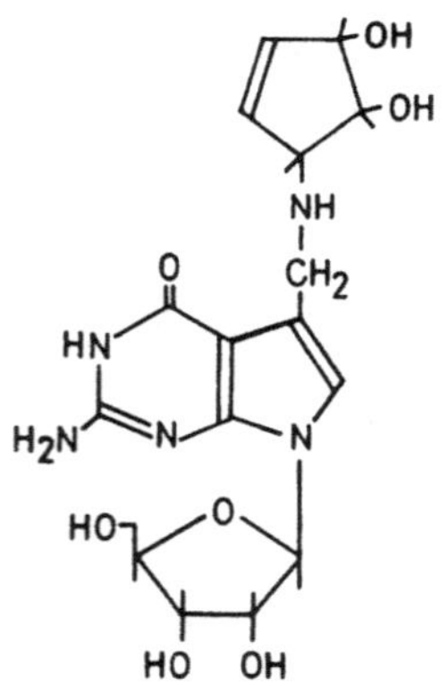

Fig. 1. Queuosine

In general the modified nucleosides are synthesized by modification of a nucleoside originally present in the polynucleotide chain. However, deazaguanine-derivatives are inserted into the polynucleotide chain by specific tRNA: guanine-transglycosylases (E.C. 2.4.2.) upon excision of a guanine residue at the first position of the anticodon

1 Institut für Physiologische Chemie der Universität Erlangen-Nürnberg, Fahrstrasse 17, 8520 Erlangen, FRG

33. Colloquium - Mosbach 1982
Biochemistry of Differentiation and Morphogenesis

(position 34) of tRNA. The eukaryotic enzyme replaces, in homologous tRNAs, guanine by queuine in an almost irreversible reaction [6]. The prokaryotic enzyme inserts precursors of queuine which finally become modified in the polynucleotide chain [7]. tRNA from mice, kept germ-free over a year and fed on a queuine-free diet, have no Q in their tRNA, suggesting that mammals obtain queuine, by the intestinal flora and/or by nutrition [8].

The purified tRNA transglycosylase from E. coli exchanges in the Q-family of tRNAs, irrespective of their origin, an unmodified guanine residue by guanine. This reaction can be used to insert labeled guanine at position 34 into corresponding Q-lacking tRNAs [9]

With this method, variable amounts of Q-lacking tRNAs have been determined in eukaryotic cells of different origin. Whereas only traces of Q-lacking tRNAs are present in differentiated cells and tissues, tRNAs with submolar amounts of queuine accumulate at distinct stages during development, differentiation, or morphogenesis of cells and organisms. This has been shown in *Dictyostelium discoideum* during development to spores [3] in *Drosophila melanogaster* during morphogenesis [10] and in *murine erythroleukemia cells* during differentiation [11].

Considerable amounts of the Q-lacking tRNAs were found in various experimental tumors of mice and rats [9]. In Ehrlich ascites tumor cells of mice we observed that the degree of Q modification in tRNA depends on cell age. Furthermore tRNAs obtained from two cell lines of solid mouse plasmacytomas show a positive correlation between the growth rate of the tumors and the amount of Q-lacking tRNA [12]. Since both cell lines originate from clones of distinct stages of differentiation the degree of Q-modification in these tRNAs might be correlated to cell-differentiation.

The melanophoric system of the platyfish *Xiphophorus* was suggested to be an ideal biological tool to evaluate Q-modification of tRNAs in relation to differentiation and neoplastic transformation. In a co-operative work with the group of Prof. Anders we discovered that tRNAs from benign melanomas and from malignant melanomas of hybrids of *Xiphophorus helleri* and *Xiphophorus maculatus* differ: Only traces of Q-lacking tRNAs are found in the benign melanomas, whereas considerable amounts are present in tRNAs of malignant melanomas.

Further results from our laboratories show that the skin of the healthy offspring of *Xiphophorus* hybrids contain an amount of Q-lacking tRNA intermediate between that of tRNAs of benign and of malignant melanomas. It is therefore possible, that the increase in the amount of Q-lacking tRNAs in malignant melanomas is correlated with the absence of the regulatory gene R_{diff}. If this can be proven in further experiments, the product of this regulatory gene should be involved in the conversion of corresponding Q-lacking tRNA to the fully modified Q-type tRNA.

To identify the gene product and its functional role we have to consider the following questions:

1. What are the causes for the occurrence of undermodified tRNAs of the Q-family!
2. What are the differences in the functional properties of Q-containing and Q-lacking tRNAs.

Possible answers to question number 1:

An incomplete modification with respect to Q might be caused by (1) limiting supply of cells with queuine, or (2) the transition of an active state of the enzyme to a less active state, or (3) the synthesis of an inhibitor of the tRNA transglycosylase.

Two types of inhibitors of tRNA transglycosylase have been found, pteridines [13] and 7-methylguanine [14]. Increased levels of pteridines and altered pteridine patterns are observable in malignant melanomas of *Xiphophorus* hybrids (Anders, personal communication). In this context it is of interest that the modified nucleoside 7-methylguanine causes hypomodification of Q tRNA and induces neoplastic transformation in *Chinese hamster embryo cells* [15].

Answers to question number 2:

So far the following functional differences of Q-containing and Q-lacking tRNAs have been observed. In a rat liver system with $tRNA^{Asp}$ as substrate the apparent V_{max} of the aminoacylation is about tenfold higher for the Q tRNA than for the unmodified counterpart [16].

In *Dictyostelium discoideum* during the onset of development we observed a transiently uncharged Q-lacking $tRNA^{Asn}$-isoacceptor. Uncharged tRNAs in eukaryotes have been suggested to be involved in regulatory mechanisms at the translational level ([17] and ref. therein).

The Q tRNAs are suggested to prefer codons NAU over NAC, whereas the Q-lacking rRNAs read NAC and NAU equally well [18]. This can be an important mechanism in the selection of specific mRNAs e.g. an adaptation of tRNA anticodons to mRNA codons.

For $tRNA^{Tyr}$ it has been shown that the Q-lacking species reads a terminator codon, probably UAG [19]. Thus the Q-lacking tRNAs might change gene-expression by termination.

In conclusion: Some particular biological functions of incompletely modified tRNAs of the Q-family have been found. The Q-deficient tRNAs can play a role as modulators of gene-expression at the transcriptional as well as at the translational level. Beside Q-lacking tRNA, free queuine may also be taken into consideration as being involved in the regulation of gene expression.

References

1. Clark BFC (1978) In: Altman S (ed) Transfer RNA. MIT Press, Cambridge, Mass.
2. Morgan SD, Söll D (1978) In: Cohn WE (ed) Nucl Acid Res Mol Biol. Academic Press, London New York, pp 181-209
3. Kersten H (1982) Transmethylation. In: Usdin E, Borchardt R Creveling B (eds) Elsevier/North Holland Biomedical Press, Amsterdam New York, pp 357-369
4. Sprinzl M, Gauss DH (1982) Nucleic Acids Res 10:57-81
5. Kasai H, Kuchino Y, Nihei K, Nishimura S (1975) Nucleic Acids Res 2:1931-1939
6. Howes NK, Farkas WR (1978) J Biol Chem 253: 9082-9087
7. Okada N, Noguchi S, Kasai H, Shindo-Okada N, Ohgi T, Goto T, Nishimura S (1979) J Biol Chem 254:3067-3073
8. Reyniers JP, Pleasants JR, Wostmann BS, Katze JR, Farkas WR (1981) J Biol Chem 256:11591-11594
9. Okada N, Shindo-Okada N, Sato S, Itoh YH, Oda K-I, Nishimura S (1978) Proc Natl Acad Sci USA 75:4247-4251

10. White BN, Tener GM, Holden J, Suzuki DT (1973) J Mol Biol 74:635-651
11. Shindo-Okada N, Terada M, Nishimura S (1981) Eur J Biochem 115:423-428
12. Kersten H (1982) J Cancer Res Clin Oncol (in press)
13. Jacobson KB, Farkas WR, Katze JR (1981) Nucleic Acids Res 9:2351-2366
14. Elliott MS, Trewyn RW (1982) Biochem Biophys Res Commun 104:326-332
15. Trewyn RW, Kerr SJ (1978) Cancer Res 38:2285-2289
16. Singhal RP, Kooper RH (1981) Fed Proc Abstr 615:1646
17. Dingermann Th, Ogilvie A, Pistel F, Mühlhofer W, Kersten H (1981) Hoppe-Seyler's Z Physiol Chem 362:763-773
18. Grosjean HJ, de Henau S, Crothers DM (1978) Proc Natl Acad Sci USA 75:610-614
19. Bienz B, Kubli E (1981) Nature (London) 294:188-190

Cell Differentiation

Regulation of Bacterial Sporogenesis

J. Szulmajster[1]

Previous speakers have stressed some important features of gene expression and regulation in different eukaryotic systems. My communication will deal with similar problems but revealed in a prokaryotic system, endowed with the properties to undergo differentiation.

What is the Problem? It has been known for about 140 years, due to the pioneering observation of Ehrenberg [1], Perty [2], Cohn [3], and Koch [4] in Germany and Pasteur in France [5], of the existence, inside some bacterial aerobic and anaerobic species, of brilliant refractile bodies that are resistant to toxic chemicals. Soon afterwards it was found that these refractile bodies, when placed in a fresh medium, are capable of being converted into the same organism from which they were derived. Thus the existence of the bacterium-spore cycle was demonstrated (Fig. 1).

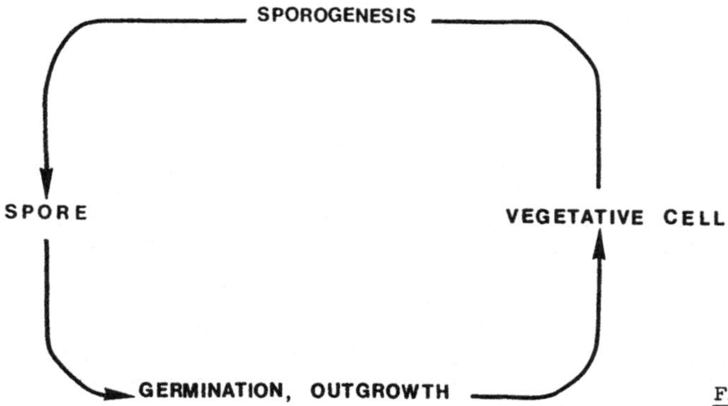

Fig. 1. Spore-bacterium cycle

Bacterial spore formation is an ordered developmental process initiated by the exhaustion of an essential nutrient in the environment. This nutritional deficiency provides a starting signal for the sequential morphological and biochemical changes accompanying the metamorphosis of a vegetative cell into a highly resistant dormant spore. The bacterial spore is known to be metabolically inert and is about 10,000-fold more resistant to heat than the vegetative cell and 10- to 100-fold more resistant to radiation [6]. These are the most striking properties of bacterial spores and it was the effort to understand and control

1 Laboratoire d'Enzymologie du CNRS, 91190 Gif-sur-Yvette, France

33. Colloquium - Mosbach 1982
Biochemistry of Differentiation and Morphogenesis

these properties, in relation to food processing, which was probably one of the initiating forces for research concerning bacterial spores over a long period.

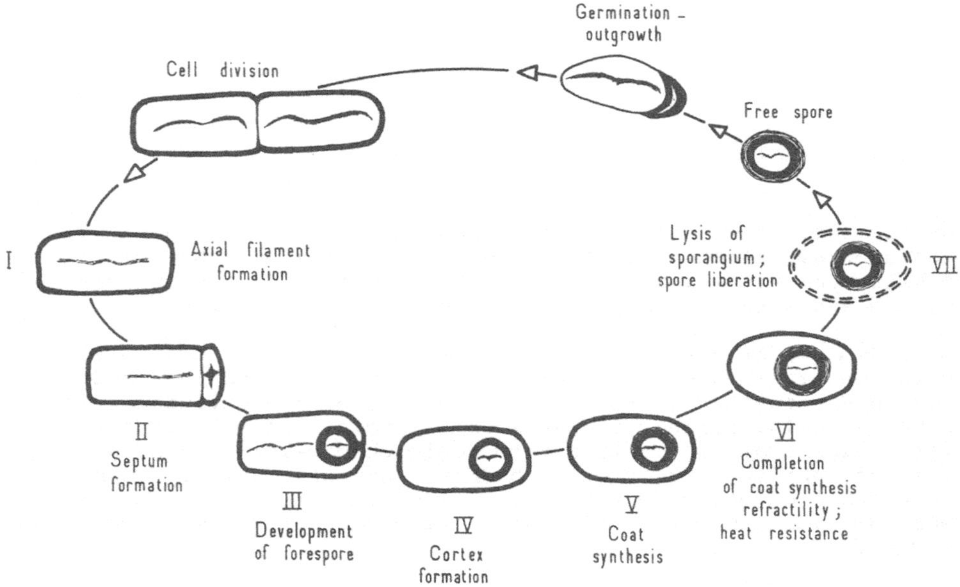

Fig. 2. Diagrammatic representation of the life cycle of a sporogenic bacterium with the seven characteristic structural steps recognizable on the electron microscope

Morphology. The morphology of spore formation is now well established in many sporogenic bacteria. Figure 2 shows a schematic presentation of a sporogenic bacterium with the seven characteristics and distinctive structural steps recognizable on the electron microscope during growth cycle (reviewed in [7]). One of the earliest events in sporogenesis is a condensation of the DNA along the axis of the cell, then a compartmentalized cell is created by a transverse septum formed on one pole of the rod by invagination of the cell membrane enclosing a part of the nuclear material and the surrounding cytoplasm, forming the forespore compartment. The latter is then enveloped by a second membrane layer and completely surrounded by the mother cell. The forespore goes through several distinct phases of development, such as deposition of a cortical layer of mucopeptide between the forspore membranes, accumulation of Ca^{++}, dipicolinic acid, appearance of refractility and assembly of a multilayered spore coat. The larger portion of the cell, the sporangium, undergoes enzymatic lysis, and the mature spore is finally freed (see Fig. 3).

Bacterial sporulation is thus regarded as a unique model for studying differentiation at a unicellular level.

Genetics. The biochemical events associated with the formation of a mature spore are under the control of genes or operon-like groups of genes, randomly distributed along the *B. subtilis* chromosome. A great number of morphological mutants (spo^-) blocked at different stages of spore development, but non affected in their vegetative growth, have been isolated and their mutations localized on the *B. subtilis* chromo-

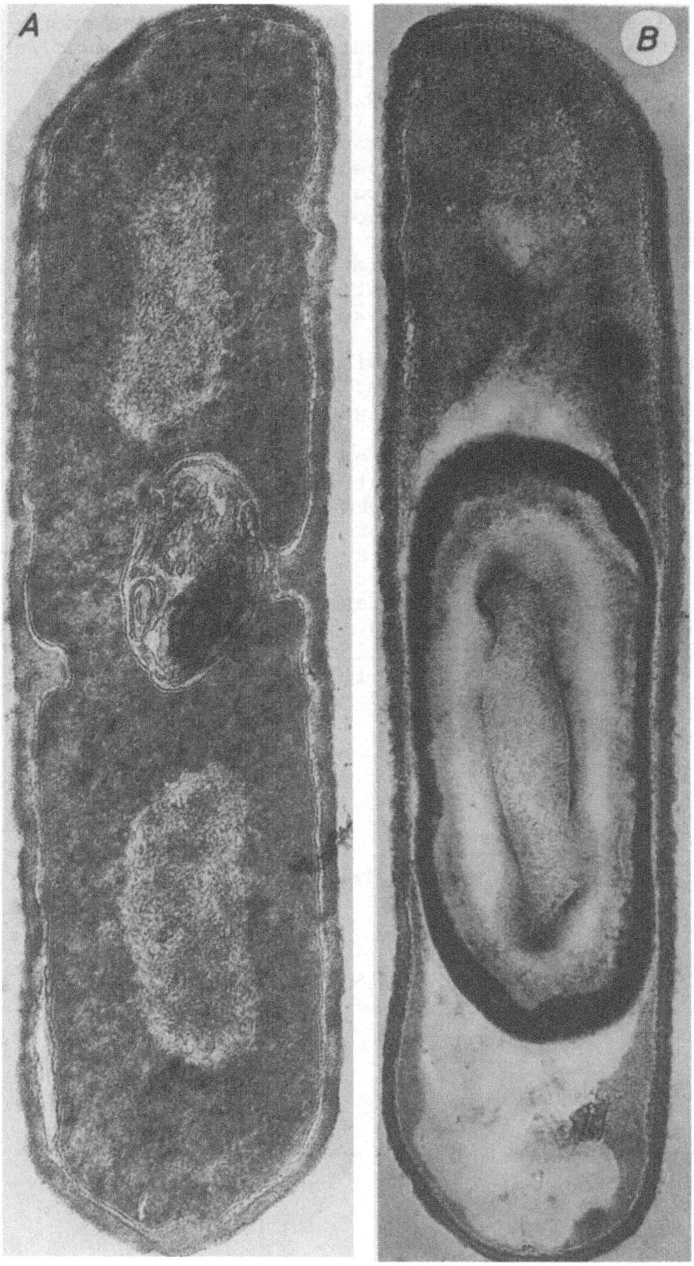

Fig. 3. (A) An exponential growing cell of *B. subtilis*. (B) Spore stage VI

some. From genetic studies it was calculated that the possible number of loci or gene clusters involved in sporulation is about 35-42, which means that sporulation might be under the control of about 200 genes (reviewed in [8]). It is quite possible that the major part of these genes are regulatory genes in addition to the structural ones.

At least ten genetic loci are involved in the initiation of the sporulation process (spoO genes) and are widely scattered on the *B. subtilis* chromosome. Mutations in these genes are of pleiotropic character. Such mutations block spore formation prior to the completion of any morphological changes and also alter several vegetative functions. Mutations in the spoOA locus, for example, prevent the synthesis of NADH oxidase, the excretion of sporulation associated antibiotics and proteases, and interfere also with the acquisition of genetic competence. It was suggested [9] that some, if not all, of the spoO genes are expressed in growing cells and that their products probably play an essential role in the initiation of sporulation. Although some of the spoO genes have recently been cloned, little is known about the nature and function of their product [10,11,12,13].

Biochemical Events. The biochemical events associated with spore development have been reviewed earlier by Kornberg et al. [14], Hanson et al. [15], Szulmajster [16,17] and Young [18]. In general these events can be divided into two main categories depending on their degree of relevance to the sporulation process. The events of the first category are those that are essential for spore development (Fig. 4), such as the spore specific component, dipicolinic acid (DPA), the cortical peptidoglycan, the uptake of Ca^{++}, etc. The events of the second category are those involved in the production of substances which are synthesized at the initiation of sporulation but play no part in this process. Examples of the products of this category are exoproteases, antibiotics, ribonuclease, alkaline phosphatase, etc. (see Fig. 3). Although the events of this category are dispensable for sporulation, their times of appearance are precisely programmed and are utilized by the geneticists as useful markers for classifying asporogenic mutants.

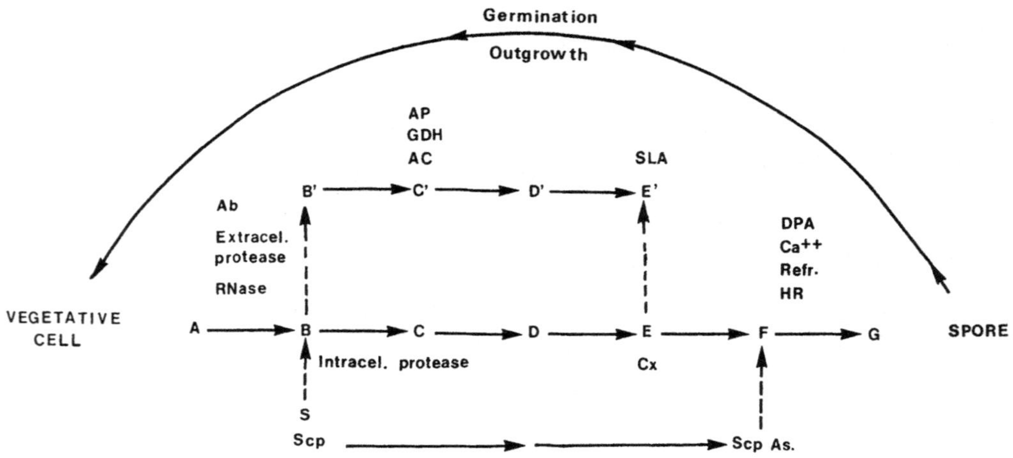

Ab : ANTIBIOTIC
GDH : GLUCOSE DEHYDROGENASE
AP : ALKALINE PHOSPHATASE
AC : ACONITASE
SLA : SULFOLACTIC ACID
Scp : SPORE COAT PROTEIN
Scp As. : SPORE COAT PROTEIN ASSEMBLY
Cx : CORTEX
HR : HEAT RESISTANCE
Refr. : REFRACTILITY

Fig. 4. Biochemical events associated to sporulation process

It is now generally assumed that the biochemical as well as the morphological events display a dependent sequence in the sense that if any event is blocked by an inhibitor, or as a result of a mutation, the expression of the later events will not occur [19]. This implies that during sporulation there is sequential gene expression. The expression of the first locus is likely to be the subject of catabolite repression [19], although it differs from the normal catabolite-sensitive functions of the cell [8]. When the repression is released by exhaustion of a nutrient in the medium, the first gene (or gene cluster) is activated and one of the products of its expression is activating the next locus, and so on [19]. How all these events are regulated in time and space is the major challenge in sporulation studies.

Control Mechanisms. During growth of prokaryotic cells there is a set of controls operating mainly at the level of transcription of DNA into RNA, the translation of RNA into proteins, and the activity of proteins as enzymes. In bacterial sporogenesis, however, an inventory of research has to be performed to understand the controls involved. There are cues and signals which turn on and off the potential information and there is a need for regulation of the expression of this information. Microbial sporulation thus represents a unique biological system whose regulation is intermediate in complexity between the regulation of the production and activity of bacterial enzymes on the one hand, and differentiation of specialized tissues in higher organisms, on the other hand. Fig. 5 shows some examples, by no means exhaustive, of the control mechanisms operating during sporulation. All these mechanisms operate inside the cell at different levels but, as mentioned before, are initiated by a signal or stimulus from outside the cell, that is starvation.

Starvation Signal and Catabolite Repression. The starvation signal implies that the levels of various catabolites in the environment provide the starting signal for the entire process and leads to the hypothesis that

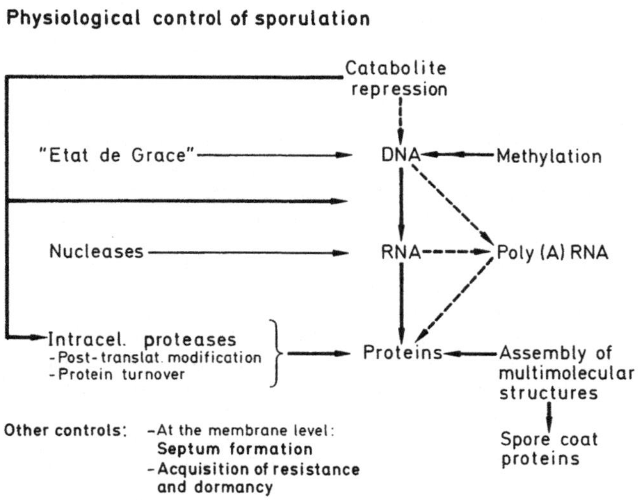

Fig. 5. Possible control mechanisms of bacterial sporogenesis

during vegetative growth, glucose (or another carbon source) causes inhibition of sporulation in the presence of an assimilable nitrogen source by catabolite repression [20]. What seems to be clear now is that the mechanisms which operate to overcome the repression of sporulation are different in several respects from those involved in the relief of catabolite repression of inducible enzymes [8]. Zero stage asporogenic mutants have been isolated that are hyper-repressed as regards all sporulation events, but not impaired in their ability to sporulate [21]. On the other hand, it was found that mutants insensitive to glucose or NH_4^+ repression of sporulation, behave normally with respect to catabolite repression of inducible enzymes [22]. Furthermore sporulation could only be initiated while DNA was being replicated, whereas the induction of enzyme synthesis was independent of DNA replication [23].

A number of metabolite effectors, that under certain conditions repress sporulation [24], and compounds that appear at the initiation of sporulation [25] have been reported. However, in both cases the mode of action still remains to be established.

Studies on the control mechanisms at the transcription, translation and post-translation levels have recently been reviewed [26] and will not be further developed here.

I will rather focus the attention on two other possible regulation mechanisms and they are (a) the role of intracellular proteases and (b) the role of poly(A)RNA as a specific message for sporulation.

Function of Proteases During Sporulation

Extracellular Proteases

It has been known for a long time that spore-forming bacteria excrete, after the end of exponential growth, enzymes with high proteolytic activity. A correlation between extracellular protease production and spore formation was suggested by a number of studies with different spore-forming bacilli and led to the assumption that the synthesis of proteases might be essential for the initiation of spore development. However, the isolation of protease deficient mutants capable of undergoing sporulation has led to the conclusion that the production of an extracellular protease is not an absolute requirement for sporulation [27]. At the present stage of knowledge it can be assumed that protease production and initiation of sporulation are two phenomena that are only linked fortuitously and that a mutation in a regulatory gene which blocks the protease synthesis also blocks some early steps in sporulation by a pleiotropic effect. In general the studies of enzymes which appear during the initial stages of sporulation, have been complicated by the fact that sporulation is initiated at the end of exponential growth. At this same period many catabolite repressed genes are also derepressed as the result of the depletion of glucose from the medium. It is then difficult to decide whether the appearance of an enzyme was the result of derepression of a catabolite repressed gene or the expression of sporulation-specific genes. It is likely that expression of the initial sporulation genes and catabolite repressed genes are both under identical catabolite control mechanisms [28].

Intracellular Proteases

Recently more attention has been focused on a possible function of intracellular proteases in some stages of spore development.

It is now well known that many intracellular differentiation processes found in micro-organisms occur under starvation conditions unfavorable for growth. Therefore, it may be expected that proteases represent essential tools in such dvelopmental processes, since they allow the development of specific new protein patterns by providing the necessary amino acids by degradation of pre-existing proteins. Two main functions are currently attributed to the intracellular proteases in spore development: one concerns the extensive turnover of proteins occurring during sporulation, and the other, post-translational modifications of pre-existing vegetative proteins [29]

As mentioned before, bacterial sporogenesis occurs under conditions unfavorable for growth. Starvation might then produce abnormal proteins which could be harmful for sporulation. Such proteins - let's say sporulation repressors - might be destroyed by intracellular proteases which then have a scavanger function allowing thus initiation of transcription of spore genes. Intracellular proteinases might also play a role in processing of spore coat precursors [30].

However, further studies on the role of intracellular proteases in spore development have been hampered by the lack of mutants deficient in these enzymes. This gap is now partially filled up by the isolation of sporulation mutants altered in the intracellular protease activity [30-32].

Thermosensitive Sporulation Mutants of Bacillus subtilis Affected in Their Intracellular Protease Functions. In pursuing this aspect of investigation we have isolated thermosensitive sporulation mutants of *Bacillus subtilis* that present the characteristics shown in Table 1, as compared to the wild type strain. It can be seen that the mutation in ts-15 grown at 42°C is affected in the intracellular protease activity and in spore formation, although the levels of the extracellular protease activity remained practically unchanged, independent of the growth temperature. It can also be seen in this table that the two thermosensitive traits of the ts-15 mutant, loss of sporulation capacity and of intracellular protease activity, were transmitted together into the wild-type strain 168M by transformation using saturating concentrations of ts-15 DNA as donor. Furthermore, two revertants (ts-15R_1 and ts-15R_2) were isolated and their properties became comparable to those of the wild type strain i.e. when grown at 42°C they have regained completely, or a major part of, their intracellular protease activity and have also recovered their entire sporulation capacity. The extracellular protease activity in these two revertants was rather increased at both temperatures of growth.

Since the altered properties of the mutant ts-15 are restored in the revertants and can also be transmitted together by transformation, it can therefore be assumed that ts-15 is most likely carrying a point mutation.

Turnover. A correlation between protein turnover with intracellular protease activity and sporulation capacity was previously established with thermosensitive sporulation mutants of *B. subtilis* [31] and *B. cereus* [30]. This relationship was further confirmed with the ts-15 mutants as shown in Fig. 6. The rate of turnover in the stationary phase of the wild-type *B. subtilis* strain varies generally between 7% and 12%

Table 1. Levels of extracellular and intracellular protease activity and sporulation capacity of parental strain SMY-512, arg⁻, of its ts-15 mutant, revertant and transformant [32]

Strain	Growth temperature	Protease activity		Spores
		Extra-cellular	Intra-cellular	
	°C	Units/mg protein		%
SMY-512, arg⁻	30	8200	660	95-100
	42	6350	550	95-100
ts-15 (arg⁻)	30	8000	525	90-95
	42	5000	n.d.	± 0
Transformant	30	-	350	90-95
	42	-	34	3

in 1 h [32]. It can be seen on this figure that in ts-15 grown at 30°C the rate of turnover was indeed about 12% in 1 h while in ts-15 grown at 42°C this rate was reduced to a negligible value. On the contrary, in the revertant (ts-15R_1) grown at 42°C the turnover rate was almost completely restored.

Immunological Studies. Although, in the assay conditions used, no intracellular protease activity was detected in the crude extracts of the ts-15 mutant grown at the restrictive temperature, it was observed, by means of antibodies made against the intracellular protease of *B. subtilis*, that an antigen crossreacting with these antibodies is present in these enzymatically inactive extracts. Using the "rocket" electrophoresis procedure [33] a quantitative estimation of the amount of antigen protein present in the extracts of ts-15 grown at 42°C and 30°C was possible. It was thus calculated that about 30%-35% of the protein was made in the ts-15 when grown at 42°C compared to that made at 30°C, while the specific activity, measured in a fivefold concentrated extract from ts-15 grown at 42°C, is about 400-500 times lower than when grown at 30°C.

These experiments strongly suggested that the mutation in ts-15 resides in a structural gene of an intracellular protease. This was further substantiated by the inactivation kinetics of the purified ts-15 enzyme[2] as shown in Fig. 7. It can be seen that the purified protease from the ts-15 mutant is considerably more sensitive to heat at 65°C than the wild type; the half-life of the ts-15 enzyme is about 7 min compared to 70 min of the wild type. No such difference was observed with the crude extract preparations where the enzyme is probably protected against heat inactivation.

---

2 The purification procedure, molecular weights, substrate specificity and inhibition pattern were described in [32]

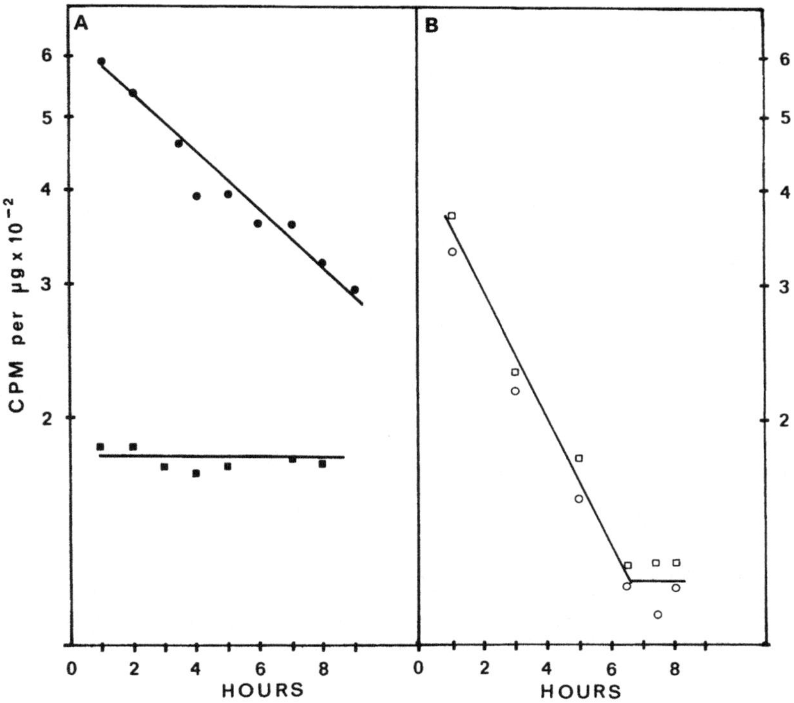

Fig. 6A,B. Protein turnover. Cells were grown in nutrient broth (Difco) and incubated, one culture at 30°C and another one at 42°C; 2 h after the end of the exponential growth (t_2) a sample of each culture was supplemented with o.6 µCi L-(^{14}C)leucine (330 Ci/mol) and incubation continued at the corresponding temperatures. After 1 h 1mg/ml of unlabeled L-leucine was added and samples of 0.5 ml were withdrawn at 1-h intervals, precipitated in 1 ml of cold 20% trichloroacetic acid and the proteins collected by a 20-min centrifugation at 15000 *g*. The pellets were further washed by 2 ml of 0.2 M NaOH, reprecipitated with trichloroacetic acid and then dissolved in 1 ml of 0.2 M NaOH. Proteins were determined by the Lowry method and the radioactivity counted in a scintillation counter. Time zero on the figure designates the time of dilution of the L-(^{14}C)leucine by unlabeled leucine. (A) ■ ts-15, ● ts-15R grown at 42°C; (B) □ ts-15, o ts-15R grown at 30°C (From [32])

Subunits Structure. Studies on the molecular weight in normal and dissociating conditions (in presence of sodium dodecylsulfate) have suggested that the active intracellular protease of the wild type or of the ts-15 mutant grown at the permissive temperature (30°C) is formed from two subunits of similar or identical size (33,000-34,000) whereas at the restrictive temperature (42°C) the mutant is also synthesizing the two subunits but is unable to assemble them to the active form.

On the basis of these results it appeared to us permissible to conclude that the mutation in ts-15 is most likely a point mutation in a structural gene encoding an intracellular protease, and results in an inability to assemble the two subunits into an active form.

Biological Functions. The results I have discussed so far show that the intracellular protease described here plays an important role in sporu-

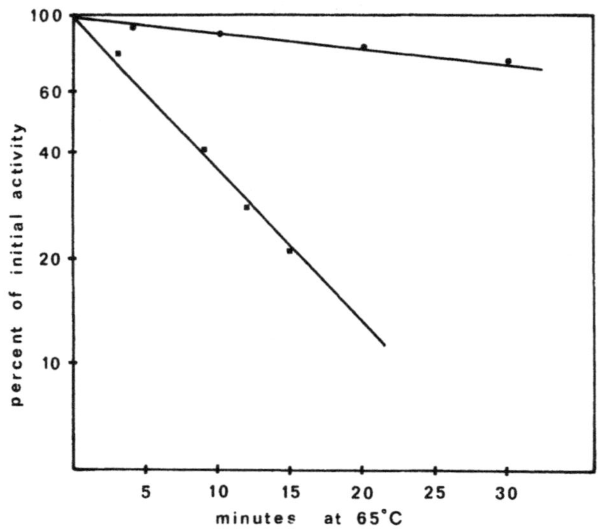

Fig. 7. Heat inactivation. Samples of purified enzyme (wild type and ts-15 grown at 30°C) were heated at 65°C for periods indicated on the figure. The reaction was stopped by immersing the tube in ice and the enzyme activity was measured in the standard conditions. ■——■ ts-15 grown at 30°C, ●——● wild type

lation process of *B. subtilis* and that one of its functions is most likely protein turnover. But we know now that the rate of proteolytic degradation is specific for different proteins. Thus protein turnover during sporulation must be selective for some proteins, which still remains to be characterized.

The intracellular protease could also be involved in modification of vegetative enzymes, in derepression of proteins specific for sporulation or in processing of precursors of spore coat proteins [30]. Therefore to gain more inside knowledge of the precise function(s) of the intracellular protease in spore development the specific target of this protease in a sporulating cell has to be discovered.

Earlier in vitro studies using the DNA dependent RNA polymerase isolated from *B. subtilis* have shown that the intracellular protease purified from *B. megaterium* or *B. subtilis* preferentially degraded one of the β subunits of this enzyme, while the extracellular protease, from the same micro-organisms, completely destroys this enzyme in 1-2 min [34]. However, no such modification was observed in vivo, i.e. when the RNA polymerase was extracted from sporulating cells in presence of protease inhibitors [35].

Another natural substrate tried for the intracellular protease was the aspartyl-transcarbamylase (ATCase) purified from *E. coli* (a gift from G. Hervé). This enzyme from *E. coli* is a multimeric one, built up from six regulatory subunits and six catalytic subunits. The molecular weight of each catalytic subunit is 33,000. We observed that the regulatory subunits are rapidly and completely destroyed by the intracellular protease from *B. subtilis* while the catalytic subunits undergo

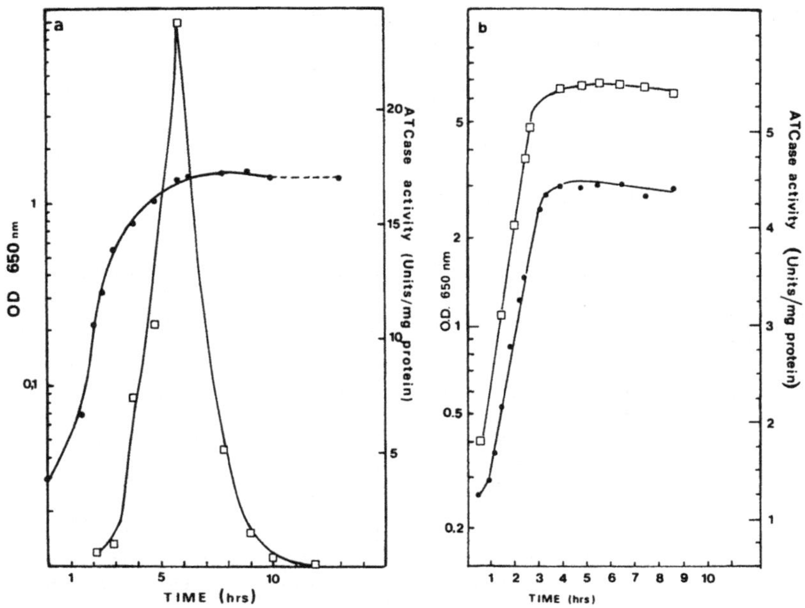

Fig. 8a,b. ATCase activity during growth of *B. subtilis* (a) wild type strain and (b) of the 12A zero stage asporogenic mutant

only a limited proteolysis (Kerjan, P., Szulmajster, J. and Hervé, G., unpublished results). The susceptible site of the protease action seems to be located at a position which is cleaving a fragment of 22,000. Since the secondary structure of this enzyme is well established it is now possible to determine the nature of the site of action of this protease at least on the *E. coli* enzyme.

The ATCase in *B. subtilis* is one of the enzymes which is completely inactivated and degraded in starving cells prior to sporulation [36]. Although it was assumed that this inactivation-degradation is carried out by an energy dependent mechanism [36], it was interesting to find out if the same phenomenon is observed in a protease deficient strain of *B. subtilis*. Figure 8a,b shows that compared to the wild type there is no degradation of the ATCase in this mutant. This would indicate that the intracellular protease might interfer with the degradation of the ATCase in vivo. Although the physiological significance of this phenomenon remains unknown, it nevertheless presents a good example of a participation of the intracellular proteases in the metabolism of the cells at the onset of sporulation (Legouar, M. and Hervé, G., unpublished results).

Proteases in Germinating Spore. Proteolysis also plays a role in spore germination. Approximately 20% of the proteins in dormant spores of *B. megaterium* is degraded to free amino acids in the first 20 min of spore germination. These free amino acids are essential for protein synthesis during germination and outgrowth [37]. The proteolytic enzymes involved in this degradation have the unique specificity to degrade only the storage proteins and not other spore proteins. As the cells progress toward vegetative growth, the protease activities rapidly disappear. The question that still remains to be answered is how these proteases are regulated so that they degrade only some specific proteins and only during germination and not in the developing forespore [37].

Role of Polyadenylated RNA in Sporulation Process

Another appealing feature in the regulation of gene expression during sporulation emerged recently from work on *B. polymyxa* and *B. subtilis* and concerns a possible role of polyadenylated RNA (poly(A)RNA) in this developmental process.

Evidence has been gathered during the last few years indicating the existence of poly(A) sequences in the RNA of prokaryotic cells including bacillus species [38-44]. However a correlation between the appearance of poly(A)RNA species and sporulation was until now only observed in *B. polymyxa* [43].

It was then of great interest to generalize this observation to other spore forming bacilli and to investigate the function of poly(A)RNA in the sporulation process of *B. subtilis*. The advantage of this organism is that it has been the subject of extensive genetic studies, providing thus a comprehensive map of the chromosomal location of genes affecting sporulation [8].

Table 2 shows the binding capacity of pulse labeled RNA isolated from wild type cells of *B. subtilis* at their vegetative and sporulating stages to oligo(dT)-cellulose. It can be seen that during vegetative growth only about 0.6% of the RNA was bound to the oligo(dT)-cellulose, while the RNA isolated from t_3 and t_6[3] stages of sporulation contained re-

Table 2. Binding of poly(A) containing RNA to oligo(dT)-cellulose. Cells were labeled with (2,8-^{3}H)adenosine for 3 min in vegetative phase and for 30 min during sporulation stages (t_3 and t_6). At the end of each labeling the RNA was extracted and samples were adsorbed on oligo(dT)-cellulose at high salt conentration and eluted by water

Strain	Time of harvest	RNA samples (cpm)	Radioactivity bound to oligo(dT)-cellulose	%
168 M	Vegetative	950,000	6,000	0.6
	t_3	675,000	81,000	12
	t_6	735,000	152,000	20.5
12A	Vegetative	755,000	± 0	-
	t_3	868,000	± 0	-
	t_6	900,000	± 0	-

3 The end of exponential growth is designated as t_0; t_1, t_2, etc. is the time, in hours, after t_0

spectively 12% and 20% poly(A)RNA species bound to oligo(dT)-cellulose. This sharp increase of oligo(dT) binding RNA appears to be sporulation specific, since the RNA extracted from stationary phase cells of the zero stage asporogenic mutants (12A) contains no such species (Table 2).

When the poly(A)RNA from sporulating cells was further analysed by sucrose gradient centrifugation and the fractions hybridized to (^{3}H)-poly(U) it was observed that poly(A)RNAs sediment with a peak at about 6 s with a broad shoulder, reflecting heterogenous poly(A)RNA populations of larger sizes up to about 13-14 s (Fig. 9).

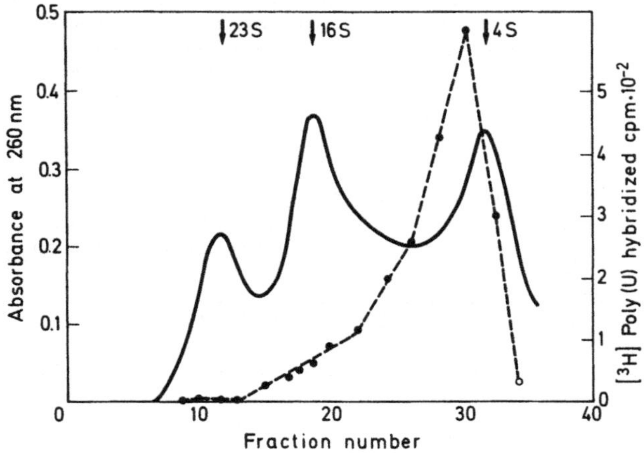

Fig. 9. Sucrose gradient centrifugation profile of (^{3}H)adenosine labeled RNA isolated from t_3 sporulating cells of *B. subtilis*. Fractions were collected and analysed by hybridization to (^{3}H)poly(U) filters

Location of the Poly(A) Sequences in the RNA Chains as Determined by Post-Labeling of the 3'OH Terminal with (^{3}H) Na-Borohydride. The procedure involves oxidation of the ribose moiety with sodium metaperiodate and reduction of the resulting dialdehyde with tritiated borohydride [45]. By combining this procedure with the action of RNases A and T_1 and alkaline phosphatase we have been able to determine the relative amounts of the poly(A) stretches located at the 3' terminus and those located inside the RNA chains. After treatment by ribonuclease A and T_1 only the poly(A) sequences which are resistant to these enzymes remain intact. The rationale for further treatment by alkaline phosphatase is to remove the phosphate groups from the internal poly(A) stretches freed by the RNases action, making these sequences also accessible to (^{3}H)-borohydride treatment. Thus the increase in the radioactivity bound to oligo(dT)-cellulose, observed afterwards, corresponds to the internal sequences in the RNA. By this method we have been able to show that about 66% of the total poly(A)RNA sequences are localized at the 3'-terminus of the RNA chains whereas about 34% are in the internal region of the RNA chains [44].

The Size of the Terminal Poly(A) Sequences was estimated by polyacrylamide gel electrophoresis with RNA samples either post-labeled by (^{3}H) Na-borohydride or pulse labeled with (^{3}H)-adenosine followed by digestion with RNases A and T_1. It was revealed that about 75%-80% of the label was migrating with the synthetic poly(A) marker of 160-180 nucleotides length. Only 20%-25% are of about 10-15 nucleotides length and probably not retained on oligo(dT)-cellulose [44].

Poly(A)RNA-cDNA Hybridization. To gain insight on the nature and role of poly(A)RNA in the sporulation process we resorted to the synthesis of the corresponding complementary DNA (cDNA) probes for poly(A)RNA in determining their hybridization specificity with RNAs isolated from various sporulation stages, or from vegetative cells, or from stationary phase cells from a zero stage sporulation mutant. As shown in Fig. 9, no hybridization was observed when total RNA isolated from either vegetative or from t_4 stationary phase cells of a zero stage asporogenic mutants of *B. subtilis* [12A] was used. However, when purified poly(A)RNA from t_6-sporulating cells was hybridized to its homologous cDNA probe in presence of an excess of vegetative RNA three distinct kinetic classes were observed (Fig. 10) and their Rot values are shown in Table 3. These results, in addition to those reported in [44], clearly indicate a close association of poly(A)RNA and sporulation.

The hybridization efficiency of t_4 cDNA with homologous and hetrologous poly(A)RNA in presence of an excess of vegetative RNA is shown in Table 4. It can be seen that compared to the hybridization efficiency of t_4-cDNA with t_4-poly(A)RNA taken as 100%, the hybridization of the same cDNA with the poly(A)RNA isolated from t_2 cells or t_6 cells was 28% and 79% respectively. These figures most likely reflect the differences in the population of poly(A)RNA present at each of the sporulation stages. Thus, for instance, at stage t_2 of sporulation 72% of the

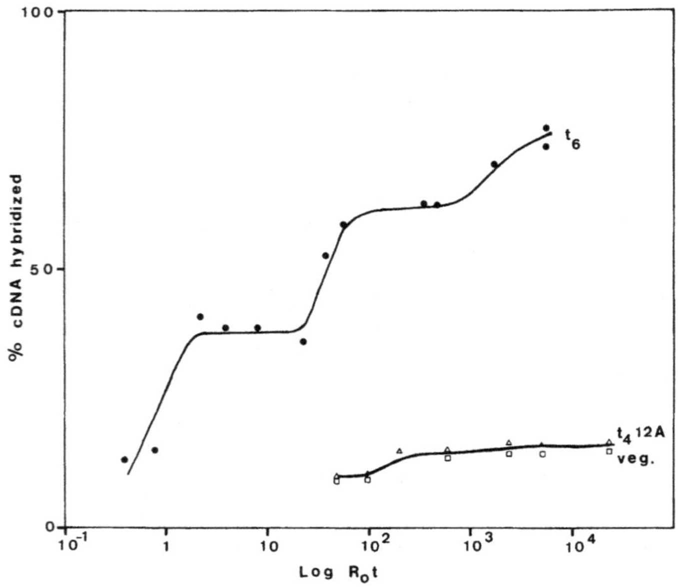

Fig. 10. Hybridization of single stranded cDNA probes with different poly(A)RNA preparations: ●—● from t_6 purified poly(A)-RNA, Δ—Δ from t_4 total RNA from the zero stage asporogenic mutant 12A, □—□ from vegetative cells

Table 3. Analysis of hybridization of t_6 cDNA to t_6 poly(A) RNA shown in Fig. 10

	% Hybridizable cDNA	Rot 1/2 (moles x sec/l)
Component I	40.7	0.54
Component II	26	30
Component III	14	973

Table 4. Hybridization of t_4-^{3}H-cDNA with poly(A)$^+$RNA isolated from different stages of sporulation. Single stranded ^{3}H-cDNA (3000 cpm) synthesized from t_4 poly(A)RNA was hybridized to either t_2 or t_4 poly(A)RNA (200 µg/ml) in presence of an excess of vegetative RNA (5 mg/ml) or to t_6 poly(A) containing total RNA (15 mg/ml) and incubated for 60 h

cDNA from stage	poly(A)RNA from stage	cpm hybridized	%
t_4	t_2	675	28
t_4	t_4	2336	100
t_4	t_6	1859	79

poly(A)RNA species are different from those present at stage t_4, while this figure decreases to 28% at stage t_6. This is in agreement with the generally accepted idea that at each stage of sporulation different classes of mRNA are expressed in a sequential manner [46-48].

Furthermore, based on saturation-hybridization experiments using nick-translated total *B. subtilis* DNA, we have calculated that the different poly(A)RNA species are coded by about 10% of the genome. Since the complexity of *B. subtilis* genome is about 5×10^6 nucleotides [49] this means that the base sequence complexity of the polyadenylated RNA is about 0.5×10^6 nucleotides. Assuming an average size of the poly(A)RNA molecules of about 1000 nucleotides we arrive at the conclusion that the population of the poly(A)RNA is composed of about 500 different species.

It is worthwhile to note that kinetic classes of poly(A)RNA with different base sequence complexities are usually found in eukaryotic cells. However, in the present case the base sequence complexity of each kinetic class could not be estimated, even approximately, due to the fact that the purity of the poly(A)RNA used was not determined.

Detection of Sequences Complementary to Poly(A)RNA in *B. subtilis* Chromosome

Southern Blotting. To investigate further the association of poly(A)RNA with sporulation process the DNA of *B. subtilis* was restricted by BamHI endonuclease and the resulting fragments separated by agarose gel electrophoresis and after denaturation transferred by blotting on strips of nitrocellulose paper [50]. Poly(A)RNA isolated from t_4 cells was labeled with ^{32}P-γ-ATP in presence of polynucleotide kinase and used as a radioactive hybridization probe.

Figure 11 shows the microdensitometer tracing of the radioautograph of the hybrids detected by the ^{32}P-labeled poly(A)RNA. It can be seen that there are five major sets of fragments of 2.3 to 6.1 Kb which retained the highest radioactivity.

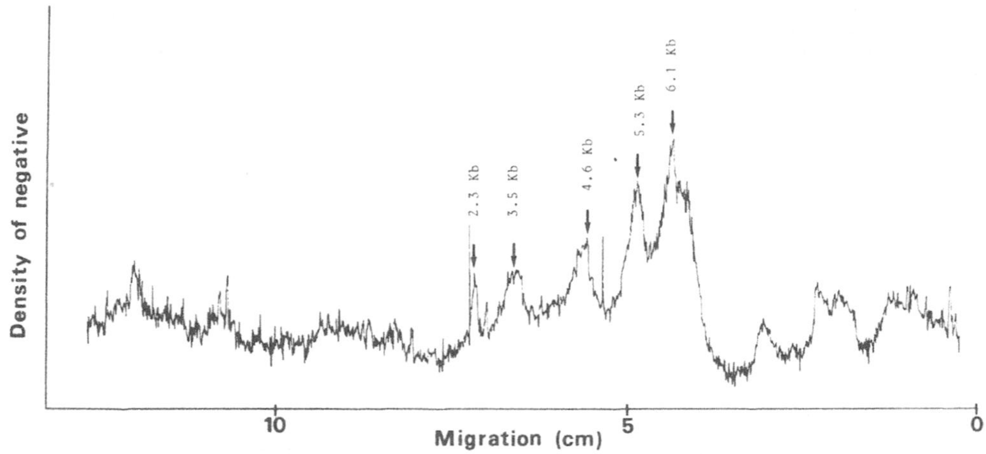

Fig. 11. Microdensitometer tracing of a radioautograph obtained from a Southern blot of BamHI endonuclease restricted fragments of *B. subtilis* DNA detected by ^{32}P-labeled poly(A)RNA

Molecular Cloning of cDNA Fragments Hybridizable to Poly(A)RNA. Five BamHI endonuclease restriction fragments which were found to hybridize efficiently to poly(A)RNA were ligated to the dephosphorylated plasmid vector pHV33 [51]. This plasmid (MW 4,7 Md) is a bifunctional vector which can replicate and express its resistance marker in *E. coli* (Ap^r, Tc^r, Cm^r) and *B. subtilis* (Cm^r). Among the Amp^r Tc^s clones (insertional inactivation of the Tc^r gene) about twenty were found to hybridize in situ (colony hybridization) with labeled poly(A)RNA. Analysis of clear lysates [52] of these clones by agarose gel electrophoresis revealed that some of these clones were carrying plasmids with DNA inserts of 1.5-2.0 Kb, some others contained inserts of smaller size, most likely generated by spontaneous deletions [53].

Clones Hybridizing with Poly(A)RNA Contain Hybrid Plasmids Bearing SpoOB Genes. When DNA from hybrid clones selected for their efficient hybridization to poly(A)RNA was used for transformation of several asporogenic mutants blocked at early stages of sporulation, it was found that the hybrid plasmid 11A5 containing an insert of 2.1 Kb was able to transform for the SpoOB marker in a rec^- recipient strain (Table 5). Simi-

Table 5. Transforming activity of hybrid plasmids. The Cm^r spo^+ transformants were selected on Donnelan agar plates (containing 3 μg/ml of chloramphenicol) after chloroform treatment

Donor DNA	Insert size	SpoOB recipients	Cm^r spo^+ transformants/μg DNA
pBs-11A5	2.5 Kb	Bs80 (rec^-)	4×10^4
pBs-2	1.5 Kb	JH648 (rec^+)	2.4×10^6
pBs-2	1.5 Kb	Bs80 (rec^-)	9×10^5
pBs-14	1.5 Kb	JH648 (rec^+)	7×10^5
pBs-14	1.5 Kb	Bs80 (rec^-)	2.4×10^5

larly, two hybrid plasmids, pBs2 and pBs14 carrying inserts of 1.5 Kb originally selected for SpoOB genes were also found to hybridize efficiently to poly(A)RNA.

Taken together these experiments favour the idea that poly(A)RNA, appearing during sporulation of *B. subtilis*, might be transcribed from DNA sequences enriched in spo genes. It is interesting that, in a recent study on the role of poly(A)RNA in sporulation of *B. polymyxa* the authors arrived at a similar conclusion, although using a different approach [54]. Indeed these authors have shown that the antibiotic netropsin, which is known to inhibit sporulation without affecting vegetative growth [55] also inhibits poly(A)RNA synthesis. This selective inhibition of netropsin was traced to the inhibition of sporulation by specific RNA polymerase under in vitro conditions. The inhibition of transcription concerns mainly the membrane-bound DNA (mDNA) enriched in d(AT) sequences. Based on these findings it was proposed by the authors that early spore genes might be enriched in d(AT) sequences and are located in the membrane bound DNA.

As to the function of poly(A)RNA in sporulation process there is now direct evidence in support of the assumption that these RNA species act as mRNA, similarly to that known in eukaryotic systems. Thus, the isolation from *B. brevis* of highly active protein synthesizing polysomes containing polyadenylated sequences has been recently reported [57]. A concentration dependent stimulation by poly(A)RNA of protein synthesis in an in vitro *B. subtilis* protein synthesizing system has recently been observed in this laboratory (unpublished results). However, the precise role of poly(A)RNA in sporulation process still remains to be elucidated. But this is also true for eukaryotic cells. In the latter case it was suggested that polyadenylation is an essential component of the process of mRNA maturation in the nucleus prior to its transport to the cytoplasm (reviewed in [57]). However, recent evidence supports a specific role of poly(A) in the initial selection of mRNA molecules for processing [58].

In the case of prokaryotes there is no evidence available for mRNA maturation unless such a process does take place to some extent during spore differentiation. But this is at present highly speculative.

Another more realistic possibility for a role of polyadenylation of RNA during sporulation stems from our earlier finding that sporulating cells of *B. subtilis* contain an exonuclease which degrades polynucleotides from the 3' end in the direction of the 5' terminus [59]. It is therefore tempting to speculate that polyadenylation might be necessary to protect the newly synthesized mRNA during sporulation. This is in agreement with the current knowledge that prokaryotes as eukaryotes may regulate the synthesis of certain proteins by regulating the levels of their respective mRNAs. This can be achieved not only by affecting their rate of production but also by affecting the rate of destruction or protection of these mRNAs.

Acknowledgements. This work was supported by grants from the Centre National de la Recherche Scientifique (ATP "Microbiologie 1979") and the Fondation pour la Recherche Médicale Française.

References

1. Ehrenberg CG (1838) Die Infusionstierchen als vollkommene Organismen, Leipzig
2. Perty M (1852) Zur Kenntnis kleinster Lebensformen, Bern
3. Cohn F (1876) Beitr Biol Pflanz 2:249
4. Koch R (1876) Beitr Biol Pflanz 2:277
5. Pasteur L (1870) Etude sur la maladie des vers à soie. Gauthier-Villars, Paris
6. Gould GW (1977) J Appl Bacteriol 42:279-309
7. Fitz-James P, Young E (1969) In: Gould GW, Hurst A (eds) The bacterial spore. Academic Press, London New York, p 39-71
8. Piggot PJ, Coote JG (1976) Bacteriol Rev 40:908-962
9. Losick R (1981) In: Levinson HS, Sonenshein AL, Tipper DJ (eds) Sporulation in germination. Am Soc Microbiol, pp 49-56
10. Hirochika H, Kobayashi Y, Kawamura F, Saito H (1981) J Bacteriol 146:594-505
11. Dubnau E, Ramakrishna N, Cabane K, Smith I (1981) J Bacteriol 147:622-632
12. Jayaraman K, Keryer E, Szulmajster J (1981) FEBS Lett 10:273-276
13. Rhaese HJ, Groscourth R, Amann E, Kühne H, Vetter R (1981) In: Levinson HS, Sonenshein AL, Tipper DJ (eds) Sporulation and germination. Am Soc Microbiol, pp 134-137
14. Kornberg A, Spudich JA, Nelson DL, Deutscher MP (1968) Annu Rev Biochem 37:51-78
15. Hanson RS, Peterson JA, Yousten AA (1970) Annu Rev Microbiol 24:53-90
16. Szulmajster J (1973) Symp Soc Gen Microbiol XXIII, pp 45-83
17. Szulmajster J (1979) Trends Biochem Sci 4:18-22
18. Young M (1978) Trends Biochem Sci 3:55-58
19. Mandelstam J (1976) Proc R Soc London 193:89-106
20. Schaeffer P, Millet J, Aubert JP (1965) Proc Natl Acad Sci USA 54:704-711
21. Brehm SP, Staat SP, Hoch JA (1973) J Bacteriol 115:1063-1070
22. Ito J, Spizizen J (1973) In: Aubert JP, Schaeffer P, Szulmajster J (eds) Régulation de la sporulation microbienne. Coll Int CNRS 227:81-82
23. Coote JG (1972) J Gen Microbiol 71:17-27
24. Freese E (1970) In: Barker, Wolf, Dring, Gould (eds) Spore research, vol I. Academic Press, London New York, pp 1-31
25. Rhaese HJ, Dichtelmuller H, Grade R, Troscurth R (1975) In: Gerhardt P, Cortilow RN, Sadoff HL (eds) Spore, vol VI. Am Soc Microbiol, Washington DC, pp 335-340
26. Sonenshein AL, Campbell KM (1975) In: Gerhardt P, Cortilow RN, Sadoff HL (eds) Spore, vol VI. Am Soc Microbiol, Washington DC, pp 179-192
27. Aronson AI (1973) In: Aubert JP, Schaeffer P, Szulmajster J (eds) Régulation de la sporulation microbienne. Coll Int CNRS 227:33-37
28. Aronson AI, Angelo N, Holt SC (1971) J Bacteriol 106:1016-1025
29. Sadoff HL Celikkol E, Engelbrecht HL (1970) Proc Natl Acad Sci USA 66:844-849
30. Cheng YE, Aronson AI (1977) Proc Natl Acad Sci USA 74:1254-1258

31. Szulmajster J, Keryer E (1975) In: Gerhardt P, Cortilow RN, Sadoff HL (eds) Spore, vol VI. Am Soc Microbiol, Washington DC, pp 271-278
32. Kerjan P, Keryer E, Szulmajster J (1979) Eur J Biochem 98:353-362
33. Laurell CB (1966) An Biochem 15:45-52
34. Millet J, Kerjan P, Aubert PJ, Szulmajster J (1972) FEBS-Lett 23:47-50
35. Orrego C, Kerjan P, Manca de Nadra MC, Szulmajster J (1973) J Bacteriol 116:636-647
36. Waindle L, Switzer RL (1973) J Bacteriol 114:517-527
37. Setlow P (1978) In: Cohen GN, Holzer H (eds) Limited proteolysis in microorganisms. DHEV Publ n° (NIH)79-1591, pp 109-113
38. Srinivasan, PR, Ramanarayanan M, Rabbini E (1975) Proc Natl Acad Sci USA 72:2910-2914
39. Nakazato H, Venkatesan S, Edmonds M (1975) Nature (London) 256:144-146
40. Ohta N, Sanders M, Newton A (1975) Proc Natl Acad Sci USA 72:2343-2346
41. Sarkar N, Langley D, Paulus H (1978) Biochemistry 17:3468-3474
42. Graef-Doods E, Chambliss GH (1978) In: Chambliss G, Vary JC (eds) Spores, vol VII. Am Soc Microbiol, Washington DC, pp 237-241
43. Kaur S, Jayaraman K (1979) Biochem Biophys Res Commun 86:331-339
44. Kerjan P, Szulmajster J (1980) Biochem Biophys Res Commun 93:201-208
45. Randerath K, Randerath E (1971) Proc Nucleic Acid Res 2:796-812
46. DiCioccio RA, Straus N (1973) J Mol Biol 77:325-336
47. Sumida-Yasumoto C, Doi RH (1974) J Bacteriol 117:775-782
48. Bonamy C, Manca de Nadra MC, Szulmajster J (1976) Eur J Biochem 63:53-63
49. Gillis M, De Ley J (1975) J Mol Biol 98:447-464
50. Southern EM (1975) J Mol Biol 98:503-517
51. Ehrlich SD (1978) Proc Natl Acad Sci USA 75:1433-1436
52. Birnboim HC, Doly J (1979) Nucleic Acid Res 7:1513-1523
53. Primrose SB, Ehrlich SD (1981) Plasmid 6:193-201
54. Jayaraman K, Murthy S (1981) Mol Gen Genet (in press)
55. Keilman GR, Tanimoto B, Doi RH (1975) Biochem Biophys Res Commun 67:414-420
56. Käufer N, Altmann M, Döhren NV (1981) FEMS Microbiol Lett 12:71-75
57. Brawaman G (1976) Progr Nucleic Acid Res 17:118-147
58. Balmain A, Minty AJ, Birnie GD (1980) Nucleic Acid Res 8:1643-1660
59. Kerjan P, Szulmajster J (1976) Biochimie 58:533-541

The Human Epsilon Globin Gene – A Paradigm for Erythroid Differentiation

J. Paul, M. Allan, J. Grindlay, and D. Spandidos[1]

The fundamental unsolved problem of cell differentiation is how a common genome can be used to give a wide spectrum of cell types. Accumulated information has clearly shown that no simple mechanism is responsible. In individual cases differences in expression of a gene have been found to be due to gene loss, gene amplification, transposition of DNA, modification of DNA, transcription, processing, transport and stability of RNA or translation, modification or turnover of protein. It is not difficult to understand why teleological arguments, though not scientifically respectable, have an appeal because it would almost seem that a cell will exploit any available mechanism to achieve the best levels of expression of individual genes to meet the demands of the environment. In searching for some order in this process theories have varied from those which ascribe a predominant role to transcriptional control [1] to those like the cascade hypothesis [2] which envisage controls at all possible stages of expression.

In order to acquire information about patterns of gene expression most workers have concentrated their efforts on studying individual systems and, among those, haemoglobin synthesis has attracted particular attention.

Two distinct aspects of differential gene expression occur in the formation of erythrocytes. All erythrocytes, megakaryocytes, granulocytes and macrophages are derived from a common precursor stem-cell population which is self-perpetuating throughout life from quite early in embryogenesis. These precursors can be demonstrated as Spleen Colony-Forming Units (CFU-S) by injecting haemopoietic cells into irradiated mice [3]. If low numbers of cells are injected discrete colonies derived from them develop in the spleen of the recipient. They can contain any of the mature cells referred to above. By analysis of the colonies and their progeny following reinoculation the presence of a common precursor can be inferred. Hence, one aspect of erythroid differentiation involves the expression of alternative sets of genes in the progeny of a cell population which proliferates at a rate sufficient to maintain itself and to give rise to maturing cells from early embryonic life to old age.

The second aspect of differential gene expression is that the erythrocytes derived from these stem cells are not the same throughout life; the differences are manifested in the haemoglobins they contain. In mammals the haemoglobin molecule is a tetramer comprising two α-type chains and two β-type chains. Several different α-type genes and sever-

1 The Beatson Institute for Cancer Research, Garscube Estate, Switchback Road, Bearsden, Glasgow, G61 1BD, Great Britain

33. Colloquium - Mosbach 1982
Biochemistry of Differentiation and Morphogenesis

al different β-type genes are expressed at different times during development. In the human the α-type gene expressed in early embryonic development is that for the ζ-chain which, between the 3rd and 4th week of development is replaced by the α-chains (two genes) which are expressed throughout all subsequent development. The first β-type chain expressed is the ε-chain which appears between the 3rd and 4th week and then is progressively replaced by γ-chains between the 6th and 12th week. Finally the γ-chains are replaced by the adult β- and δ-chains during the two weeks preceding and the two weeks following birth (Fig.1).

The appearance of these chains gives rise to a series of haemoglobins Gower 1 ($\zeta_2\varepsilon_2$) Gower 2 ($\alpha_2\varepsilon_2$) HbF ($\alpha_2\gamma_2$) HbA ($\alpha_2\beta_2$) and HbA_2 ($\alpha_2\delta_2$) (Fig.1). There are several important features of development of the cells which produce these haemoglobins.

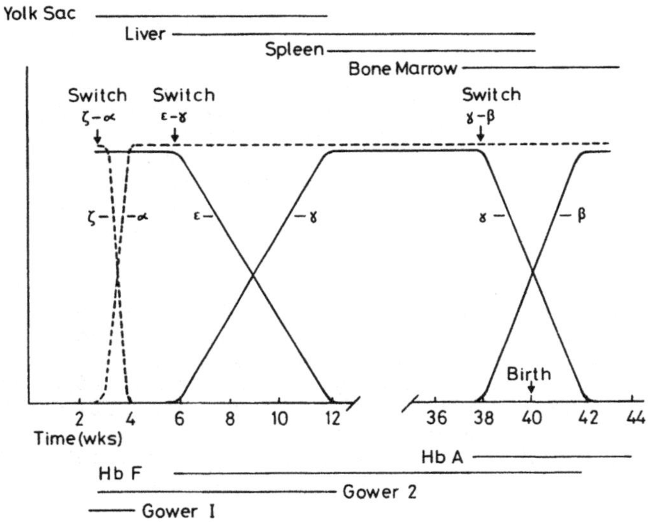

Fig. 1. The appearance of different globin chains during human embryonic, fetal, and adult life. Major "switches" are shown

First, maturing erythroblasts also produce other proteins in characteristic amounts, e.g. spectrin, red blood cell glycoproteins and carbonic anhydrase. Therefore many functions are co-ordinately regulated during cell differentiation.

Secondly, the extent of differentiation in a mature reticulocyte can be estimated from the facts that each reticulocyte may contain 150,000 molecules of globin mRNA, comprising about 90% of all mRNAs present, whereas in non-erythroid tissue it is a matter of contention whether a single copy of globin mRNA is present or not. Hence the quantitative differences in gene products can be enormous. Nevertheless no difference has been found between the DNA of globin genes in erythroid and non-erythroid cells.

Thirdly, although the production of α-type chains and β-type chains is co-ordinated, the genes for these are on different chromosomes and therefore completely unlinked. However, the β-type genes are clustered in one locus on one chromosome and the α-type genes in one locus on another. Moreover, in most species, including the human they are arranged in the order in which they are expressed in ontogeny (Fig.2). Hence, close linking of these genes is not correlated with co-expression but with structural and functional similarity and there is an intriguing relationship between the order in which they are arranged and the sequence in which they are expressed.

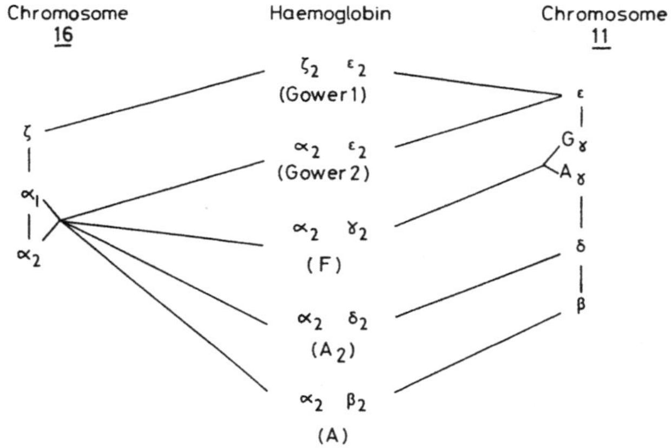

Fig. 2. The normal human haemoglobins. All are formed from two α-type and two β-type chains. The genes for β-type chains are clustered on chromosome 11 and those for α-type chains on chromosome 16

Finally, three clearcut gene switches can be recognised. The first at about 3 weeks of development involves inactivation of the ζ-gene and activation of the α-genes, the second at about 6 weeks involves inactivation of the ε-gene and activation of the γ-genes and the third at about the 38th week involves inactivation of the γ-genes and activation of the β-and δ-genes. There must also be important switches earlier in development when the ζ-and ε-genes are first activated but little is known of them.

Hence, two simultaneous sets of gene controls operate in erythropoiesis. One involves the activation of genes which result in formation of a haemoglobin-synthesising erythroid cell. The other is concerned with the kind of erythrocyte formed - whether it contains embryonic, fetal or adult haemoglobins.

Regulation of Globin Gene Expression

In attempts to obtain some understanding of the ways in which genes are regulated during differentiation, studies have been undertaken on the behaviour of individual genes which have a specific association with a developmental pathway. In erythropoiesis the globin genes have been regarded as the most suitable markers. To study factors involved in their control four approaches have been used.

The first is comparison of the genes in erythroid and non-erythroid cells. Studies of this kind have revealed no differences between the restriction patterns of human β-genes in erythroid and non-erythroid cells [4] but they have shown that the conformation of the globin genes is different in the two circumstances. In particular, the globin genes appear to be more accessible in erythroid cells and the most convincing evidence for this comes from the DNAase digestion studies referred to below.

The second approach is to compare the genes in erythroid cells in which they are being actively expressed and erythroid cells in which they are not actively expressed. In normal human erythroid tissue by far the greatest number of erythroid cells represent stages in which globin genes are already active and no one has successfully isolated precursor cells in which the genes are inactive in sufficient amounts to make this comparison. It has proved necessary, therefore, to rely on artificial systems such as the mouse Friend cell [5] and the human K562 cell [6] in which the level of expression of the gene can be greatly increased by inducers. This again is not the best approach because, even in the uninduced cells, there is usually a detectable level of globin messenger RNA.

A third approach is to study the effects of natural mutations. In the human these result in haemoglobinopathies, notably thalassaemia, but although a large number of different genetic abnormalities producing this disease have been studied, the information bearing on regulation has been minimal. However, one particularly interesting piece of information has emerged. This is that in some cases deletion of the β- and δ-genes and regions adjacent to them results in persistent synthesis of fetal haemoglobin even in the adult. This observation implies that the deletion somehow disturbs the normal switch which would result in no further fetal haemoglobin being synthesised shortly after birth [7].

A fourth approach is to engineer recombinants containing globin genes which can be inserted into foreign cells (e.g. cells from other species) by means of recombinant DNA technology. By observing the behaviour of these transferred genes both before and after inducing mutations by biochemical modification, information can be gained about the importance of certain regions of the gene and neighbouring sequences.

Finally, it is hoped that eventually synthetic systems may by created whereby the regulation can be studied completely in vitro and factors involved in it identified and characterised.

We have taken a particular interest in the behaviour of the ε-globin gene because of its involvement in globin gene switches and because it is expressed in a readily available cell-line, the K562 cell.

The Human Epsilon Globin Gene

The ε-globin gene was first cloned by Proudfoot and Baralle [8] and the same group determined the primary structure [9]. For the present discussion some particular features should be noted. Like other β-type globin genes it has three exons and two introns and transcribed but untranslated leader and tail sequences which appear in mature messenger RNA. It bears considerable resemblance to the γ-, β- and δ-genes, particularly at the 5' end (in relation to transcription). At the 3' end it diverges more and the Bam HI(2) to EcoRI(2) fragment (Fig.3) has so little homology with the corresponding regions in other β-type genes that it can be used as a gene-specific probe. At the 5' side,

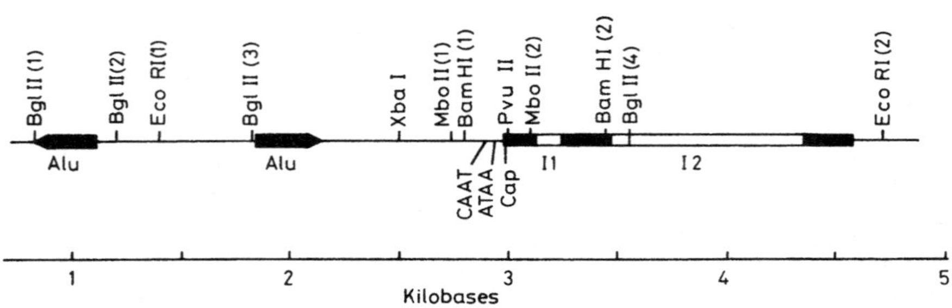

Fig. 3. Restriction map of the ε-globin gene and adjacent regions. Alu = Alu sequences. I1 and I2 = intervening sequences (introns)

53 to 55 bp upstream of the initiating codon (ATG) there is a recognisable putative cap sequence from which transcription might be predicted on the basis of homology with other genes [9]. Moreover this is preceded about 20 bp upstream by an ATAA sequence and 80 bp upstream by a CAAT sequence. Those sequences are commonly found in the same relationship to other transcribed genes. Still further upstream, about 1-2 kb from the putative cap site, there occur two members of the Alu family of highly repeated sequences [10]. These are in opposite orientations, forming an inverted repeat. Members of the Alu family are interspersed among the β-type genes but the significance of this, if any, is not known. However, an interesting feature of the Alu sequences is that they commonly contain an RNA polymerase III initiation sequence.

The objectives of the present study were to define the transcriptional unit of the ε-gene and to investigate ways in which transcription is regulated, particularly during the onset of erythropoiesis and at the ε-γ-switch.

Transcription of the Epsilon Globin Gene

From the study of the nucleotide sequence of the ε-globin gene Baralle et al. [9] predicted two possible cap sites 53 and 55 base pairs respectively upstream of the initiating codon ATG. Proudfoot et al. [11] undertook in vitro transcription experiments of a cloned ε-globin gene fragment. Using the Manley system and studying the origin of the transcripts by the primer extension method they identified a transcript which originated in this region. For the mouse β-globin gene there is good evidence that the precursor originates in a corresponding position and is transcribed through to a polyadenylation site downstream of the 3' end of the coding sequence [12]. This 15S precursor is processed to mature messenger RNA by splicing out the intron sequences and the findings quoted would be in accord with the existence of a similar situation in the human ε-gene. However there are some uncertainties about the relevance of in vitro transcription to the in vivo situation and, moreover, there have been persistent claims to have identified even larger precursors of the mouse β-globin mRNA in erythroid cells [13-15]. We therefore undertook to study ε-globin gene transcripts in K562 cells and in human embryos using the S1 mapping procedure [16]. In this technique an isolated DNA segment of the gene is hybridised to RNA and the DNA protected by the hybrid is studied by S1 digestion of unhybridised DNA followed by analysis. In our experiments we used DNA fragments labelled with ^{32}P at the 5' end. The length of DNA from the labelled end which was colinear with complementary RNA could be measured by determining the size of the labelled protected fragments. Figure 4 shows the results of an experiment in which a 370

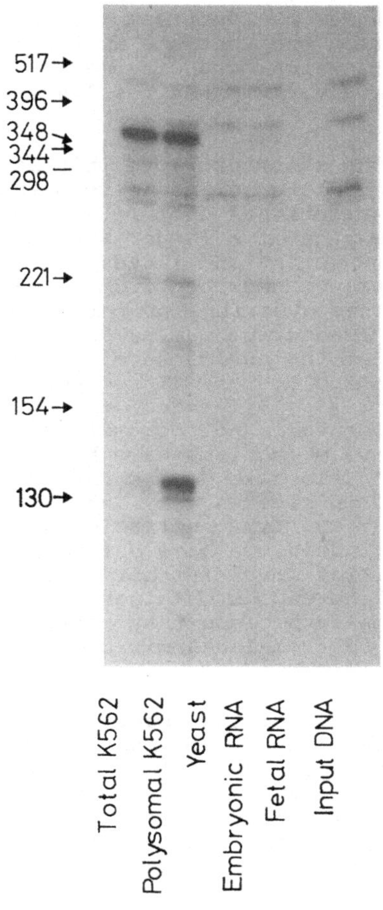

Fig. 4. S1 analysis of RNA transcribed from the ε-globin gene. A DNA fragment represented the region between Mbo II (1) and Mbo II (2) in Fig.3 was labelled at the 5' end with ^{32}P. After hybridising to RNA the unpaired RNA and DNA were digested with S1 nuclease. The sizes of the protected DNA fragments were then determined by electrophoresis in a polyacrylamide gel followed by autoradiography. The lengths in nucleotides are indicated at the side

base pair fragment between two Mbo II sites (marked Mbo II (1) and Mbo II (2) in Fig.3) was used. Two sizes of protected fragments were seen, one extending from the Mbo II (2) site upstream to a position corresponding approximately to the putative cap site, and the other extending about 220 bases further. In this figure it can also be noted that there may be double bands at each site. The analyses were extended and the precise sites of initiation mapped by running sequencing gels in parallel. These experiments showed that the smaller molecules mapped at exactly 53 and 55 base pairs upstream of the initiating codon ATG, i.e. in the region of the cap positions predicted by Baralle et al. [9]. However the larger molecules mapped at about 272 bases upstream of the initiation codon.

In subsequent experiments, RNA molecules have been identified by Northern blotting, using appropriate probes, which correspond to transcripts extending from both points, but the transcript originating from the upstream initiation site is present in much lower concentrations than the other. It appears prominently in the S1 experiments possibly because hybridization of molecules originating at the downstream site face competition from γ-globin mRNA. A canonical cap site cannot be recognised in the DNA sequences corresponding to the upstream initiation site. It follows that either transcription can initiate at this site in which event the requirements for a canonical cap site may be less rigorous than is thought, or, alternatively, this RNA results from processing of a larger molecule.

These findings are in accord with observations of transcripts initiating upstream of the canonical cap site in the mouse β-gene and leave open the question of the actual size of the transcriptional unit [17]. If it were possible to cap an RNA formed by cleavage of a larger RNA, one could envisage a very much larger initial transcript and indeed, the existence of molecules which quite clearly are transcribed from a region upstream of the CAAT and TATA boxes raises questions of this kind, especially the suggestion that the upstream transcript may be derived from an even larger molecule. However, most of the published evidence concerning initiation of transcription by RNA polymerase II suggests that RNA transcription starts at the cap site [18,19]. Indeed the normal capping mechanism which is thought to operate is not compatible with any other mechanisms although alternative mechanisms have been suggested [20]. The claim by Shaul et al. [15] to have demonstrated precursors to mouse β-globin mRNA about 5 kb in size is very interesting and there is just the possibility that the true initiation site may be about 1 kb upstream of the cap site. Indeed, we have obtained preliminary evidence for molecules which initiate much further upstream than the sites reported here. These findings have been obtained not only in K562 cells which might behave abnormally because they represent a continuously cultured cell line derived from leukaemic blood, but also in normal human embryonic material.

The alternative explanation for our findings and for the observation of very large precursor molecules is that the fragment we observe extending from the structural gene to 272 bases upstream of the initiation codon is only part of a larger molecule and that the protected DNA ends at a splice junction. Such a hypothesis would be compatible with the observations on synthesis of α-amylase mRNA in which alternative cap sites and adjacent sequences may be spliced to an otherwise identical precursor RNA [21,22]

The Transcriptional Unit

If we try to define the transcriptional unit of the mouse globin β-gene, or indeed of any of the mammalian globin genes, on the basis of the RNA transcripts isolated biochemically, it becomes apparent that we do not have any clear idea of its limits. Is the normal primary transcript the 15-16S precursor molecule isolated by Curtis and Weissmann [12], or do the larger molecules reported by Aviv's group [13,15] and Scherrer's group [14] give an indication of the true initial transcript. If the former is true we have to assume that the large transcripts which have certainly been demonstrated are artefacts occurring possibly by read-through from upstream cap sites in a small percentage of cells. If the latter is true, we have to postulate very rapid processing of the initial transcript and a capping mechanism different from the one which has been most intensively studied. Some observations on other genes extend these possibilities.

Electron microscopic studies on chromosomes have brought out some particularly intriguing points. Nascent transcripts can be revealed as ribonucleoprotein fibres attached to actively transcribed genes which tend to present a typical "Christmas tree" structure of the well-known type demonstrated for ribosomal genes. However in non-ribosomal genes the "Christmas trees" are very much less dense, reflecting a lower rate of initiation, and moreover the structure is not always regular. In particular, complex structures have been described in which cleavage of the nascent transcript clearly occurs [23-25]. This phenomenon of "instantaneous processing" may explain why, in general, EM estimates of the size of primary transcripts are considerably larger than biochemical estimates of the same material.

An indirect estimate of the length of transcriptional units may also be indicated by the extent of DNAase I sensitivity in the region of individual genes. The observation by Weintraub and Groudine [26] that pancreatic deoxyribonuclease (DNAase I) preferentially digests transcriptionally active genes has now been extended. It has emerged that the altered conformation of chromatin indicated by DNAase I sensitivity occurs, not only in genes which are being transcribed, but also in genes which are destined to be transcribed. In the chick globin genes the peptide coding regions themselves are in a particularly sensitive state but relative sensitivity extends to some kilobases outside both α- and β-genes [27,28] and highly sensitive sites extend some hundreds of bases to the 5' side of the gene [29]. Moreover, hypersensitivity to DNAase I cleavage has been shown in individual sites near genes which are being actively transcribed.

There is evidence that the active genes are associated with the nonhistone proteins HMG14 and 17 and that these in some way are responsible for DNAase I sensitivity [30]. Other correlative pieces of information are less convincing. Globin genes in sperm DNA are highly methylated whereas in DNA from erythroid tissue the extend of methylation is much less [4]. Similar observations have been made for some other genes. However, Van der Ploeg and Flavell found that the extent of methylation of sites near the β-globin genes in a human erythroid cell line (K562) was exactly the same as in DNA from a non-erythroid cell line (KB) or placenta, so care has to be exercised in interpreting this relationship [4].

At the present time these data are not sufficiently complete to permit any precise correlation between DNAase I sensitivity, methylation and the transcriptional unit.

Transfer of Human Globin Genes to Cells of Other Species

The transfer of globin genes to cells of other species has been reported by a number of authors but in the experiments conducted in other laboratories fibroblasts were used as recipients [31,32]. Since these are non-erythroid cells the physiological significance of the observation is dubious. We therefore undertook to transfer human globin genes to mouse erythroleukaemic cells as well as to fibroblasts. As recipients we used mouse fibroblasts ($LMTK^-$), Chinese hamster fibroblasts ($BHKTK^-$), and mouse erythroleukaemic cells ($F4\text{-}12B2TK^-$) which were deficient in thymidine kinase (TK) activity. As donor molecules we constructed recombinants between the plasmid pTK-1 and the 8 kilobase human DNA fragment from a recombinant phage λ788 which contains the human ε-globin gene [8]. pTK-1 itself was made by inserting the thymidine kinase gene from herpes simplex virus into the bacterial plasmid pAT 153 [33]. The resulting recombinants (Fig. 5) have the ε-gene-containing fragment in one orientation or the other. The TK gene provides the opportunity for selection of transformed TK⁻ cells by converting to cells which can grow in the presence of a selective medium (HAT). The identification of a suitable recipient Friend cell has been described [34].

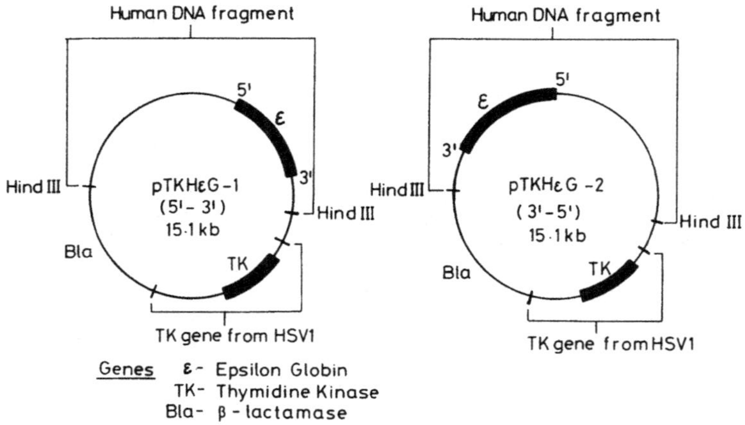

Fig. 5. The two recombinant molecules used in the gene transfer experiments. The plasmid pTK1 was formed by inserting the thymidine kinase gene (TK) from herpes simplex virus (HSV-1) into the bacterial plasmid pAT 153. An 8 kb fragment containing the ε-globin gene was then inserted into the Hind III site in pTK1 in both orientations

LMTK⁻ cells, BHKTK⁻ cells and F4-12B2TK⁻ cells were treated with the two recombinant plasmids containing the gene. As shown in Fig.6 RNA extracted from all these cells contained a species which migrated at the same rate as 9S RNA and which hybridised to an isotopically labelled probe specific for ε-gene sequences. Hence, all three cell lines apparently were able to synthesise RNA derived from the ε-gene and process it to a size similar to that of genuine messenger RNA. It is noteworthy that in this experiment more 9S ε-RNA was detected in induced cells transformed with the ε-gene than in fibroblasts. When the number of plasmid copies per cell in individual transformants was measured, it was found to vary considerably in all types of cell, and the number of copies of ε-specific RNA was also found to vary greatly. However, we have observed in other experiments [34] that more messenger RNA may be present in induced transformed Friend cells than in uninduced cells. If this is a regularly demonstrable phenomenon, then it offers an excellent system for investigating some aspects of globin gene control. To determine whether the effect was specific for the globin gene, the experiment illustrated in Fig.7 was undertaken. The RNA from three separate Friend cell lines which had been transformed with pTKHεG-1 was spotted on to nitrocellulose paper and probed with four different probes. In all cases a probe for the

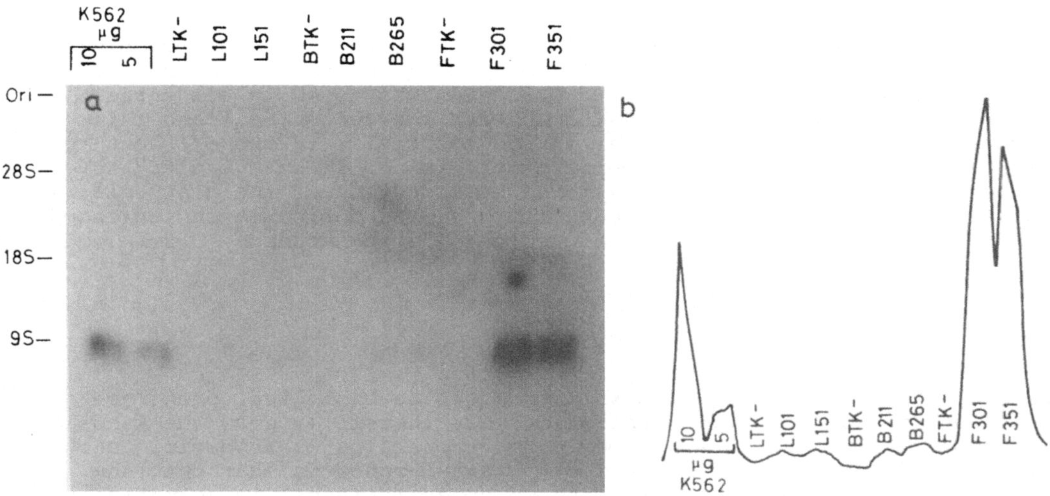

Fig. 6a,b. Northern blot hybridization of RNA isolated from transformed LMTK^{+}, BHKTK^{+} and F4-12B2TK^{+} cells. Lanes A and B represent 10 and 5 μg respectively of total RNA isolated from non-induced K562 cells. All other lanes represent polyA^{+} RNA from 200 μg total RNA from the cell lines indicated. LTK^{-}, BTK^{-} and FTK^{-} were the untransformed cell lines LMTK^{-}, BHKTK^{-} and F4-12B2TK^{-}. L101 and L151 were derived from LMTK^{-} cells, B211 and B265 from BHKTK^{-} cells and F301 and F351 from induced F4-12B2TK^{-} cells. L101, B211 and F301 were transformed with pTKHεG-1 DNA and L151, B265 and F351 were transformed with pTKHεG-2 DNA. The 1.3 kb long human 3'-ε-globin DNA fragment from the internal Bam H1 (2) to the external (3') EcoRI (2) site (see Fig.3), labelled with ^{32}P, was used as a probe. (a) The audoradiogram. (b) Densitometer scanning

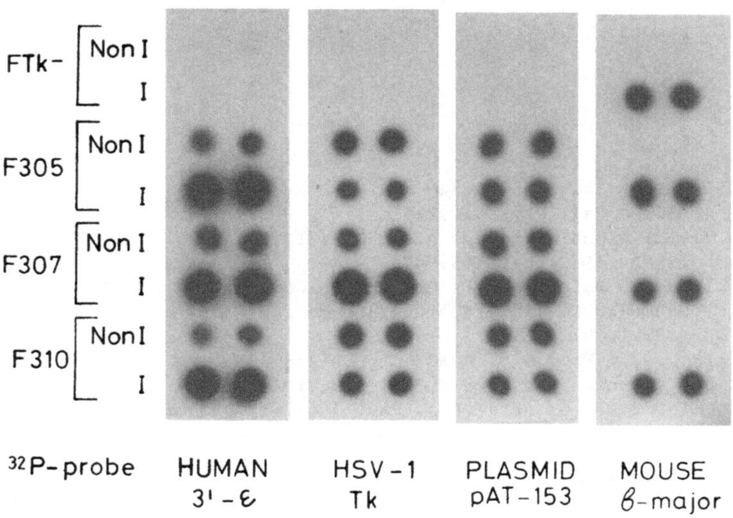

Fig. 7. Spot Hybridization reactions of RNA samples from non-induced and induced F4-12BTK^{-} and F4-12BTK^{+} cells transformed with pTKHεG-1 DNA. Twenty μg of each sample of total RNA were applied per spot in duplicate to four separate pieces of nitrocellulose paper [35]. Each of the following nick translated probes was hybridized to a separate paper: the '3-ε-gene probe described in the legend to Fig.6; a 0.6 kb long HSV-1 TK DNA fragment from a Bgl II site near the cap site to an Sst I site within the TK gene [33]; the plasmid pAT 153 and a 1.0 kb long mouse β-major globin DNA fragment from a Hind III site close to the 5' end of the β-gene to a Hind III site inside the gene. FTK^{-} represents the F4-12B2TK^{-} cells and F305, F307 and F310 three different F4-12B2TK^{+} cell lines transformed with pTKH G-1 DNA

mouse β major mRNA showed barely detectable amounts of mouse globin mRNA in the uninduced cells but large amounts in the induced cells. A significant amount of human ε-globin mRNA was found in all three cell lines which had been transformed with the recombinant and this amount was significantly greater in induced cells. In two of the three cell lines the amount of RNA complementry to the TK gene and to the bacterial plasmid DNA was not altered by induction of the host cell, but in the third case (F307) induction appeared to result in increased transcription of both the TK gene and pAT 153 sequences.

Conclusions

We now have a fairly detailed knowledge of the biological phenomena involved in erythroid differentiation. We, therefore, know the kinds of questions we wish to ask. How is the expression of a specific set of genes, including those for globin chains, spectrin, carbonic anhydrase and haem enzymes, co-ordinated so as to result in formation of an erythrocyte, rather than a megakaryocyte, granulocyte or macrophage from a common precursor? How are the genes for a specific haemoglobin selected at different stages of development? These are fundamental biological questions with very extensive implications. The globin genes provide a particularly interesting system for study because they have been characterised, in many cases, down to the details of nucleotide sequences. Moreover, the genes can be manipulated by gene transfer and by artificial mutation and they can be studied in vitro. The biochemical challenges are obvious. Yet, despite a remarkably extensive accumulation of background information, our knowledge, in biochemical terms, of the processes which control the biological switches involved in erythroid differentiation is quite rudimentary.

Acknowledgements. The Beatson Institute for Cancer Research is supported by grants from the Cancer Research Campaign.

References

1. Britten RJ, Davidson EH (1969) Science 165:349
2. Scherrer K, Marcaud L (1968) J Cell Physiol 72:181
3. Till JE, McCulloch EA (1980) Biochim Biophys Acta 605:431
4. Ploeg van der LHT, Flavell RA (1980) Cell 19:947
5. Friend C, Scher W, Holland JG, Sato T (1971) Proc Natl Acad Sci USA 68:378
6. Anderson LC, Jokinen M, Gahmberg CG (1979) Nature (London) 278:364
7. Weatherall DJ, Clegg JB (1979) Cell 16:467
8. Proudfoot NJ, Baralle FE (1979) Proc Natl Acad Sci USA 76:5435
9. Baralle FE, Shoulders CC, Proudfoot NJ (1980) Cell 21:621
10. Coggins LW, Lanyon WG, Slater AA, Grindlay G.J, Paul J (1981) Biosci Rep 1:309
11. Proudfoot NJ, Shander MHM, Manley JL, Gefter ML, Maniatis T (1980) Science 209:1329
12. Curtis PJ, Weissmann C (1977) J Mol Biol 106:1061
13. Bastos R, Aviv H (1977) Cell 11:641
14. Reynaud C, Imaizumi-Scherrer M, Scherrer K (1980) J Mol Biol 140:481
15. Shaul Y, Kaminchik J, Aviv H (1981) Eur J Biochem 116:461
16. Berk AJ, Sharp PA (1978) Proc Natl Acad Sci USA 75:1274
17. Hofer E, Darnell JE Jr (1981) Cell 20:585
18. Shatkin AJ (1976) Cell 9:645
19. Contreras R, Fiers W (1981) Nucleic Acids Res 9:215
20. Abraham G, Rhodes DP, Bannerjee AJ (1975) Cell 5:518

21. Hagenbuchle O, Tosi M, Schibler U, Bovey R, Wellauer PK, Young RA (1981) Nature (London) 289:643
22. Young RA, Hagenbuchle O, Schibler U (1981) Cell 23:451
23. Beyer AL, Bouton AH, Miller OL Jr (1981) Cell 26:155
24. Laird CD, Chooi WY (1976) Chromosoma 58:193
25. McKnight SL, Miller OL Jr (1976) Cell 8:305
26. Weintraub H, Groudine M (1976) Science 193:848
27. Stalder J, Groudine M, Dogson JB, Engel JD, Weintraub H (1980) Cell 19:973
28. Weintraub H, Larsen A, Groudine M (1981) Cell 24:333
29. McGhee JD, Wood WI, Dolan M, Engel JD, Felsenfeld G (1981) Cell 27:45
30. Weisbrod S, Weintraub H (1979) Proc Natl Acad Sci USA 76:630
31. Wold B, Wigler M, Lacy E, Maniatis T, Silverstein S, Axel R (1979) Proc Natl Acad Sci USA 76:5684
32. Mantei N, Boll W, Weissmann C (1979) Nature (London) 281:40
33. Wilkie NM, Clements JB, Boll W, Mantei N, Lonsdale D, Weissmann C (1979) Nucleic Acids Res 7:859
34. Spandidos D, Paul J (1982) EMBO Journal 1:15
35. Spandidos DA, Harrison PR, Paul J (1981) Biosci Rep 1:911

Erythrocyte Heterokaryons – A Model System for the Analysis of Cell Determination and Gene Regulation

N. R. Ringertz, S. Linder, S. Zuckerman, and G. Lanfranchi[1]

Introduction

Embryological studies using cell and nuclear transplantation techniques have shown that during early development, cells undergo a stepwise restriction in their potential to undergo all forms of cell differentiation. This process has been termed *determination*. Genetic evidence suggests that the determined state is heritable and may occur several cell generations before it results in altered gene expression and altered cellular phenotypes, i.e. in *differentiation*. The formation of terminally differentiated cells such as erythrocytes and nerve cells, most likely consists of a well-ordered series of determination events followed by sequential phenotypic changes. While it is now well established that differentiation is associated with altered patterns of transcription and mRNA translation, very little is known about the molecular basis of determination. Among the mechanisms which have been considered are rearrangements of nucleotide sequences (recombination, amplification, deletion etc.), DNA methylation and altered DNA-protein interactions.

We have used somatic cell hybridization to study cell determination and to examine the role of nucleocytoplasmic interactions in the control of gene expression (review, see [1]). Using Sendai virus-induced fusion of cells or cell fragments, nuclei from one type of cell have been confronted with cytoplasms from other types of cells. The fusion products have then been analysed with respect to their phenotypes in order to obtain information about the stability of the determined state and about mechanisms controlling tissue specific gene expression.

In this presentation we would like to focus on one of these systems, namely heterokaryons, produced by Sendai virus-induced fusion of nucleated chick erythrocytes with mammalian, avian and amphibian cells.

Avian, fish and reptile erythrocytes are terminally differentiated cells containing condensed, inactive cell nuclei (Fig.1). Inactivation occurs gradually during erythropoiesis and results in a cessation of RNA and DNA synthesis, a disappearance of nucleoli and a compaction of the chromatin. Compared to the nuclei of active cells, erythrocyte nuclei are small and very low in non-histone proteins. They do, however, contain a normal quantity of DNA and histones.

1 Department of Medical Cell Genetics, Medical Nobel Institute, Karolinska Institutet, S-104 01 Stockholm, Sweden

33. Colloquium - Mosbach 1982
Biochemistry of Differentiation and Morphogenesis

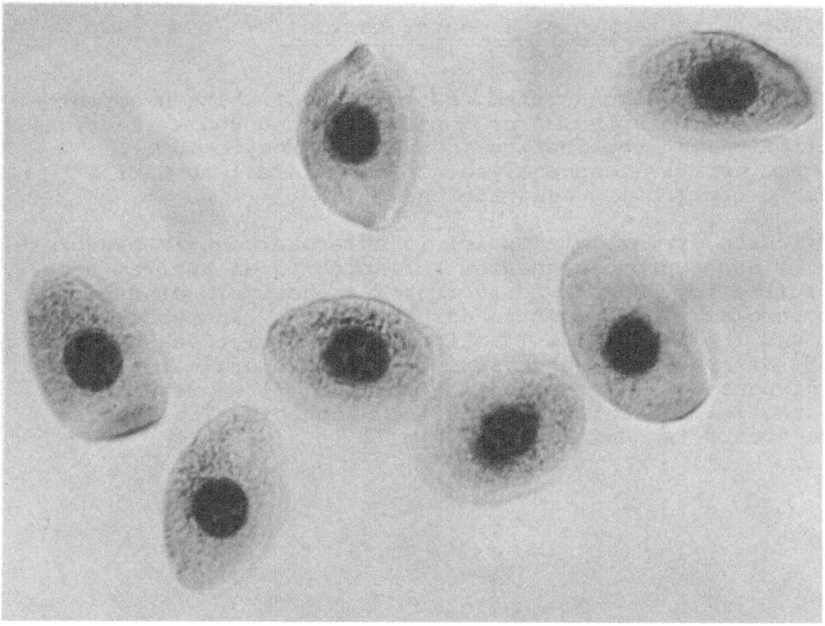

Fig. 1. Nucleated erythrocytes from 13-day hen eggs. The nucleus is in a condensed, inactive state. UV-microphotograph at 260 nm

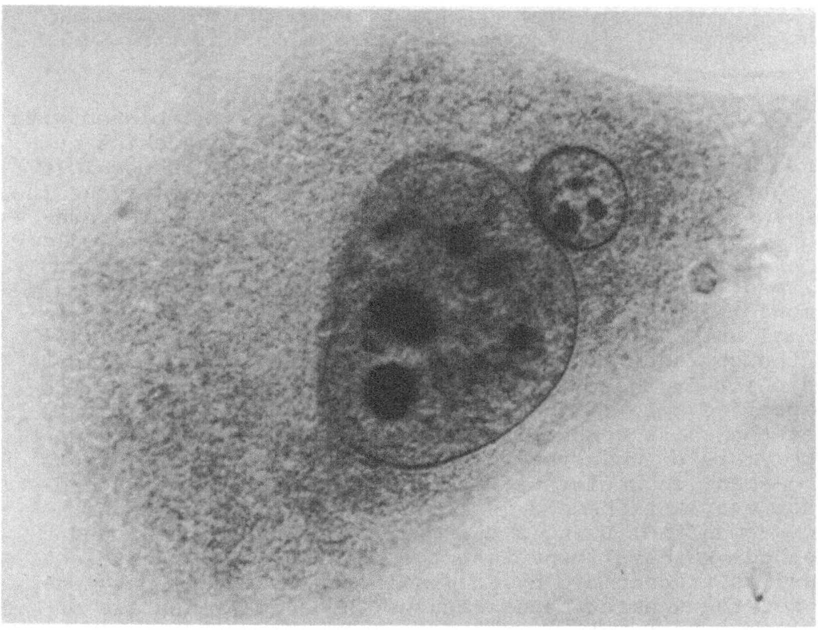

Fig. 2. Heterokaryon formed by Sendai virus-induced fusion of a human HeLa cell with a nucleated chick erythrocyte. The erythrocyte nucleus has undergone chromatin dispersion and enlargement but is still smaller than the HeLa nucleus. As a result of the reactivation, nucleoli have formed in the chick erythrocyte nucleus. UV-microphotograph at 260 nm

Reactivation of Erythrocyte Nuclei in Heterokaryons

In 1965 Henry Harris [2] demonstrated that fusion of chicken erythrocytes with actively dividing human or mouse cells induced a reactivation of transcription and replication in the erythrocyte nuclei (Fig. 2). This process was then analysed in some detail by Harris' group and by our group (review see reference [1]).

Reactivation of chick erythrocyte nuclei in heterokaryons is associated with (a) early nucleoprotein changes reflected in an altered heat denaturation profile for DNA and an increased capacity to bind intercalating fluorochromes, (b) chromatin dispersion, (c) increase in nuclear volume and dry mass, (d) initiation of transcription and an acceleration of RNA synthesis which runs parallel to nuclear growth and chromatin dispersion, (e) DNA replication and (f) expression of chick genes in the form of chick polypeptide synthesis. The time sequence of these events is summarized in Fig. 3.

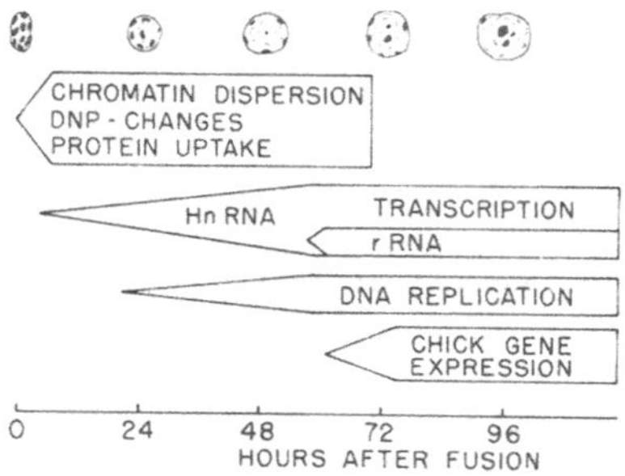

Fig. 3. Schematic summary of the reactivation of chick erythrocyte nuclei in chick-mammalian heterokaryons

Nuclear growth during the first 2 days after fusion is associated with a massive migration of mammalian nuclear proteins into the chick nucleus. Using human autoantibodies to nuclear antigens, it is possible to demonstrate that in chick erythrocyte x HeLa cell heterokaryons the chick nuclei specifically concentrate antigens typical of human nuclei, whereas antigens characteristic of human cytoplasm are excluded from the chick nuclear compartment. These immunological observations have been confirmed in biochemical and cytochemical studies of specific proteins. Among the proteins taken up by the reactivating chick nucleus are mammalian DNA repair enzymes, SV40 T-antigen, human H1 histone etc. Although these proteins may have a functional role in the reactivated chick erythrocyte nucleus, it is difficult to attribute to them a specific role in the reactivation process. In order to find out whether the chick genome is transcribed by a chick or by a mammalian RNA polymerase, we have recently [3] studied heterokaryons made by fusing α-amanitin resistant rat myoblasts and chick erythrocytes. The mutant myoblasts have a modified RNA polymerase II which is resistant to α-amanitin. Using a combination of biochemical and cytochemical methods we have found that the chick genome, including its globin genes, is transcribed by an α-amanitin-resistant polymerase. Thus in these rat x chick heterokaryons the chick genes are effectively transcribed by a mutant mammalian polymerase.

We have also tried to examine if some protein components of the chick erythrocyte chromatin are lost during the reactivation process in heterokaryons. Chick erythrocyte nuclei contain a tissue-specific linker histone called H5. Using immune fluorescence staining with a rabbit H5 antibody kindly provided by Dr. Thomas Graf, we have shown that H5 antigen is lost from the chick nucleus during reactivation. Some of this antigen is taken up by the mammalian nuclei in the heterokaryons [4].

What we have so far learnt about the mechanisms involved in the reactivation of chick erythrocyte nuclei in heterokaryons can be summarized in the following points:

1. The earliest changes so far detected are altered dye-binding properties of the chromatin and decreased stability of DNA to thermal denaturation. The molecular basis for these changes has not yet been defined.
2. The growth of the erythrocyte nucleus is parallelled by a selective uptake of nucleospecific antigens and proteins typical of the mammalian cell nucleus.
3. Among the mammalian proteins taken up by the erythrocyte nucleus are proteins which play a functional role.
4. There is also a loss of some components of the chick erythrocyte chromatin. Thus mammalian H1 histone appears to replace chick H5 histone.
5. The reactivation of the chick erythrocyte nucleus results in the synthesis of chick specific polypeptides. The polypeptides demonstrated so far do not, however, provide any information about tissue specific gene expression.

Different Types of Erythrocyte Heterokaryons

Most of our work has been carried out with chick erythrocytes from 12-17-day hen eggs. At this stage of development, the erythrocytes are of the definitive type and contain the adult types of globin chains. Erythrocytes have been fused to monolayers of mammalian cells using UV-inactivated Sendai virus. A number of mammalian cell lines have been used as partners in these fusions e.g. rat L6 myoblasts, mouse A9 fibroblasts, mouse C1300 neuroblastoma cells, human HeLa and K562 erythroleukemia cells and baby hamster kidney cells (BHK). In most of these combinations the fusion yield has been very high, resulting in the introduction of erythrocyte nuclei into most of the mammalian cells. For biochemical work we have used cultures containing 50-100 chick erythrocyte nuclei per 100 mammalian nuclei. Because the mammalian cells remain attached to the culture dishes during fusion, very few mammalian cells fuse with each other to generate homokaryons. Unfused erythrocytes are effectively removed by rinsing the cultures after fusion. In order to prevent overgrowth of unfused mammalian cells these cells have been pretreated prior to fusion with mitomycin C. This drug effectively blocks DNA replication and cell division after fusion but has little or no effect on the rate at which the chick erythrocyte nuclei are reactivated. The number of heterokaryons and of unfused mammalian cells respectively, remains constant for at least 12 days post fusion. In some cases heterokaryons have been maintained for more than 3 weeks. If mitomycin is not used, unfused mammalian cells overgrow the heterokaryons. Some heterokaryons do, however, enter mitosis and give rise to mononucleate hybrid cells.

Because nuclear reactivation takes several days and because of the difficulties in obtaining mononucleate hybrids, we have also examined heterokaryons between cells from species that are more closely related than are the chick x mammalian cell combinations. To this end

Xenopus erythrocytes have been fused with frog cell lines. Heterokaryons have been obtained and preliminary evidence suggests that the *Xenopus* erythrocyte nuclei undergo reactivation. Since *Xenopus* erythrocytes are unable to multiply and because there are mutant frog cell lines, in principle, it should be possible to isolate mononucleate hybrids on semiselective media. This possibility is presently being examined.

Because most work on erythrocyte heterokaryons has been done with established mammalian cell lines we have also examined the possibilities to fuse chick erythrocytes to primary cells from quail, a closely related avian species. We have found that chick erythrocytes can easily be fused to quail myoblasts. Reactivation in chick-quail and chick-chick heterokaryons was found to occur at approximately the same rate as in chick-mammalian heterokaryons.

Pattern of Gene Expression in Erythrocyte Heterokaryons

The fact that nucleated erythrocytes can be fused to a number of different cell types makes it possible to analyse whether the cytoplasmic environment influences the pattern of genes which becomes active. An interesting question in this context is whether the chick nucleus retains its determination for erythroid-type gene expression or whether it undergoes a reprogramming so as to express genes analogous to those which are active in the mammalian partner in the fusion. In what follows we will summarize some recent experiments [5,6] in which we have tried to analyse these questions by (a) examining chick erythrocyte heterokaryons for globin gene expression and (b) analysing the pattern of chick polypeptides synthesized in heterokaryons where the mammalian partner represents different forms of tissue-specific differentiation.

Globin Gene Expression

Heterokaryons obtained by Sendai virus induced fusion of mitomycin blocked mammalian cells; rat L6J1 myoblasts, mouse C1300 neuroblastoma cells and mouse A9 fibroblasts, have been examined for globin gene expression by RNA blot hybridization analysis using fetal and adult chicken globin genes cloned in pBR322 and by 2D gel separations of chick polypeptides isolated by immunoprecipitation from ^{35}S-methionine or ^{14}C-leucine labelled heterokaryon lysates.

Globin mRNA was not detected in chick erythroyte x rat L6 myoblast heterokaryons one day after fusion but appeared 10 days post fusion (Fig. 4). The globin mRNA was polyadenylated since it bound to oligo-(dT) cellulose. No synthesis of fetal globin mRNA could be detected by blot hybridization of fetal globin DNA probes to mRNA from chick erythrocyte x L6 heterokaryons.

Chick α^A, α^D and β-globins of the adult type were detected in several types of heterokaryons. The levels of globin expression varied in different types of heterokaryons, however (Table 1). Strong globin expression was found in heterokaryons made with L6 rat myoblasts (Fig. 5) and an α-amanitin resistant L6 subline. Heterokaryons made with mouse C1300 neuroblastoma and A9 fibroblasts synthesized globins at a level which was some 20 times lower than that of L6 heterokaryons. Furthermore, the level of globin synthesis correlated with the level of globin mRNA. Heterokaryons made by fusing chick erythrocytes to a mononucleate L6 x A9 hybrid cell line synthesized globins at the same low level as the original A9 line. It appears, therefore, that the ability to suppress globin expression in heterokaryons is a dominant property and suggests that the A9 parental cell may produce some negative regulator that inhibits globin expression. In order to further examine the basis for the variable expression of chick globin expression we also examined the general level of chick gene expression by 2D gel

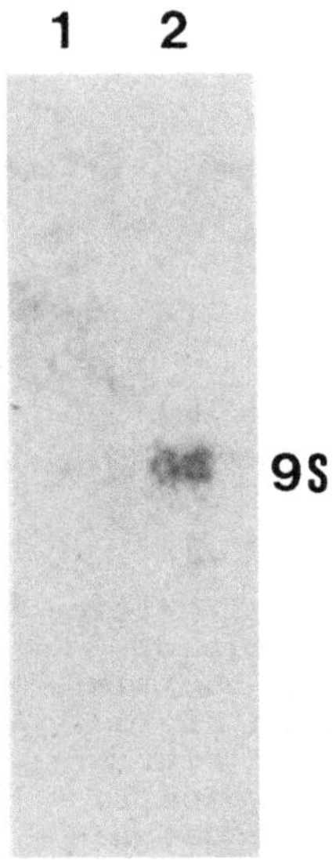

Fig. 4. Alpha-globin transcripts in rat myoblast - adult chick erythrocyte heterokaryons. RNA isolated from heterokaryons was electrophoresed, blotted, and hybridized with ^{32}P-labelled nick-translated pBR322 plasmids containing an adult alpha-globin chicken genomic insert. *Lane 1:* 1 day after fusion, no globin RNA, *Lane 2:* 10 days after fusion, 9S globin RNA present [5]

Table 1. Globin expression chick erythrocyte heterokaryons [6]

Parental cell		Globin expression
L6J1	Rat myoblast, diploid, myogenic	100
420, 431	Rat myoblast, tetraploid, non-myogenic	100
A9	Mouse L-cell	15
BB2	L6J1 x A9 hybrid	10
NA	Mouse neuroblastoma	5
BHK	Hamster	5
3T3	Mouse	5
NRK	Rat kidney	5
Chicken primary fibroblasts		5

Globin expression is relative to L6 heterokaryons which is defined as 100. Globin was quantitated by measuring the maximal peak intensity of α-globin using a Joyce-Loebl densitometer and normalized by dividing by the fusion frequency times the total cpm subjected to immunoprecipitation.

The levels indicated for the low globin producers are maximum values. In many experiments these values were 2-3-fold lower.

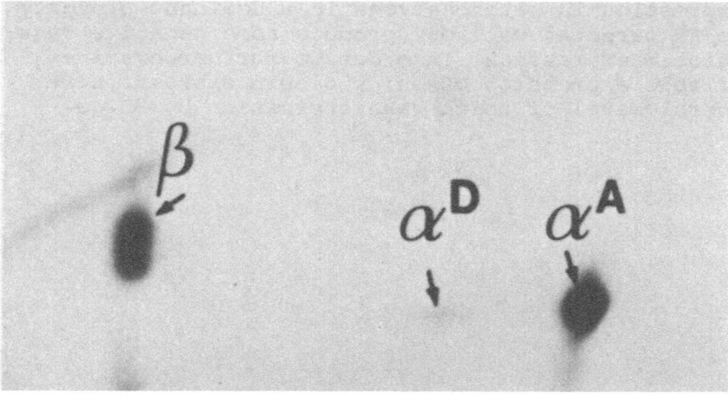

Fig. 5. Two-dimensional gel analysis of immunoprecipitated globin polypeptides from rat myoblast-17-day embryonic erythrocyte heterokaryons labelled with ^{14}C-leucine. [5]

analysis of chick polypeptides (see below), the sensitivity of heterokaryon globin expression to known globin inducers, and the role of gene dosage effects caused by the different ploidy levels in the mammalian parental cells. It was found that at comparable fusion yields, the different heterokaryons synthesized chick polypeptides at similar levels. Hemin, DMSO and sodium butyrate induced a slight increase in globin expression but did not significantly alter the differences recorded between different types of heterokaryons. Heterokaryons made with diploid and tetraploid L6 cells produced chick globins at similar levels.

Conformation and Methylation of Globin Genes

The observation of differential globin expression in a series of heterokaryons made with different types of mammalian cells suggested that the cytoplasmic environment influenced the pattern of chick gene expression. The finding of globin mRNA in those heterokaryons which produced significant quantities of chick globin and the failure to find globin mRNA in heterokaryons which showed low or non-detectable levels of globin synthesis suggested that there were differences in chick globin gene transcription. In order to further explore this aspect at the chromatin level we have examined the conformation of the globin gene by nuclease digestions using DNAase I and micrococcal nuclease. Furthermore, attempts have been made to detect changes in DNA methylation within the globin genes by digestion with isoschizomeric restriction enzymes.

Previous work by other groups has shown that in erythrocyte nuclei, the globin genes are in an "open" configuration that is sensitive to nuclease attack. In other tissues where the globin gene is never expressed, the globin genes are resistant to these nucleases. It was of interest, therefore, to examine if after fusion, the globin genes retained their "open" nuclease sensitive conformation. Preliminary evidence suggests that the globin genes in reactivated erythrocyte nuclei show the same overall nuclease sensitivity and the same "hypersensitive" sites as they do in the inactive nuclei of unfused erythrocytes.

Analogous findings have been made concerning the DNA methylation pattern of the globin gene. Electrophoretic analysis of the type of DNA restriction fragments obtained after digestion of erythrocyte DNA with HpaII and MspI has failed to detect changes during the reactivation process. The variable expression of the globin gene cannot be explained by methylation changes in the globin genes in some heterokaryons. The globin genes are hypomethylated in chick erythrocyte and remain so in the types of heterokaryons that we have analyzed so far. The analysis of chromatin conformation in specific genes by nuclease digestion and nucleic acid hybridization with specific probes is presently being extended to other genes that are of interest in describing the pattern of gene activation.

Expression of Other Chick Polypeptides

In order to obtain some information about the pattern of chick genes expressed in different types of erythrocyte heterokaryons, cultures of fused cells have been labelled with ^{35}S-methionine. Chick polypeptides have then been purified from heterokaryon lysates by immunological methods using an antiserum against chicken cells. The labelled chick polypeptides were then eluted from the antibody and analysed by 2-D gel electrophoresis and autoradiography. Chicken polypeptide synthesis increased several-fold during the reactivation of the erythrocyte nucleus. About 40 different chicken polypeptides were detected. The pattern of chicken polypeptide synthesis was very similar in heterokaryons made by fusing chicken erythrocytes with rat L6 myoblasts, hamster BHK cells and mouse neuroblastoma cells. Three polypeptides synthesized by non-erythroid cells but less so by embryonic erythrocytes were conspicuous in heterokaryons. Two abundant erythrocyte polypeptides were insignificant in non-erythroid chicken cells and in heterokaryons. Thus there are undoubtedly differences in the pattern of chick polypeptides synthesized after reactivation of chick erythrocyte nuclei in different mammalian cells. It is difficult, however, to interpret the data in terms of the stability of the determined state of the erythrocyte nuclei. Most of the chick polypeptides detected are likely to be "household" proteins synthesized by most chick cells irrespective of their differentiation. The differences observed may well reflect phenotypic modulation rather than stable alterations in the gene expression pattern.

Summary

Techniques have been developed to obtain high yields of heterokaryons by Sendai virus-induced fusion of chick erythrocytes with mammalian cells representing different forms of cell specialization. By pretreating the mammalian cell with mitomycin before fusion it is possible to maintain heterokaryons for 2-3 weeks while at the same time preventing overgrowth of unfused mammalian cells. This procedure allows for a more complete reactivation of the chick erythrocyte nuclei than has so far been possible and makes it possible to accumulate enough chick gene products to be analysed by biochemical methods.

Mammalian nuclear proteins taken up by the chick erythrocyte nuclei appear to play an important part in the reactivation process. Thus chick genes are transcribed with mammalian RNA polymerase II. Different chick genes are expressed in different types of heterokaryons. Globin gene expression occurs in some heterokaryons showing that the erythrocyte genome can retain its programming for erythroid gene expression. Many of the non-globin genes expressed in heterokaryons appear to be such which are expressed in non-erythroid chick cells.

References

1. Ringertz NR, Savage RE (1976) Cell Hybrids. Academic Press, London New York
2. Harris H (1965) Behaviour of differentiated nuclei in heterokaryons of animal cells from different species. Nature (London) 206:583-588
3. Zuckerman SH, Linder S, Ringertz NR (1982) Transcription of chick genes by mammalian RNA polymerase II in chick erythrocyte - mammalian cell heterokaryons. J Cell Physiol 113:99-104
4. Linder S, Zuckerman SH, Ringertz NR (1982) Distribution of histone H5 in chicken erythrocyte - mammalian cell heterokaryons. Exp Cell Res 140:464-468
5. Linder S, Zuckerman SH, Ringertz NR (1981) Reactivation of chicken erythrocyte nuclei in heterokaryons results in the expression of adult chicken globin genes. Proc Natl Acad Sci USA 78:6286-6289
6. Linder S, Zuckerman SH, Ringertz NR (submitted) Pattern of chick gene activation in chick erythrocyte heterokaryons. J Cell Biol

Regulation of Development of Salivary Glands in 3rd Instar *Drosophila melanogaster* Larvae by Ecdysone

B. Dworniczak, S. Kobus, K. Eiteljörge, and O. Pongs[1]

Introduction

There is ample evidence that growth and development in *Drosophila melanogaster* is controlled and coordinated by hormones. Molting of larvae, puparium formation, pupal and adult development are triggered and regulated by steroid hormones, the ecdysteroids, and by sesquiterpenoids, the juvenile hormones [1]. In this paper we will not discuss the importance of juvenile hormones for ecdysteroid action during molting processes. Instead, we will concentrate on that developmental stage of *Drosophila melanogaster*, where 3rd instar larvae develop into prepupae, i.e. the beginning of metamorphosis. It is very fortunate that salivary glands of 3rd instar larvae possess polytene interphase chromosomes, which are so gigantic that they can be analyzed under a phase contrast microscope [2].

Polytene chromosomes contain a characteristic banded pattern. The bands, which vary in thickness, contain a high concentration of DNA and protein. The spaces between the bands, known as interbands, contain a low concentration of these substances. Reversible structural modifcations of the polytene chromosome bands are puffs, which originate from a single band [3,4]. The molecular events that occur at the chromosomal level when a band is transformed into a puff, and vice versa, are unknown. Puffing activity is tissue specific and changes during development. Based on this important obversation, a great body of experimental evidence has been accumulated to show that puffs are uncoiled sections of the chromatin where transcription occurs [3-5].

Control of Puffing by 20-hydroxyecdysone

The importance of ecdysone in the control of puffing in larval tissues was first suggested by Becker [6] and soon directly demonstrated by Clever and Karlson [7]. Puffing activity dramatically increases towards the end of the third instar. This correlates well with the release of ecdysone into the blood from the ring gland of third instar larvae of *Drosophila melanogaster*, as first shown from ligature experiments by Becker [6]. The time of ecdysone release is between 3 1/2 and 5 h before puparium formation.

---

1 Lehrstuhl für Biochemie, Universität Bochum, Postfach 102148, D-4630 Bochum, FRG

33. Colloquium - Mosbach 1982
Biochemistry of Differentiation and Morphogenesis

Since the salivary glands histolyse during metamorphosis, the study of the ecdysone induced developmental changes in puffing in *Drosophila* is restricted to that short period in pre-imaginal development, where polytene chromosomes are large enough and, hence, accessible for analysis [8]. This period encompasses the last 5-6 h before puparium formation, where a large number of puffs become active for short periods of time. Changes in puffing represent a temporal sequence which has been dissected into "Puff Stages" (PS) [9]. Individual puff stages are landmarks in this temporal sequence, where the puffing pattern is sufficiently different from that at other times to be readily recognized. The puffing pattern moves from PS1 to PS2 at 5-6 h before puparium formation, i.e. the time where the hormone titre increased in the blood and, as a consequence, ecdysone induced puffs appear. Thus, the pattern of puffs, described as PS1, is relatively undefined and refers to early instar development before the increase in blood hormone titre. Puff stages are, of course, correlated with particular morphological events, e.g. secretion of the salivary gland's glue into the lumen occurs after PS4 and puparium formation itself at PS10/11 [9].

The larval response to ecdysone during the development, which occurs between PS1 and PS10/11, can be studied with cultured salivary glands in defined media [10]. This has several advantages over the intact animal since not only do the isolated glands not metabolize the hormone, but especially the ability to use one lobe of the paired gland as a control, makes many experiments possible that would otherwise either be impossible or very difficult.

Since Ashburner's studies [9-11] on the ecdysone control of puffing in *D. melanogaster* are the basis of our own studies, they will be briefly summarized here. Ecdysone induces puffs either very rapidly (within a few minutes) or after a lag of one and more hours. Accordingly, puffs are described as "early" or as "late" puffs. In general, protein synthesis is not necessary for the induction of early puffs (with the exception of the puff at 23E), whereas induction of late puffs requires protein synthesis [12]. The hormone dose-puff-response relationship is sigmoidal for early puffs, which respond over a 500-fold ecdysone concentration range. Late puffs show an almost all or nothing dose-response [13]. In addition to the induction of early and of late puffs, another response to ecdysone is the regression of some "intermolt puffs" (e.g. the puff at 68C), which are active earlier in the instar [11].

Localization of 20-hydroxyecdysone on Polytene Chromosomes

These data have been interpreted in the light of the common paradigm of steroid-hormone action, which postulates that 20-hydroxyecdysone acts directly, presumably via hormone-receptor complexes, at genetic loci, which as a result become puffed [14]. Early attempts to localize radioactive 20-hydroxyecdysone, by autoradiography on polytene chromosomes, however, were not successful [15]. We recognized that 20-hydroxyecdysone carries in its B-ring an α, β-unsaturated ketone-moiety, which is photo-activatable with near-visible light [16]. Irradiation of 20-hydroxyecdysone with light of ~320 nm induces an exited triplet state at the ketone which then, in turn, can react with surrounding molecules, presumably via proton-abstraction and subsequent C-H-insertion [17]. This photochemical property of 20-hydroxyecdysone was exploited to localize the steroid-hormone at genetic loci on polytene chromosomes of various puff stages during development of third instar salivary glands [16,18]. The methods used are illustrated in Fig. 1. Salivary glands of 3rd instar larvae of *Drosophila melanogaster* were dissected and incubated in vitro. At suitable times, one lobe of a gland at a time was irradiated for 5 min with light of > 320 nm. The light source was a high pressure 500 W mercury lamp. Then chromosomes were prepared by conventional squashing procedures. Bonded 20-hydroxyecdysone on these chromosomes was detected by incubating the chromosome preparations successively with anti 20-hydroxyecdysone-rabbit IgG and with fluoresceine-labelled goat-anti-rabbit IgG. Chromosomes were then analyzed in the fluorescence microscope and, subsequently, in the phase contrast microscope after orcein staining. Thus, we could correlate the fluorescence, caused by bonded hormone, with a particular site on the respective chromosome.

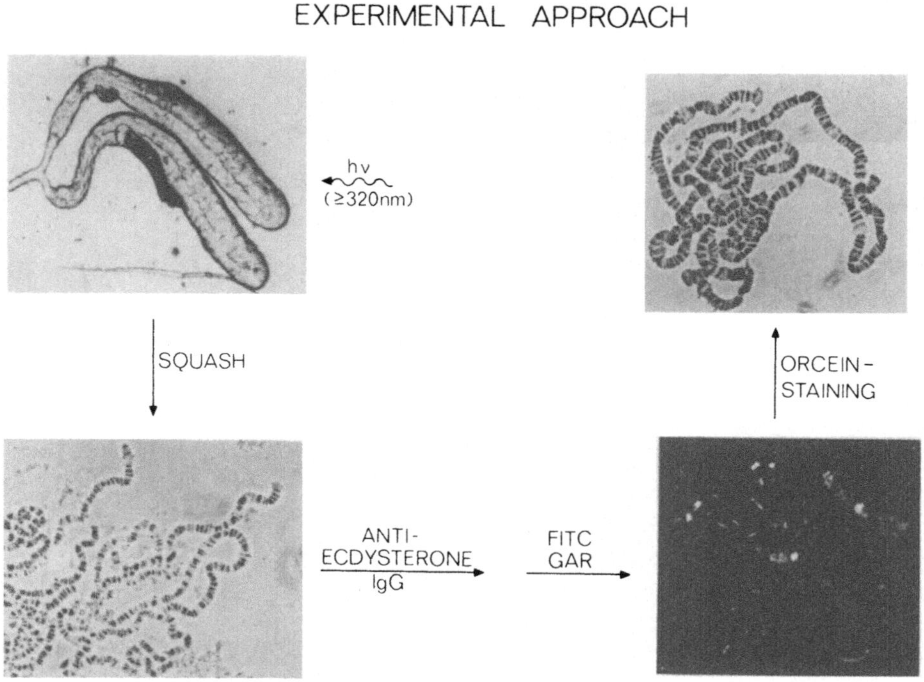

Fig. 1. Illustration of the indirect immunofluorescent microscopic approach to localize bonded 20-hydroxyecdysone on polytene chromosomes of 3rd instar larval salivary glands. Glands were dissected and irradiated as described [16]. Squashed preparations of the gland chromosomes were incubated with antiecdysterone-antiserum/ fluorescein isothiocyanate-coupled goat antirabbit immunoglobulins [18]. Specimens were photographed by incident UV illumination using a Zeiss standard microscope with an incident UV-attachment and fluorescence optics. Phase-contrast photographs were taken after staining the preparations with lacto-acetic acid/orcein [35]

This technique permitted us to ask several pertinent questions for the action of steroid hormones in general and of 20-hydroxyecdysone in particular.

We wanted to know (a) whether 20-hydroxyecdysone directly interacts with the polytene chromosome and whether this interaction is specific; (b) how long the steroid hormone is necessary for the active state of a transcribed gene; (c) how the binding of a hormone-receptor complex to chromatin is regulated, especially terminated; (d) whether the temporal change in puffing is exerted by a temporal binding of 20-hydroxyecdysone to the relevant chromosomal loci. Before going into the results of the indirect immunofluorescence microscopy, the specificity of the photoreaction and of the antibodies should be briefly discussed. The specificity of the photoreaction was analyzed in salivary glands as well as in a 20-hydroxyecdysone-sensitive *Drosophila* tissue culture cell line [19,20]. After irradiation, proteins were extracted from the tissues and separated on PAGE-gels. They were then blotted onto nitrocellulose sheets, which were incubated with hormone-antibody and ^{125}I-Protein A.

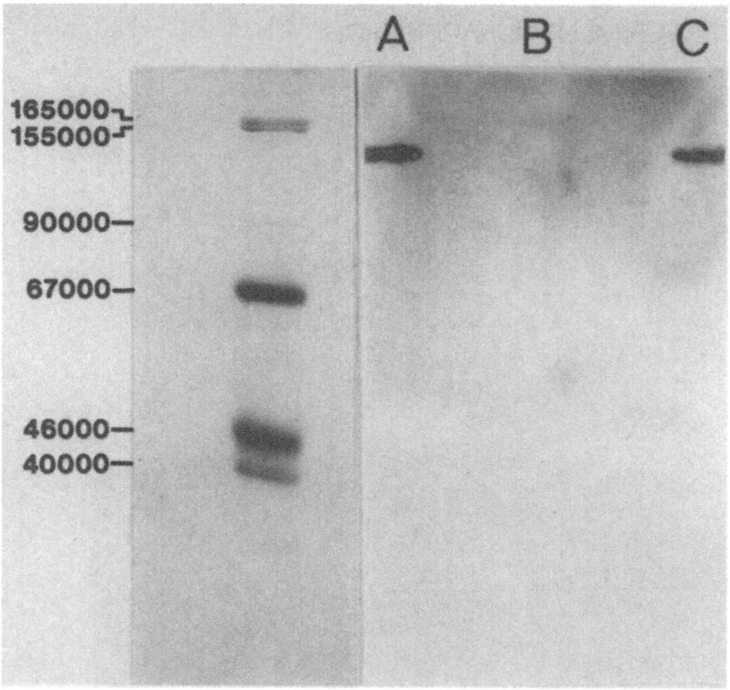

Fig. 2. Identification of ecdysterone-binding proteins in salivary glands and in tissue culture cells of *D. melanogaster*. Protein extracts of salivary glands of third-instar larvae and of K_c cells containing covalently linked ecdysterone were prepared. After polyacrylamide gel electrophoresis according to Laemmli [36], proteins were blotted onto nitrocellulose membranes and incubated with ecdysterone antiserum and ^{125}I-protein A as described [19]. Molecular weight standards (*left lane*) were the subunits of *Escherichia coli* RNA polymerase. The autoradiogram was obtained with proteins extracted from 5×10^8 irradiated K_c cells that had been incubated either with 0.1 μM ecdysterone (*lane A*) or with 10 μM 2-acetylecdysterone (*lane B*) for 15 min. *Lane C* shows the result with proteins that were extracted from 2000 irradiated salivary glands

Only one protein-hormone crosslink was detected in the resulting autoradiograms (Fig. 2). The denatured protein has an apparent molecular weight of 130,000, on the basis of electrophoretic mobilities in the gel. Since the protein, to which 20-hydroxyecdysone became bonded, is very similar, if not identical, in *Drosophila* larval salivary glands and in *Drosophila* tissue culture cells of the Kc line, further photoaffinity labeling experiments were carried out with Kc cells to characterize the 20-hydroxyecdysone binding protein. The distribution of this protein in Kc-cells before and after hormone addition, its sedimendation behaviour and its high affinity steroid binding properties, demonstrated that this protein, which reacts with photoactivated 20-hydroxyecdysone, is in fact the ecdysteroid-receptor [20]. This receptor forms a complex with 20-hydroxyecdysone in the cytoplasm and than migrates to the nucleus, where it becomes associated with chromatin (Figs. 3 and 4). This result has important consequences with respect to our immunofluorescence work, which we will

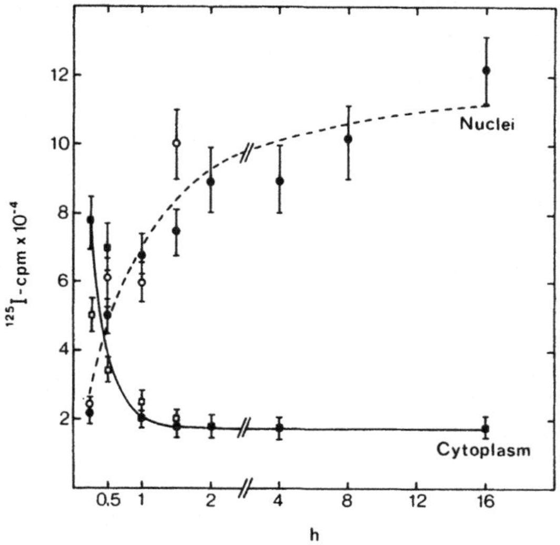

Fig. 3. Distribution of ecdysterone binding protein in cytoplasm and nuclei of K_C cells. Cells were incubated with 0.1 μM ecdysterone and irradiated at the times indicated. The amount of covalently linked ecdysterone in protein extracts of cytoplasmic (■) or nuclear (●) fractions was determined immunologically with ecdysterone antiserum and ^{125}I-protein A. In a second experiment cells were irradiated immediately after 0.1 μM ecdysterone had been added. Cytoplasmic (□) and nuclear (○) protein fractions were prepared after continuing the incubation for the times indicated. Error bars indicate SD

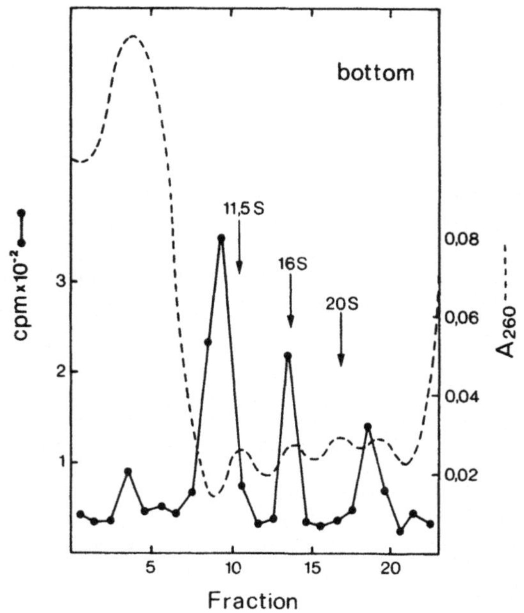

Fig. 4. Association of ponasterone A with chromatin of K_C cells. Nuclei of K_C cells, which had been incubated with [^{3}H]-ponasterone A (120 Ci/mMol), were prepared as described [37]. 10^9 nuclei were irradiated and then digested with micrococcal nuclease. Nucleosomes were separated on 5%-31% isokinetic sucrose gradients [38]. Each fraction was made with 7.5% trichloroacetic acid. Trichloroacetic acid precipitable material was collected on Whatman GFC-filters, which were washed and then counted. --- A_{260} of chromatin materials, —●— cpm of bonded [^{3}H]-ponasterone A

discuss below, because it suggests that the fluorescence, which we observed at genetic loci of polytene chromosomes, represents hormone-receptor complex, and not only hormone. Having shown the specificity of the photoinduced reaction of 20-hydroxyecdysone with hormone receptor, the problem remains whether the antibody, employed in the immunofluorescence work, binds only to 20-hydroxyecdysone or also recognizes other determinants on polytene chromosomes which are not related to hormone. First of all, non-irradiated salivary glands do not yield any fluorescence in our system. This demonstrates that the determinants for antibody binding are only created by the photoreaction. Also rabbit-IgG can be substituted with monoclonal antiecdysterone-mouse IgG. Non-immune serum did not give any fluorescence at all. There are two physiological controls which can be exploited in the *Drosophila* system. When larvae are transferred from 25°C to 35°C, they experience a "heat-shock" [21]. This results in inactivation of almost all transcribed genes except for heat-shock genes, which are turned on. Chromosomes prepared from heat-shocked larvae do not yield

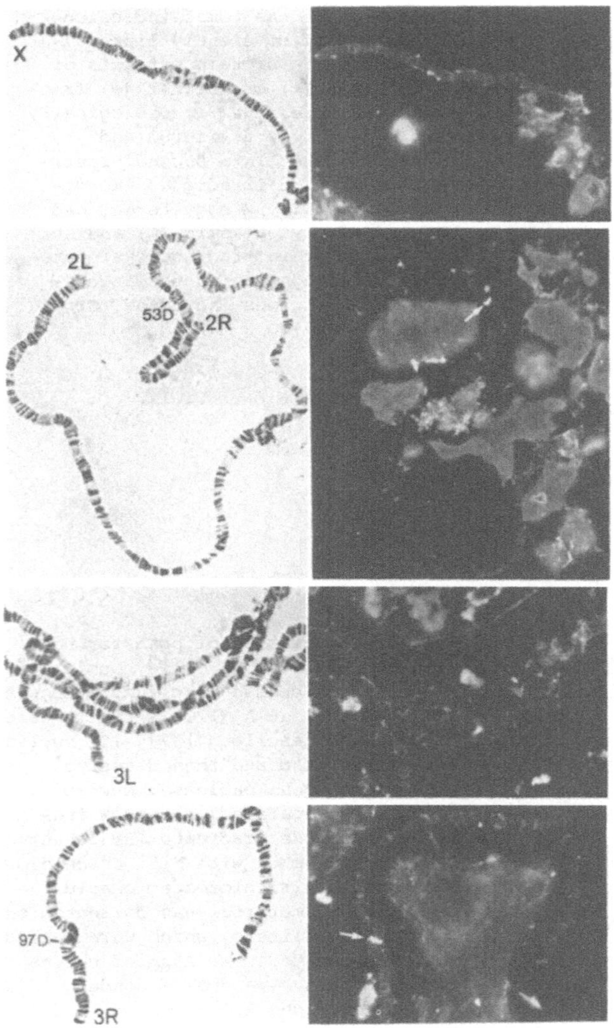

Fig. 5. Indirect immunofluorescence of chromosome squashes obtained from salivary glands of ecd[1]-*Drosophila melanogaster* 3rd instar larvae after UV irradiation for 5 min. Glands were fixed and analyzed as described [16]. *Left panels* show phase-contrast pattern of the chromosome after orcein staining, and *right panels* corresponding images obtained by indirect immunofluorescence microscopy to visualize bonded 20-hydroxyecdysone

ecdysone-mediated fluorescence after irradiation. A temperature-sensitive lethal mutant of *Drosophila melanogaster* called ecd^1 becomes deficient in ecdysone, when there is a shift in temperature from 20° to 29°. When the shift occurs early in third-instar stage, the mutant larvae grow to full size but fail to pupariate [22]. The puparation block in the ecd^1 mutant is associated with a reduced level of ecdysone, i.e. 22 pg as compared to 500 pg per larvae in wild type. Ecd^1 mutants were shifted to 29°. Salivary glands were dissected from these larvae after 24 h, irradiated and analyzed for bonded ecdysone as described in Fig. 1. Not surprisingly, Fig. 5 demonstrates that all the chromosomes appeared dark in the fluorescence microscope except for two sites, 53D on chromosome 2R and 97D on chromosome 3R. A low level of ecdysone did not give rise to fluorescence on chromosomes X, 2L and 3L, which could explain the observed block in pupariation of the ecd^1 mutant. In summary, the photoreaction induces in salivary glands a covalent

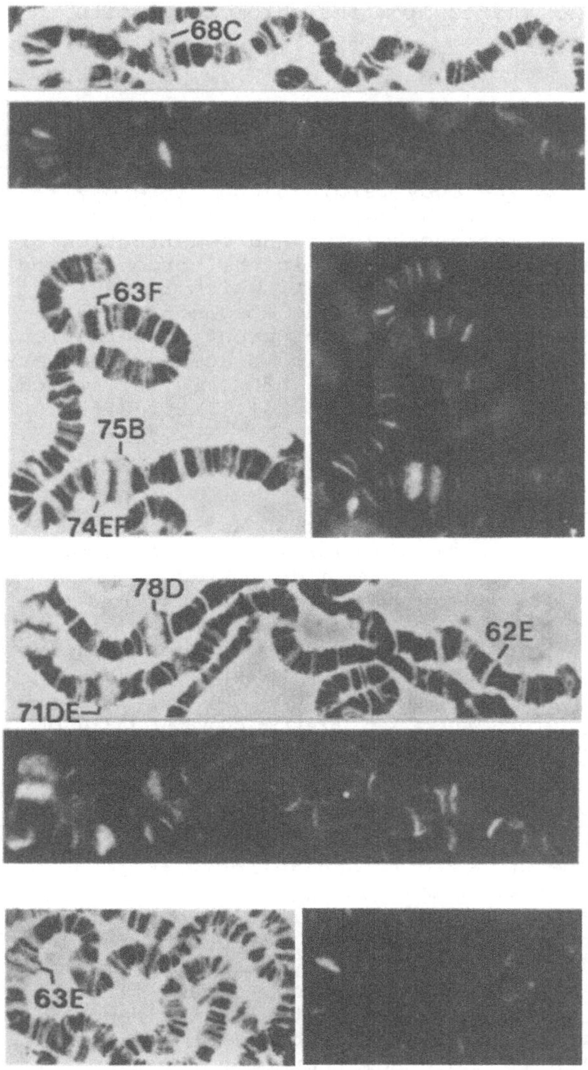

Fig. 6. Changes in puffing patterns and hormone-binding patterns on chromosomes of salivary glands of 3rd instar larvae. Puff stages are from top to bottom: PS1, PS4, PS7, PS10 [9,11]. Indirect immunofluorescence microscopy was carried out as illustrated in Fig. 1 for localizing 20-hydroxyecdysone on polytene chromosomes

specific bond between hormone and receptor. The binding of this complex to polytene chromosomes can be detected and analyzed by indirect immunofluorescence microscopy.

Figure 6 shows the left arm of chromosome 3 at various puff stages. A comparison of puffing pattern and fluorescence pattern demonstrates that the 20-hydroxyecdysone-receptor complex binds specifically to a certain number of genetic loci. This binding is temporal and is correlated to the temporal puffing activity at a respective chromosomal locus. With few exceptions our data show that in early puff stages, 20-hydroxyecdysone is localized e.g. 63F/74EF/75B, whereas in later puff stages (PS6/7) it shifts to e.g. 62E/71DE/78D and in still later puff stages (PS10/11), 20-hydroxyecdysone is localized on for e.g. 63E, 66B. The data, which are summarized in Fig. 6, already answer some of the questions, raised in the introduction. They demonstrate that the continous presence of hormone-receptor complex is necessary for active gene-transcription. They also suggest that during development the hormone-receptor complex shifts from one chromosomal site to another.

Influence of Hormone Concentration on Puffing Patterns

During the life cycle of *Drosophila* the ecdysteroid blood titre oszillates [23]. This is believed to trigger the extensive cellular changes that accompany ecdysis (molting of the exoskeleton and construction of a new cuticle). In early instar, there is a burst in the hormone blood titre concomitant with the onset of the development, which leads to pupariation (Fig. 7). Therefore, 20-hydroxyecdysone does not only influence cellular functions, which are important for puparium formation, but among many others, also cellular functions which precede the intermolt. A typical intermolt puff is observed at site 68C (Fig. 6), which is responsible for synthesis of a glue protein [24,25]. This puff is also under hormonal control.

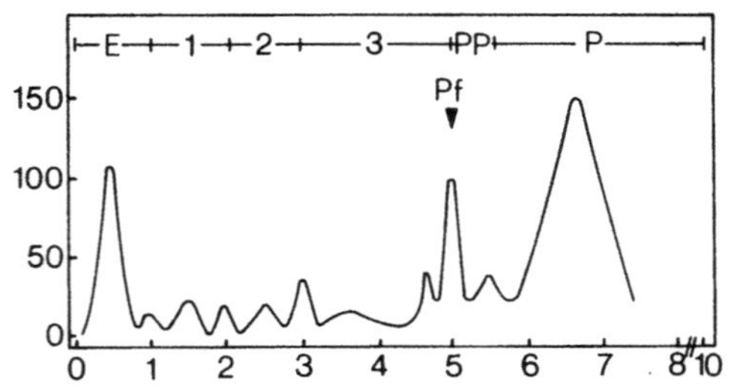

Fig. 7. Oscillations in hormone titer in the hemolymph of *Drosophila* during the development of eggs into pupae. Developmental stages are abbreviated as *E* egg; *1*, *2*, *3* first, second, third instar; *PP* prepupae; *P* pupae; *Pf* pupariation. The ordinate gives the time of development in days. The abscissa indicates the relative ecdysone titre in arbitrary units. (Data from [23])

These observations immediately raise the problem of how a rise in hormone concentration determines the specificity and selectivity of a hormone-receptor complex and how the hormone-receptor complex is able to shift from early puffs to late puffs in an apparently temporally and coordinately controlled manner. A formal model to account for the responses of early and late puffs to 20-hydroxyecdysone in *Drosophila* salivary gland chromosomes was suggested by Ashburner et al. [14]. It was proposed that the early sites respond directly to the concentration of hormone-receptor complex and that they code for two functions - to repress the activities of the genes that code for them and to induce activity of late puff sites. Recently, genotypes aneuploid for two early puffs (74EF/75B) have been constructed, being either duplicated or deficient. These mutants were used for gene-dosage-studies, which permitted two important conclusions: the activities of the 74EF and 75B early puffs are compensated for altered gene dose of the puffs themselves, i.e., they are autoregulated, and some late puffs respond to altered gene dose of these two early puffs, quantitatively as well as temporally [26]. Similarly, larvae carrying a deletion of the early puff at 2B5 die at the end of third larval instar, because they are unable to perform the activation of late puffs and also the regression of intermolt puffs such as 68C [27]. These data clearly implicate gene products for establishing temporal puffing patterns. They also suggest that the time needed to accumulate these gene products is important for the duration of the developmental program. These data also imply that the half-life of a hormone-receptor complex is of secondary importance in order to explain the duration of hormonally controlled puffing activity at a given chromosomal site, e.g. 68C. This was experimentally demonstrated in the following way: the fluorescence pattern, shown in Fig. 6, was obtained by incubating the salivary gland in vitro to reach a certain puff stage, where the glands then were irradiated for localizing bonded 20-hydroxyecdysone. In another series, glands at puff stage 1 were irradiated and then further incubated, until the puff stage to be analyzed was reached. Figure 8 shows that the same puffing pattern was obtained as above. This demonstrates that the photo-induced covalent bond of 20-hydroxyecdysone to receptor does not inactivate the activities of hormone-receptor complexes and that it still can be normally dissociated from the puff at 68C and bind to other puffs (e.g. 74EF/75B in Fig. 8)

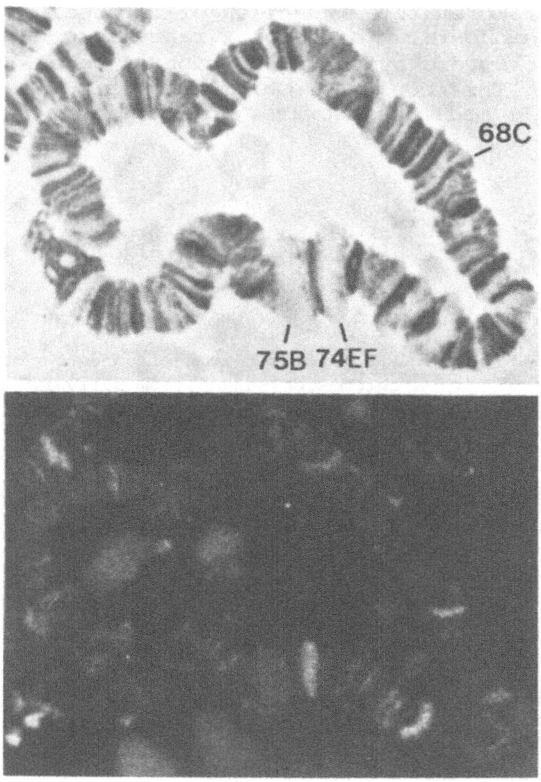

Fig. 8. Indirect immunofluorescence of a chromosome squash obtained from salivary glands of *Drosophila melanogaster* 3rd instar larvae. Salivary glands were explanted and irradiated at PS1, then incubated in vitro for further 2 h to reach PS7/8 before squashing and microscopic analysis as shown. The photograph depicts chromosome 3L. The phase-contrast pattern is at the top

at other chromosomal sites. When PS1-salivary glands are transferred into hormone-free medium and are incubated for a further 2 h, binding of 20-hydroxyecdysone to 68C is not influenced. However, a similar experiment with PS4-salivary glands shows that binding of 20-hydroxyecdysone to early puffs is very sensitive to such a wash out experiment (data not shown). The following conclusions can be drawn from these results: (a) Hormone-receptor complexes move on and off chromosomal sites without dissociation; (b) binding of hormone-receptor complexes to the intermolt puff 68C at PS1 is insensitive to a wash out of hormone; (c) binding of hormone-receptor complexes to early puffs at PS4 is very sensitive to a wash out of hormone.

Based on cytogenetic data with respect to mutations at 2B5, regression of the puff at 68C requires a gene product, presumably encoded in the 2B5 locus [27]. We investigated the effect of inhibition of protein synthesis on puff regression at 68C. One lobe of a PS1 salivary gland was incubated in the presence of 20-hydroxyecdysone until PS4 and then transferred into hormone free medium; the sister lobe was incubated in the presence of 20-hydroxyecdysone and of cycloheximid until PS4 and also transferred into hormone free medium in the continued presence of cycloheximid. After irradiation, chromosomes were analyzed for the presence of hormone-recptor complex at 68C. Figure 9 shows that in the "normal" case, hormone-receptor complex dissociated off at 68C, whereas the presence of cycloheximid inhibited this dissociation. This demonstrates that the regression of the puff at 68C is controlled by an active removal of hormone-receptor complex by protein(s), which are synthesized in early puff stages, i.e. between PS1 and PS4.

Withdrawal of 20-hydroxyecdysone from media, in which salivary glands are incubated in vitro, results in a premature induction of late puffs [28]. It was suggested that ecdysterone-receptor complex represses late puff activity, which then is overcome by the production of gene products of early puffs (EGP) [14]. This could account for the premature induction of late puffs. This model is, however, difficult to

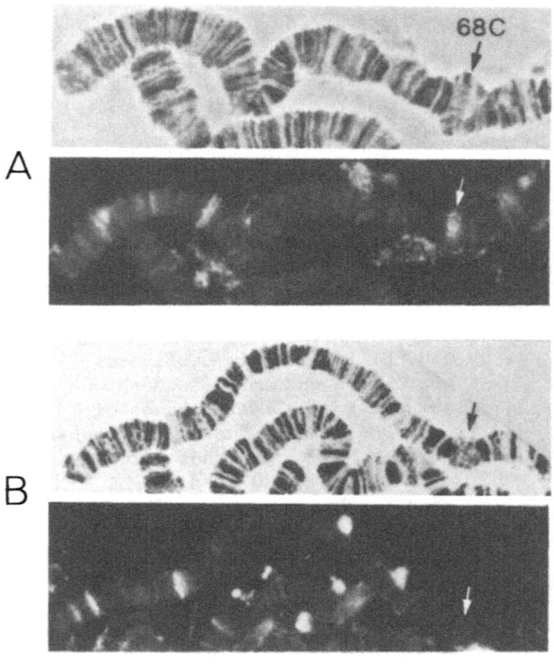

Fig.9A,B. Indirect immunofluorescence of a chromosome squash from salivary glands of *Drosophila melanogaster* 3rd instar larvae after UV irradiation for 5 min. (A) One lobe of a PS1-gland was incubated for 1 h in the presence of 20-hydroxyecdysone and cycloheximid. They were then transferred to a medium only containing cycloheximid ("wash out conditions") and incubated for another 2 h before irradiation, squashing and antibody-staining. The pattern depicts the distal half of chromosome 3 L.
(B) The other PS1-lobe was incubated similarly, but without cycloheximid. The pattern depicts again the distal half of chromosome 3L

reconcile with the observation that 20-hydroxyecdysone is also present on late puffs, and in almost all cases, first at the time of puff activation. Also, it does not explain the all-or-nothing response of late puffs to 20-hydroxyecdysone concentrations, which points to a highly cooperative mechanism of activation of late puffs. A competition between EGP and hormone-receptor complex for the same chromosomal locus might not yield such cooperativity. A direct cooperation between hormone-receptor complex and EGP's seems to be more likely. This would explain together with the observed autoregulation of early puffs how the switch in hormone-receptor binding from early to late puffs comes about.

A 2-h treatment of glands with 20-hydroxyecdysone followed by transfer into hormone-free medium results in the premature induction of many late puffs (e.g. 63E, 82F). These experiments were interpreted to show that induction of these late puffs is not under direct hormonal regulation. This interpretation was done with the tacit assumption that transfer into hormone-free medium would lead to a rapid dissociation of chromatin bound as well as unbound nuclear hormone-receptor complexes. The following experiments shows that this assumption is probably not valid. Salivary glands were incubated in vitro for 1 h in the presence of 20-hydroxyecdysone and then were transferred into hormone-free medium for 2 h. As a result, early puffs (e.g. 74EF/75B) have regressed and late puffs have been induced (e.g. 82F, 63E). However, this late puff induction is again accompanied by binding of hormone-receptor complex to the respective site. If we now re-added 20-hydroxyecdysone to the medium, early puffs (e.g. 74EF and 75B) were re-induced [28], and Dworniczak B, Pongs O in preparation). These results demonstrate that a change in the 20-hydroxyecdysone concentration of the medium changes the affinity of hormone-receptor complex to chromosomal sites and induces its migration from early to late puff sites and vice versa.

Conclusions

The general concept of steroid-hormone action is the following: the steroid-hormone enters its target cell by active or passive transport. Thereupon, it is recognized by a cytoplasmic protein, "the hormone receptor", which then initiates the signals, which lead to expression of particular "hormone regulated" genes [29]. The simplest mechanism is that steroid-hormone binds to receptor protein. The steroid-hormone complex becomes activated, translocates into the nucleus and finally binds to a chromatin site, whereupon transcription of the gene located at that site is regulated. Although our results apparently give strong support to this basic model of steroid-hormone action, since we in fact see a migration of 20-hydroxyecdysone into the nucleus and subsequent specific binding to puffing sites, where transcription eventually is activated, several observations do not fit with this model.

First of all, such a model cannot explain the temporal binding sequence of hormone-receptor complex to different genes, which underlies the developmental program studied in our experiments. This model also lacks dynamic properties (in time or space), which would be necessary to reconcile the observed effect of oscillations in hormone concentration upon hormone-receptor binding properties to chromosomes. For reasons of clarity, we will firstly discuss the mechanisms underlying the early to late puff switch and secondly, the intermolt to early puff switch.

Early puffs (e.g. at 2B5, 74EF/75B) code for gene products (EGP) which are necessary for two functions: to regress intermolt as well as early puffs and to induce late puffs [26,27]. Our data show that EGP's are important for dissociating hormone-receptor off intermolt puffs (e.g. 68C) as well as off early puffs. In addition to this feedback, EGP's are also involved in the induction of late puffs together with hormone-receptor complex as illustrated in Fig. 10. This dual function of EGP's would then explain the temporal sequence, which is observed in the puffing patterns and how the temporal puffing is related to temporal hormone-receptor binding. It could also explain the different dose-response curves for early and late puffs as a function of 20-hydroxyecdysone concentration; early puffs respond over a broad range of 20-hydroxyecdysone concentration, whereas late puffs respond to a very narrow range of 20-hydroxyecdysone concentration, meaning that there induction is a highly cooperative phenomenon [13]. The gene-dosage experiments indicated that a given amount of EGP's has to be synthesized [26]. Once this quantity has been made in the gland, the development goes on from early to late puff stages. It suggests that this quantity is necessary to compete hormone-receptor complex out of early puff sites and to direct hormone-receptor complex into late puff sites. However, it is difficult to see how the EGP-controlled removal of hormone-receptor complex from chromosomal sites should be affected by hormone-withdrawal.

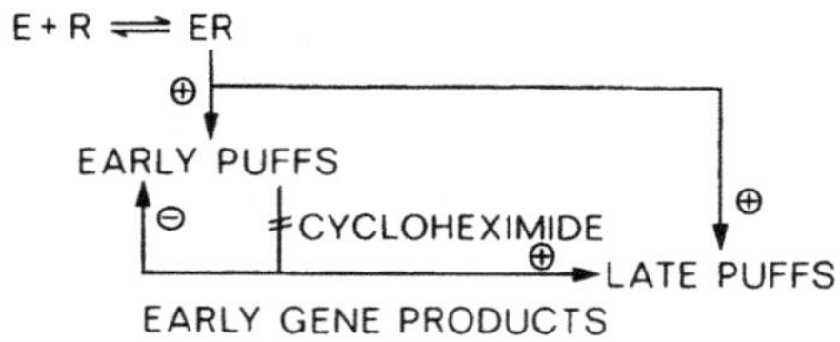

Fig. 10. Model of 20-hydroxyecdysone action during 3rd instar development to explain the temporal change in puffing patterns of polytene chromosomes of salivary glands, which is observed during puff stages 2 to 10. For further details see text

The hormone induced developmental switch from PS1 to PS4 is similarly puzzling. The observation that now more hormone gives more hormone-receptor complex and thus, *new* binding sites on chromosomes, is difficult to understand. Why does the hormone-receptor complex bind to different sites depending on hormone concentration? It could be that different cooperativities are required for one or another chromosome binding site, which would give different hormone-receptor occupancies and thus, different hormone concentration dependencies as reported in the case of estrogen regulated expression of ovalbulmin/conalbumin genes [30]. Alternatively, early puff sites might not be available for hormone-receptor binding before the onset of PS2. This could be one of the functions of juvenile hormone, which suppresses a metamorphic change at a molt. Early puff induction in 3rd instar larvae, however, is the beginning of the metamorphic change of larvae into pupae. Yet another alternative could be that 20-hydroxyecdysone exerts, at high concentrations, cellular signals which are not mediated by hormone-receptor complex formation, but lead to different chromosomal responses. For progesterone, it has been well documented that its control of oocyte development is mediated by the deposition of the steroid-hormone in the cellular membrane, whereupon the Ca^{2+}-influxes are changed with all its pleiotropic aftermath [31]. It has been reported that upon exposure to 20-hydroxyecdysone a straight increase as well as an initial decrease in junctional permeability of the plasma membrane of salivary glands was observed [32]. Measurements of membrane potentials revealed that the salivary gland cells are electrically coupled during the late third instar. If this were under hormonal control, the contradictory experiments which report the influence of ions on puffing [33,34] might well indicate that ecdysterone changes ion fluxes in the salivary gland, which in turn sensibilize those chromosomal loci, to which subsequently hormone-receptor complex binds and activates transcription. The initial step, however, which leads to the hormonal regulation of transcription, could come from enzymes or other cellular signals, which are only indirectly related to hormone concentration. Such a view, however, would easily explain the profound and rapid effects, which oscillations in hormone concentrations have on the selectivity and specificity of hormone-receptor binding to chromosomal loci, where early or late puffs are induced. Thus, the actual high rise in blood hormone titre in early 3rd instar larvae might not be to provide more hormone-receptor complex but to provide the necessary cellular environment, which now allows the hormone-receptor complex to recognize genes which were not otherwise recognizable. Only future experiments can decide which of all the alternatives might, eventually, underly the hormonal regulation of pupariation formation.

References

1. Karlson P (1966) Naturwissenschaften 53:445-452
2. Beermann W (1952) Chromosoma 5:139-198
3. Beermann W (1972) In: Beermann W (ed) Results and problems in cell differentiation, vol. IV. Springer, Berlin Heidelberg New York, pp 1-33
4. Pelling C (1972) In: Beermann W (ed) Results and problems in cell Differentiation, vol IV. Springer, Berlin Heidelberg New York, pp 87-99
5. Sass H (1982) Cell 28:269-278
6. Becker HJ (1959) Chromosoma 10:645-678
7. Clever U, Karlson P (1960) Exp Cell Res 20:623-626
8. Berendes HD, Ashburner M (1978) In: Ashburner M, Wright TRF (eds) The genetics and biology of drosophila, vol IIb. Academic Press, London New York, pp 453-498
9. Ashburner M, Berendes HD (1978) In: Ashburner M, Wright TRF (eds) The genetics and biology of drosophila, vol IIb. Academic Press, London New York, pp 315-359
10. Ashburner M (1972) Chromosoma 38:255-281
11. Ashburner M (1972) In: Beermann W (ed) Results and problems in cell differentiation, vol IV. Springer, Berlin Heidelberg New York, pp 101-151
12. Ashburner M (1974) Dev Biol 39:141-157
13. Ashburner M (1973) Dev Biol 35:47-61
14. Ashburner M, Chihara C, Meltzer P, Richards G (1975) Cold Spring Harbor Quant Biol 38:655-662
15. Emmerich H (1969) Exp Cell Res 58:261-270
16. Gronemeyer H, Pongs O (1980) Proc Natl Acad Sci USA 77:2108-2112
17. Benisek WF (1977) Methods Enzymol 46:469-479
18. Gronemeyer H, Hameister H, Pongs O (1981) Chromosoma 82:543-559
19. Schaltmann K, Pongs O (1980) Hoppe Seyler's Z Physiol Chem 361:207-210
20. Schaltmann K, Pongs O (1982) Proc Natl Acad Sci USA 79:6-10
21. Ashburner M, Bonner JJ (1979) Cell 17:241-254
22. Garen A, Kauvar L, Lepesant JA (1977) Proc Natl Acad Sci USA 74:5099-5103
23. Richards G (1981) Mol Cell Endocrinol 21:181-197
24. Korge G (1975) Proc Natl Acad Sci USA 72:4550-4554
25. Akam ME, Roberts DB, Richards G, Ashburner M (1978) Cell 13:215-225
26. Walker VK, Ashburner M (1981) Cell 26:269-277
27. Belyaeva ES, Vlassova IE, Biyasheva ZM, Kakpakov VT, Richards G, Zhimulev IF (1981) Chromosoma 84:207-219
28. Ashburner M, Richards G (1976) Dev Biol 54:241-255
29. Yamamoto K, Alberts BM (1976) Annu Rev Biochem 45:722-746
30. Palmiter RD, Mulvihill ER, Shephard JH, McKnight GS (1981) J Biol Chem 256:7910-7916

31. Baulieu EE, Godeau F, Schorderet M, Schordert-Slatkine S (1978) Nature (London) 275:593-598
32. Hax WMA, Venrooij G, Vossenberg J (1974) J Membr Biol 19:253-266
33. Kroeger H (1977) Mol Cell Endocrinol 6:105-110
34. Ashburner M, Cherbas P (1976) Mol Cell Endocrinol 5:89-107
35. Lefevre G (1976) In: Ashburner M, Novitsky A (eds) The genetics and biology of Drosophila, vol Ia. Academic Press, London New York, pp 32-66
36. Laemmli UK (1970) Nature (London) 227:680-685
37. Worcel A, Han S, Wong ML (1978) Cell 15:968-977
38. Finch JT, Noll M, Kornberg RD (1975) Proc Natl Acad Sci USA 72:3320-3322

The Gene for Chicken Lysozyme: Structure and Expression of a Steroid Regulated Gene

A. E. Sippel and J. Nowock[1]

Introduction

The function of a multicellular organism depends on the exchange of information between cells in distant parts of the body. The endocrine system is one of the systems for intercellular communication. Steroid hormones, for example, are produced in specialized gland cells, distributed in the bloodstream and are finally able to regulate the differentiation and function of whole organs.

It is generally believed that steroids act via a common molecular mechanism [1]. Target cells are characterized by the presence of a highly specific, cytoplasmic steroid binding protein, the so called receptor. Steroid receptors can be considered key molecules for the initiation of a specific program in cellular differentiation. Upon binding with the hormone the receptor is translocated to the nucleus where selective changes in gene activity are induced.

There are two central questions for understanding the molecular basis of steroid action.

1. Which processes occur between the nuclear entry of the steroid-receptor complex and the subsequent accumulation of specific mRNAs?
2. What regulates the presence and level of steroid receptors in the target cell?

In this report we are dealing with our work considering the first of these questions.

The Chicken Oviduct

The chicken oviduct is an attractive model organ with which to study steroid control of specific gene expression [2,3]. In the immature chicken, administration of estradiol causes growth and differentiation of the oviduct. In the tubular gland cells of the developed oviduct tissue the synthesis of the major or egg white proteins ovalbumin, conalbumin, ovomucoid, and lysozyme accounts for more than 75% of the total protein synthesis. The coordinated regulation of these few proteins and the fact that their synthesis can be readily induced and deinduced by application and withdrawal of estrogens and gestagens, has caused the considerable interest in this system. Previously we could show, by specific cDNA titration experiments, that application

1 Institut für Genetik der Universität Köln, Weyertal 121, D-5000 Köln 41, FRG

33. Colloquium - Mosbach 1982
Biochemistry of Differentiation and Morphogenesis

of female sex hormones to hormonally withdrawn chickens led to a dramatic increase of specific egg white protein mRNAs [4,5]. In case of lysozyme mRNA the level changed from an average of 7 molecules to 30,000 molecules per oviduct cell. In order to decide which processes are responsible for changes of individual mRNA concentrations we measured transcriptional activity of specific genes in isolated nuclei and specific mRNA degradation in vivo [5,6]. From our experiments we conclude that steroids raise mRNA levels by an increase of transcriptional activity at the gene as well as by differential mRNA stabilization in the cytoplasm.

The Structure of the Lysozyme Gene

For the detailed analysis of the molecular processes involved in steroid mediated gene activation we ultimately need a regulatable in vitro transcription system. Therefore it is our aim to isolate all components of such a system and to characterize their structure and function. Since we expect that the transcriptional specificity, at least in part, is based on the sequence organization of the DNA of the gene and its flanking regions, we have isolated the gene for lysozyme from a chicken DNA library [7].

Figure 1 shows that the lysozyme gene is a typically fragmented eukaryotic gene. The final decision about number, size and position of its 4 exons and 3 introns could only be made by comparison of the complete nucleotide sequence of the mRNA with its respective sequences

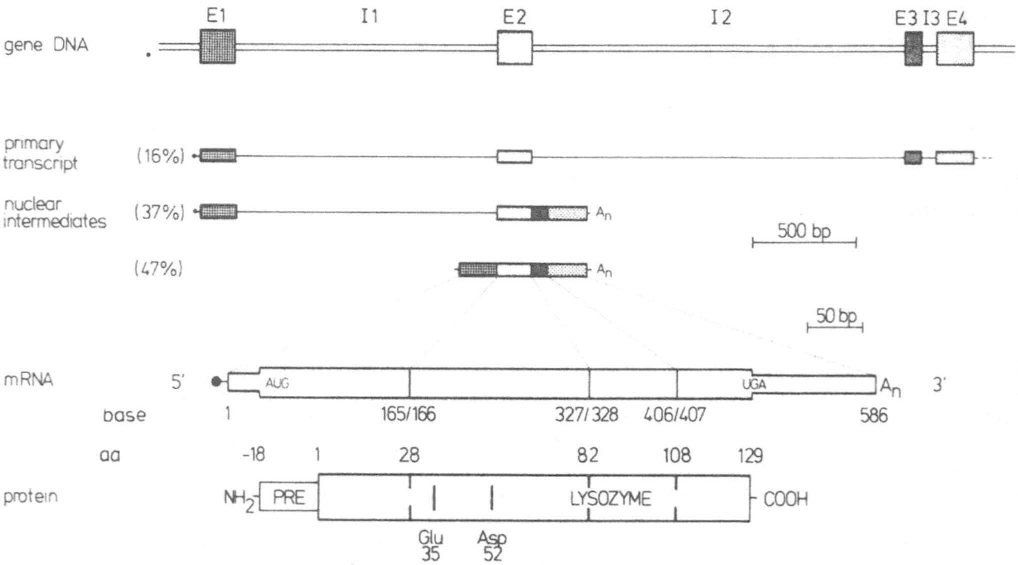

Fig. 1. Chicken lysozyme: From gene to protein. This figure summarizes data on the exon-intron pattern of the gene [9], on the structure and mass proportion of nuclear lysozyme specific hnRNA as derived by EM-analysis of hnRNA: gene DNA hybrids (Giesecke, unpublished data), on the structure of lysozyme mRNA [8,9] and on the correlation of exon borders to the amino acid sequence of the final protein product [9]

in the cloned gene DNA [8,9]. We found that introns interrupt the coding region at the codons for amino acid trp 28, ala 82 and trp 108 of mature lysozyme. Since lysozyme is one of the best understood enzyme molecules we were able to correlate the exon structure of the gene to the structural and functional subdomains of its final protein product (Fig. 1). On a structural level exons correlate to topologically compact units of the protein. In functional respect exon 3 encodes the complete information for the katalytically active center of the molecule. Exon 2 can thus be considered a primitive β,1-4 glycosidase minigene. The product encoded on exon 3 provides increasing substrate specificity for the enzyme. The products of exon 1 and exon 4 are not involved in the katalytic function. From evolutionary considerations [10] one may assign importance for exon 1 and exon 4 to provide for the necessary DNA-, RNA- and peptide signal environment for transcription, translation and protein-transport of the katalytically active core carried on exon 2 and 3.

Different Lysozyme mRNAs Are Produced from the Same Gene

In order to gain insight into the process of transcription of the gene and of processing of its primary transcript we first analyzed hybrids of oviduct nuclear RNA with cloned gene DNA by S1 mapping and by electron microscopy ([11] and Giesecke, unpublished). We found that the primary transcript, covering the complete 4 kilobases (kb) sequence from the beginning of exon 1 to the end of exon 4, is processed by a splicing pathway preferentially operating in the opposite direction to transcription (Fig. 1).

When we sequenced the 5'-end of the lysozyme gene we noticed that it contains a surprising repetition of sequence motifs similar to those

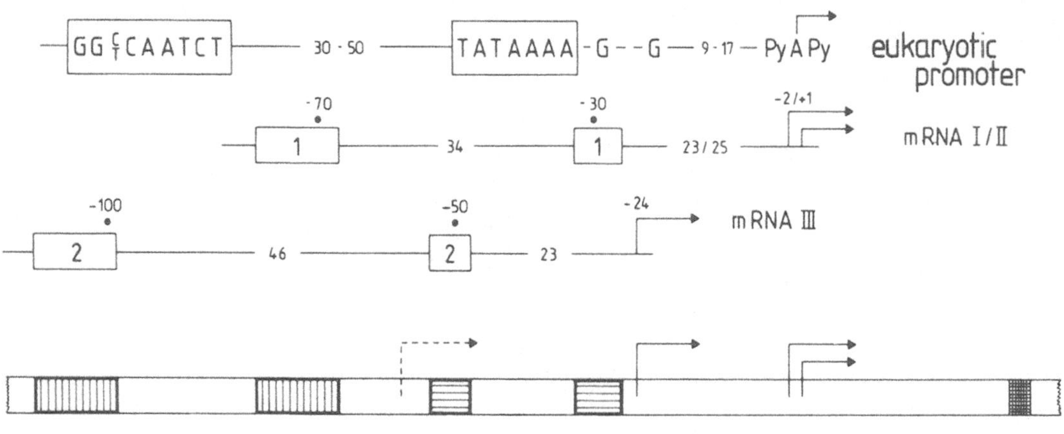

Fig. 2. Organization of the conserved sequence motifs in the lysozyme promoter region. The position of the conserved sequence motifs, as generally found in eukaryotic RNA polymerase II promoter regions (*top line*), are delineated for the chicken lysozyme promoter region (*bottom line*) according to their position in the published DNA-sequence [13]. TATA-motifs are marked as *horizontally striped boxes* and CAAT-motifs are marked as *vertically striped boxes*. The position of the translational start codon AUG is marked as a *checkered box*. The two overlapping sets of motifs are separately outlined in the middle lines with numbers giving nucleotide distances. *Arrows* indicate positions for transcriptional initiation as deduced from S1 mapping of oviduct nuclear RNA [13]

signals which are found to be used for the initiation of transcription by RNA polymerase II. The organization of the conserved sequence motifs in the lysozyme promoter region are shown in Fig. 2. There exist two overlapping sets of motifs each consisting of a "TATA"-like sequence and a CAAT sequence. By S1 mapping [12] with oviduct specific nuclear RNA and with polysomal mRNA we found that there exist multiple 5'-termini of the transcripts [13]. Since, as far as we know, the gene is only represented once per haploid genome, this means that multiple transcriptional initiation sites are used at the same lysozyme gene. Most likely the two overlapping sets of conserved sequence motifs are responsible for the production of mRNAs differing only in their lengths of 5'-noncoding region. The sequences of the 5'-terminal end of the various lysozyme mRNAs are organized in such a way that they either lack or contain a pronounced 5'-terminal hairpin structure (Fig. 3).

The significance of this finding for the tissue- and hormone-specific expression of the gene is unknown to us at the present time. This double promoter is reminiscent of the situation in the gal operon of *E. coli*. There also exist two overlapping promoters, only one of which is dependent, in its activity, on the cyclic AMP-dependent catabolite repression protein [14].

frequency of mRNA types		
type	S1 mapping	cDNA cloning
I	} ~90 %	~40 %
II		~60 %
III	~10 %	~2.5 %
IV	~1 %	

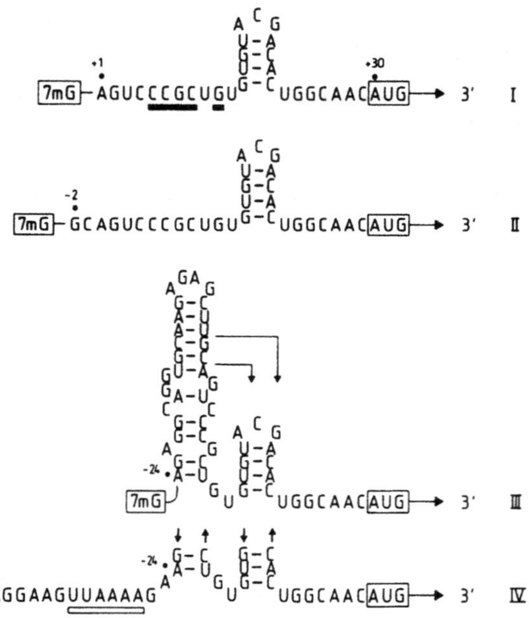

Fig. 3. Multiple 5'-ends of lysozyme mRNA in oviduct cells. The four different 5'-terminal sequences of lysozyme mRNA are shown as derived from S1-mapping and gene-DNA sequencing from primer extension sequencing on lysozyme mRNA and from cloned cDNA analysis [13]. Cap structures were only confirmed for mRNA type II and III (Rhoads, personal communication). Sequences are written in a way to demonstrate absence and presence of a possible 5'-terminal hairpin. *Closed bars* delineate a complementary-region to the 3'-end of 18S rRNA and *open bars* delineate the RNA region homologue to the TATA-motif in the promoter region of the lysozyme gene. Relative abundance of mRNA types shown in the inset are derived from S1-mapping experiments [13] and from frequencies of cDNA clones [20]

Middle Repetitive Sequence Elements in the Lysozyme Gene Region

In analogy to the situation in prokaryotes we assume that DNA signal sequences should exist in the gene region responsible for regulation of gene activity and chromatin structure. Sequence comparison of the immediate 250 bp prior to the cap-site of all four major chicken egg white protein genes shows no significant sequence homology. We therefore looked in the vast intergenic spacer region flanking the gene for sequence elements, which also occur at other sites in the genome. Using labelled total chicken DNA as a hybridization probe for Southern blots of cloned gene DNA we could map five middle repetitive sequence elements, one in intron 2 of the lysozyme gene, two in the 5'- and two in the 3'-flanking region (Fig. 5b). Some of these elements belong to a family of middle repetitive sequences which we determined to be present about 12,000 times per haploid genome, which are interspersed with single copy DNA and which also occur in the ovomucoid gene region (Sippel, unpublished). The reasons for the ubiquitous occurrence of middle repetitive sequence elements in the genome of all higher organism has not yet been determined. Work at those elements which are located in the vicinity of specific single copy genes may be the first step to finding out whether they are sites involved in gene expression, in structural organization of the chromatin, in initiation of replication, or in processes for translocation within the genome.

Specific Protein Binding Sites in the Flanking Regions of the Gene

In the chromatin of eukaryotic genes, distinct structural elements have been defined by their sensitivity to various nucleases [15-18]. Differential nuclease sensitivity of DNA in gene regions has been correlated to cellular commitment and transcriptional activity. We assumed that recognition of DNA signals by specific proteins must be an element in the regulation of chromatin structure and of gene activity. We therefore started a search for proteins which can recognize specific sequences close to the lysozyme gene.

Endlabelled, subcloned Hind III and Eco RI DNA fragments of the lysozyme gene region (Fig. 5a) were incubated with protein fractions from oviduct nuclei and filtered through nitrocellulose membranes. The analysis of the filter retained DNA fragments indicated differential protein mediated retention of certain DNA fragments. In the filter binding experiment presented in Fig. 4, two subcloned fragments were used, one (H2.1(SB)) containing the promoter region of the gene and another (E3.7) starting 1.4 kb downstream from the Poly(A)-point. Specific nuclear protein binding can only be detected for the non-promoter fragment. The result is supported by controls, which show that under the conditions used, bovine serum albumin does not retain any DNA, whereas *E. coli* lac repressor *does* differentially retain the lac operator containing vector fragment. In addition we observed that the cytosol fraction does contain a low level of the same DNA binding specificity found in nuclear extracts (Fig. 4). We assume this to be the result of nuclear leakage during cell fractionation.

By using subcloned fragments, which together cover the complete gene region of 22.2 kb, we could clearly map four distinct loci, two of which were located several kb upstream and two several kb downstream from the lysozyme gene (Fig. 5b). The arrangement of these elements flanking the gene suggests a possible function in the establishment of chromatin domains.

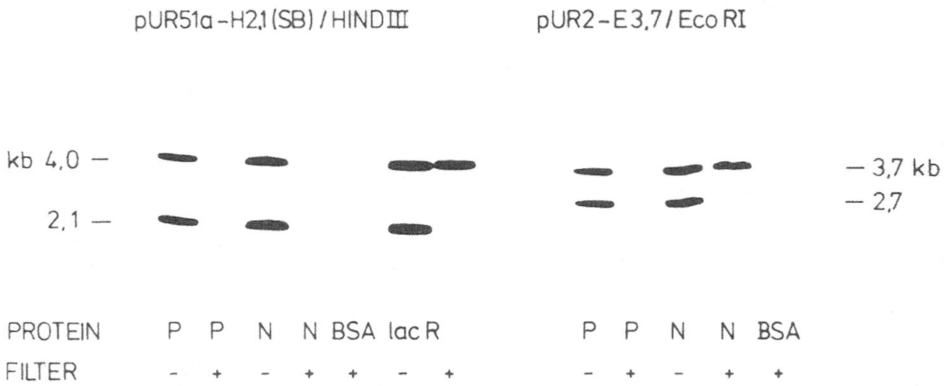

Fig. 4. Filter binding assay with DNA fragments from the lysozyme gene region. A 2.1 kb Hind III fragment (promoter fragment H6-H7 in Fig. 5a) and a 3.7 kb Eco RI fragment (E6-E7 in Fig. 5a) were subcloned in pUR51a (4.0 kb) and pUR2 (2.7 kb, see [21]). DNA of the recombinant plasmids pUR51a-H2.1(SB) and pUR2-E3-7 were digested with Hind III and Eco RI respectively, endlabelled by α-^{32}P-dATP "fill in" with DNA polymerase, Klenow fragment and used for filter binding assays [22] with nuclear (*N*) and cytoplasmic (*P*) protein fractions [19] and bovine serum albumin (*BSA*) and *E. coli* lac repressor (*Lac R*) as controls. Plus and minus signs at the bottom line indicate whether protein-DNA solutions, after incubation (30 min at 25°C), were passed through nitrocellulose filters or not. DNA retained on the filters was separated on agarose gels and analysed by autoradiography

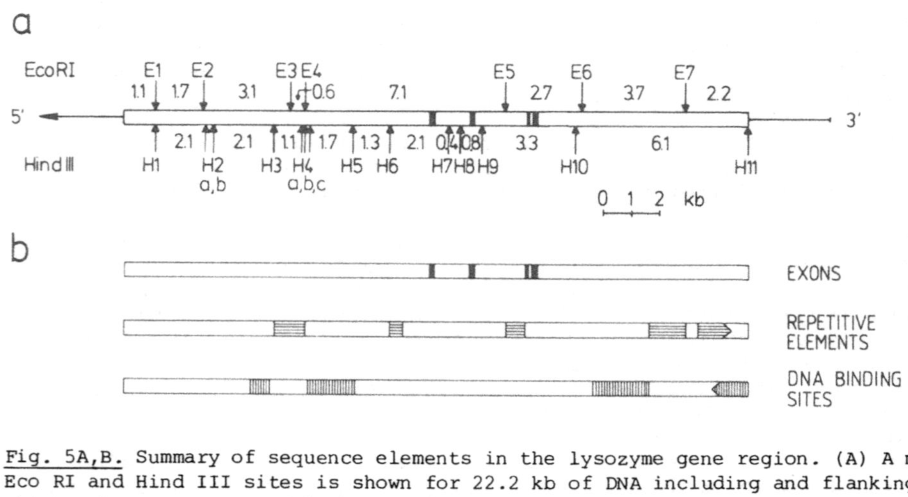

Fig. 5A,B. Summary of sequence elements in the lysozyme gene region. (A) A map of Eco RI and Hind III sites is shown for 22.2 kb of DNA including and flanking the chicken lysozyme gene. (B) The exact position of the exons of the gene are given by *black boxes*. The location of the smallest restriction fragments mapped thus far, which contain repetitive sequence elements and specific protein binding sites [19], are marked as *horizontally* and *vertically striped boxes* respectively

For further characterization of the nature of this specific protein-DNA interaction we concentrated on the most upstream of the binding sites (6.0 kb before the cap-site of the gene). Under optimal reaction conditions we found an apparent binding constant of $K_D = 6 \cdot 10^{-12}$M. This is in the range of prokaryotic repressor-operator binding. From DNA competition experiments it was evident that all four binding sites could mutually serve as efficient competitors. We thus concluded that only one protein component is involved in the recognition of all four sites [19]. Fine mapping by restriction enzyme and DNase I digestions uncovered that the first binding-site itself consists of two attachment sites 40 bp apart from each other. The size of the composite binding site makes it possible that more than one molecule of binding protein is involved.

We are currently determining the precise DNA sequence of two of the binding sites to gain insight into the nature of this new DNA signal. Isolation of the protein component, on the other hand, should clarify whether a relationship exists between this protein and steroid hormone receptors.

References

1. Yamamoto KR, Alberts BM (1976) Annu Rev Biochem 45:721-746
2. Palmiter R (1972) J Biol Chem 274:6450-6458
3. Schütz G, Nguyen-Huu MC, Giesecke K, Hynes NE, Groner B, Wurtz T, Sippel AE (1978) Cold Spring Harbor Symp Quant Biol 42:617-623
4. Hynes NE, Groner B, Sippel AE, Nguyen-Huu MC, Schütz G (1977) Cell 11:923-933
5. Hynes NE, Groner B, Sippel AE, Jeep S, Wurtz T, Nguyen-Huu MC, Giesecke K, Schütz G (1979) Biochemistry 18:616-623
6. Nguyen-Huu MC, Sippel AE, Hynes NE, Groner B, Schütz G (1978) Proc Natl Acad Sci USA 75:686-690
7. Lindenmaier W, Nguyen-Huu MC, Lurz R, Stratmann M, Blin N, Wurtz T, Hauser HJ, Sippel AE, Schütz G (1979) Proc Natl Acad Sci USA 76:6196-6200
8. Sippel AE, Land H, Lindenmaier W, Nguyen-Huu MC, Wurtz T, Timmis KN, Giesecke K, Schütz G (1978) Nucleic Acids Res 5:3275-3294
9. Jung A, Sippel AE, Grez M, Schütz G (1980) Proc Natl Acad Sci USA 77:5759-5763
10. Artymink PJ, Blake CCF, Sippel AE (1981) Nature (London) 290:287-288
11. Sippel AE, Nguyen-Huu MC, Lindenmaier W, Blin N, Lurz R, Hauser H, Giesecke K, Land H, Grez M, Schütz G (1980) In: Beato M (ed) Steroid induced uterine proteins. Elsevier/North-Holland Biomedical Press, Amsterdam New York, pp 297-314
12. Weaver RF, Weissmann C (1979) Nucleic Acids Res 7:1175-1193
13. Grez M, Land H, Giesecke K, Schütz G, Jung A, Sippel AE (1981) Cell 25:743-752
14. Musso R, DiLauro R, Rosenberg M, DeCrombrugghe B (1977) Proc Natl Acad Sci USA 74:106-110
15. Wu C, Bingham PM, Livak KJ, Holmgren R, Elgin S (1979) Cell 16:797-806
16. Stalder J, Groudine M, Dodgson J, Engel D, Weintraub H (1980) Cell 19:973-980
17. Levy A, Noll M (1981) Nature (London) 289:198-203
18. McGhee JD, Wood WI, Dolan M, Engel JD, Felsenfeld G (1981) Cell 27:45-55
19. Nowock J, Sippel AE (in press) Cell 30
20. Land H, Grez M, Hauser H, Lindenmaier W, Schütz G (1981) Nucleic Acids Res 9:2251-2266
21. Rüther U (1980) Mol Gen Genet 178:475-477
22. Jones OW, Berg P (1966) J Mol Biol 22:199-209

Regulation of Structural Sperm Protein Synthesis by Y Chromosomal Loci

T. Hulsebos, J. Hackstein, and W. Hennig[1]

Cellular differentiation is usually controlled by regulatory processes at the genomic level. In principle, two aspects of such regulatory events might be considered. External signals can determine the response of the genome of a cell to distinct environmental conditions. An example of this type has been given by Pongs (this volume). Ecdysone interacts with distinct genetic loci leading to metabolic changes in the cells which are related to molting or puparium formation. Alternatively, regulatory genes may govern the activities of other genes during development. Such a concept has been developed by Garcia-Bellido [1] who postulated the existence of "selector genes" which determine certain differentiation pathways. Although genetics has provided evidence on the existence of such regulatory genes, their exact nature is still unclear.

Spermiogenesis in *Drosophila* is a suitable system for studying cellular differentiation and morphogenetic processes [2]. In the course of our biochemical studies we revealed gene interactions during sperm development which apparently reflect the action of a regulatory system within this particular pathway of cellular differentiation.

The Y chromosome of *Drosophila* carries a series of male fertility genes which are required for spermiogenesis [3]. This chromosome is inactive in tissues other than testes. In *D. hydei* our genetic analysis has revealed 16 fertility genes on the Y chromosome, which all are indispensable for sperm development [4]. Similarly, in *D. melanogaster* 6 fertility genes could be detected on the Y chromosome [5]. In comparing the effects of the Y chromosome in *D. hydei* and *D. melanogaster*, remarkable differences are found. While spermiogenesis without Y chromosome in *D. melanogaster* proceeds into advanced postmeiotic stages, the postmeiotic differentiation of sperm is almost entirely prevented in *D. hydei*. This is also reflected by studies of the protein composition of the respective X/O testes. While in *D. melanogaster* the protein composition of X/O testes hardly differs from wild-type testes [6], the contrary is found in *D. hydei*. The absence of the Y chromosome (or parts thereof) results in drastic changes in the protein pattern (see [7]). Certain protein fractions are markedly reduced in synthesis or are even entirely absent. We studied three of these proteins (see Fig. 2A) of molecular weights 155,000 ("sph155"), 55,000 ("sph55") and 35,000 ("sph35") daltons.

With the aid of an antiserum against calf-brain tubulins, sph55 was identified as tubulin. Another antiserum was raised against sph155.

1 Department of Genetics, Katholieke Universiteit, Toernooiveld, 6525 ED Nijmegen, The Netherlands

33. Colloquium - Mosbach 1982
Biochemistry of Differentiation and Morphogenesis

By cyto-immunofluorescent techniques both proteins could be shown to be constituents of mature sperm (Fig. 1). The nature of the sph35 protein remains to be determined.

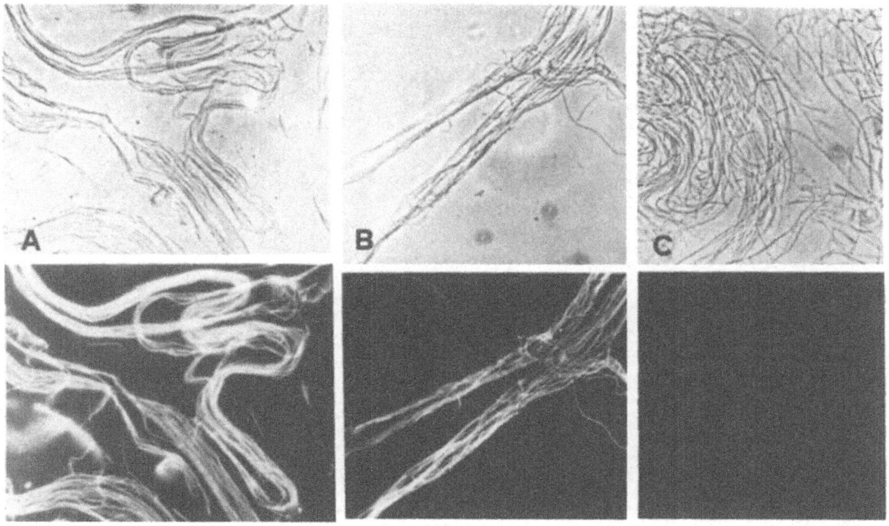

Fig. 1A-C. Cyto-immunofluorescence of *D. hydei*, wild-type, sperm with (A) sph155 antiserum, (B) tubulin antiserum, (C) preimmune serum. *Upper row*: phase contrast, *lower row*: immunofluorescence

The next step in our analysis was the establishment of the genomic origin of the three proteins. We separated protein extracts of the testes of males with various genetic constitutions on SDS-polyacrylamide gels, transferred the proteins, after separation, to nitrocellulose membranes and reacted these blots with the sph155-specific antiserum (Fig. 2B). The immune reaction with a blot of wild-type testes proteins can be seen in lane 1 of Fig. 2B. The various bands of lower molecular weight seen in this lane most probably represent immune reactions with degradation products of the sph155 protein. The low molecular weight band, also seen in the neighbouring lanes, results from a nonspecific reaction of the antiserum. It is the only band which emerges after reaction with pre-immune serum. Blots from testes proteins of X/O males show no reaction with the sph155-antiserum (not shown).

A similar absence of specific immune reactions was noted for blots of testes proteins of mutants without certain parts of the Y chromosome. Two examples are given in Fig. 2B: Lane 2 shows the absence of sph155-specific immune reaction for a testes protein blot of a mutant without gene O, while in lane 3 the Y chromosome is deleted for genes N and O. Obviously, in these mutants the sph155 protein is not synthesized. From Coomassie Brilliant Blue stained gels of the *same* testes preparations it became evident that the absence of the sph155 protein is directly correlated with the disappearance of the sph35 protein and with a dramatic reduction of the amount of protein in the tubulin region. Apparently, the synthesis of all three proteins is in parallel influenced by certain Y chromosomal loci. By analyzing, in this way, the protein patterns and blots of testes of more than 60 different Y chromosomal mutants we could assume that the expression of the

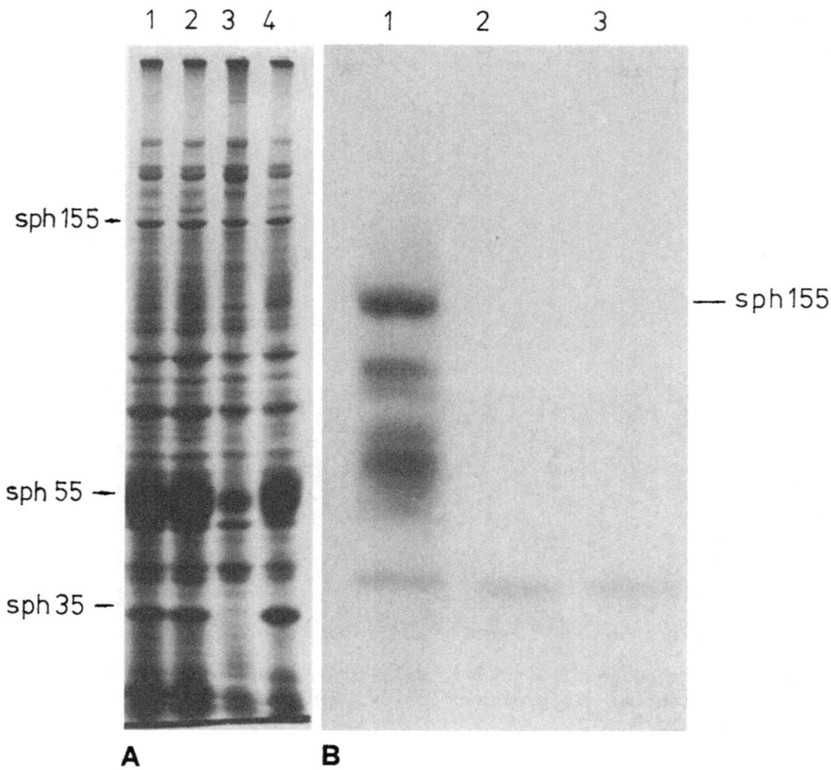

Fig. 2A,B. (A) Polyacrylamide gel patterns of testes proteins after Coomassie Brilliant Blue staining. The protein content of wild-type testes and of X/O testes is shown in *lanes 4* and *3*, respectively. *Lanes 1* and *2* display the testes protein content of mutants, without Y chromosomal genes, not required for the expression of these proteins (genes A, B and C in *lane 1*, gene A in *lane 2*). (Note: The absence of sph 155 in X/O testes (*lane 3*) cannot be seen because of the limited resolution of this gel in the high molecular weight range.) (B) Immune reaction of the sph155 antiserum with testes protein blots of wild-type males of *D. hydei* (*lane 1*) and two mutants, in which gene O (*lane 2*) and genes N and O (*lane 3*) are missing

three protein fractions depends on the activity of two genes on the Y chromosome, the genes O and Q (see [4]). This is demonstrated in Fig. 3 which contains a selection of representative genetic constitutions. In this figure, the different loci on the Y chromosome are indicated by capital letters (A - Q). The Y regions present in the respective males are indicated by bars. It is obvious that all constitutions missing the activity of genes O and Q display none of the three protein fractions.

This discovery, however, needs some further evaluation. By inspecting the testes of the various constitutions, it becomes obvious that all testes from males without functional locus Q, contain only premeiotic and few postmeiotic cells, which are maintained in an early differentiation stage hardly different from an X/O constitution. Earlier evidence [7] indicates that the sph155 protein is only synthesized relatively late in postmeiotic differentiation. The effect of locus Q in the expression of the proteins sph155, sph55, and sph35 might thus be an indirect effect simply resulting from the fact that cells do not arrive at stages where the respective proteins are usually synthesized. We therefore assume, that gene O is the only Y chromosomal gene directly involved in determining the expression of the studied protein fractions.

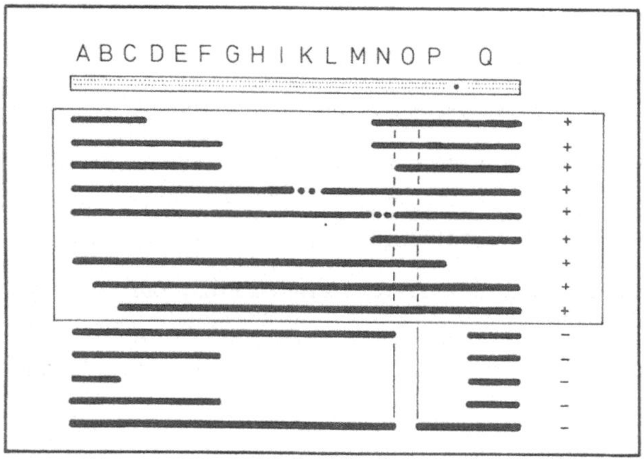

Fig. 3. Diagram of the Y chromosome and its fertility genes (*A* - *Q*). In the *lower part*, various deletion mutants are displayed and the presence (+) or absence (-) of the sph155, sph55 and sph35 proteins is indicated

The immediate question is whether gene O codes for these various proteins. In this respect it must be recalled that our genetic analysis [4] indicates the presence of only one complementation group in this region of the Y. It is, of course, uncertain what relationship exists in eukaryotes between genetic complementation groups and protein coding gene structure (cf. [8]). However, in earlier studies with species hybrids of different autosome constitutions (see [9]) we obtained evidence that sph155 is coded for by chromosome 2. For technical reasons these experiments could not yet be repeated with immunological methods.

The localization of the tubulin genes is nevertheless a clearcut example of an autosomal location of the coding sequence of a protein governed in its expression by a Y chromosomal gene. We have carried out in situ hybridization experiments with recombinant tubulin sequences. From these data it is established that a main site for tubulin genes in *D. hydei* is on chromosome 3, while there is no evidence for loci on the Y chromosome. This autosomal location of tubulin genes agrees with their location in *D. melanogaster*. While α-tubulin sites are on chromosome 3 and β-tubulin loci in chromosome 2, a testes-specific tubulin has been located in chromosome 3R of *D. melanogaster* [10,11].

These data support our interpretation that the Y chromosomal gene O controls the expression of structural sperm proteins which are encoded by autosomal loci. So far no major protein in testes could be directly correlated with any Y chromosomal site. Moreover, the effects of an inactive locus Q are of a rather general character, which indicates that this gene controls the early differentiation events after meiosis. We believe that other Y chromosomal genes might have similar functions. The Y chromosome may thus carry a number of regulatory genes responsible for controlling the postmeiotic morphogenesis of sperm.

Acknowledgements. We would like to thank R. Hochstenbach, Nijmegen, for carrying out the in situ hybridization; Dr. A. van Lammeren, Wageningen, for the tubulin antiserum; and Dr. D. Cleveland, Baltimore, for the tubulin clones.

References

1. Garcia-Bellido A (1975) In: Cell patterning. Ciba Found Symp 29:161
2. Hennig W (1978) Ent Germ 4:200
3. Tokuyasu, Lindsley D (1981) In: Ashbuner M, Wright T (eds) The genetics and biology of Drosophila, vol IId. pp 226
4. Hackstein J, Leoncini O, Beck H, Peelen G, Hennig W (1982) Genetics (in press)
5. Kennison (1981) Genetics 98:529
6. Ingman-Baker J, Candido EPM (1981) Biochem Gen 18:809
7. Hennig W, Meyer GF, Hennig I, Leoncini O (1974) Cold Spring Harbor Symp Quant Biol 38:673
8. Beermann W (1972) In: Results and problems in cell differentiation, vol IV, p 1
9. Hennig W (1977) Adv Enzyme Regul 15:363
10. Kemphues KJ, Raff EL, Raff RA, Kaufmann TC (1980) Cell 21:445
11. Sanchez F, Natzle JE, Cleveland DW, Kirschner MW, McCarthy BJ (1980) Cell 22:845

Cell Recognition

Cell Recognition by Axonal Growth Cones in Vitro

F. Bonhoeffer and B. Kern-Veits[1]

The brain of a vertebrate contains, in one cubic millimeter, about 100,000 neurons. The neurons are connected with each other by fibers, the axons and dendrites. Each cubic millimeter contains about one kilometer of such fibers. The number of synapses, the sites where information can be exchanged between neurons, found within one cubic millimeter reaches 10^9. These are rather inconceivable numbers.

From neuroanatomy it has been known for a long time that these neurons, axons, dendrites, and synapses are arranged and interconnected in a highly specific and ordered manner. How does such a complicated and specific system develop? How does an axon sent out by a neuron find its specific path to the target organ which may be thousands of cell diameters away from the neuron and how does it finally find the specific target cell within the target organ? Although we have much detailed anatomical information about development and regeneration of the CNS, almost nothing is known about molecular mechanisms which govern these steps in neurodevelopment. This is a very challenging set of questions, for biochemists and biologists, that may perhaps be successfully answered using the techniques now available in molecular biology. Here we will describe an in vitro system which, hopefully, will help us to understand some of the fundamental principles involved in the formation of topographic projections and to identify molecules which are essential for the recognition of cells by axonal growth cones.

We have chosen to work on the development of the visual system of the chick embryo, because the visual system displays the classical topographic projection of neurons from one area of the brain to another - from the retina to the surface of the tectum opticum. An axon sent out to the tectum by a ganglion cell of the retina has to find a certain area in the tectum where it forms synapses. The exact location of that target area depends on the location of the ganglion cell within the retina. It is a mirror image projection (Fig. 1) with the following rules: the more dorsal (ventral) the position of the ganglion cell in the retina, the more ventral (dorsal) the target area on the tectum and, correspondingly, the more nasal (temporal) the axon originates in the retina, the more posterior (anterior) is its field of projection on the tectum. Neighbouring relationships between different axons are retained in this projection. The projection does not seem to be a strict point-to-point projection. A retinal ganglion cell does not send its axons to a particular target cell but rather to cells of particular area.

---

1 Max-Planck-Institut für Virusforschung, Abteilung Physikalische Biologie, Spemannstrasse 35, D-7500 Tübingen, FRG

33. Colloquium - Mosbach 1982
Biochemistry of Differentiation and Morphogenesis

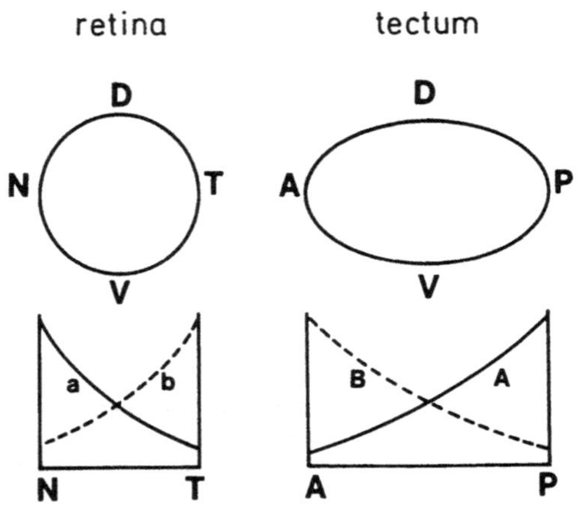

Fig. 1. Graded distribution of hypothetical positional markers along the naso-temporal or antero-posterior axis of retina and tectum. *D* dorsal, *V* ventral, *N* nasal, *T* temporal, *A* anterior, *P* posterior

Several mechanisms have been suggested to explain the formation of such orderly projections. Here we will limit ourselves to discussing only that hypothesis which in our opinion is most likely describing the correct principle of regional specific projections.

With his chemoaffinity hypothesis, Sperry [9] suggested that the information of the topographic position of each cell is expressed in some chemical way on the cell surface and also on the surface of the corresponding growth cones. According to this hypothesis not only would each retinal cell carry positional markers, but also the tectal cells would carry complementary positional markers. Thus a precise final projection could be defined. However, such a simple key-lock system where retinal cells and tectal cells carry complementary positional markers would require many different marker molecules and could not easily explain how the axonal growth cones find the target area without searching all over the tectal surface. As we know, such a search does not occur in vivo, the fibers grow more or less directly to the target area.

An extension of the chemoaffinity hypothesis, which assumes graded distributions of only a few positional markers, will however explain not only the final projection but also how axons are guided on a direct route to the target area [5,6]. Let us consider how according to a model based on differential adhesion [5] the projection in the naso-temporal or antero-posterior axes could perhaps be achieved. We assume the retinal cells contain the positional markers a and b with the graded distributions shown in Fig. 1 and that the tectal cells contain complementary positional markers A and B with a complementary distribution (Fig. 1). We further assume that the growth cones carrying a and b will tend to form as many aA and bB pairs as possible and that reversible pair formation is, so to speak, the driving force. It is then obvious that nasal axons will keep growing until they reach posterior parts of the tectum, because the number of pairs formed are there the greatest. On the other hand, temporal fibers can form most pairs in the anterior part of the tectum. Here only four different components with graded distributions are sufficient to explain a regional specific projection in one dimension. In this model the dorso-ventral dimension would require another set of four components. The model explains how axons find the target area on a rather direct route which is determined by the shape of the graded distribution. There is another model which explains axonal guidance by more general mechanisms of directional activation of growth cones [6]. Also this model requires graded distributions of positional markers.

There are two experimental methods by which positional differences of cell surface components have been detected. One method is based on cell adhesion assays [1,7]. The other approach is an immunological investigation of the cell surface. The latter has led to the discovery of a surface antigen which has a graded distribution along the dorso-ventral axis of the retina [10]. However, there is no indication that these dorso-ventral gradients have any function in the formation of the topographic retino-tectal projection.

In order to search for essential gradients on *tectum* and *retina*, we have chosen to probe the cell surfaces with the natural instrument which examines these surfaces in vivo: the axonal growth cone. The major questions we wanted to investigate in vitro are the following:

1. Can growth cones recognize different cell types as different?
2. Can growth cones recognize two cells as different if they are of the same cell type but come from different positions in the tissue, i.e. do target cells express and do growth cones detect positional differences in vitro?
3. If growth cones do detect positional differences, are these positional markers distributed in a graded fashion over the target surface?
4. Do retinal growth cones all have the same recognizing abilities or would they have different abilities, depending on the position of their origin in the retina?
5. Are the recognized positional markers membrane bound?

To investigate these questions, we have chosen an experimental design in which axonal growth cones are exposed simultaneously to two cell monolayers or a combination of a cell monolayer and a layer of membranes and in which the growth cones are forced to make the decision which of the layers they like better as a substrate [3,4]. Details of the experimental design are explained in Figure 2. Retinal explants cut into naso-temporal strips were used for these choice assays. Thus nasal and temporal axons could be tested separately for their preference of particu-

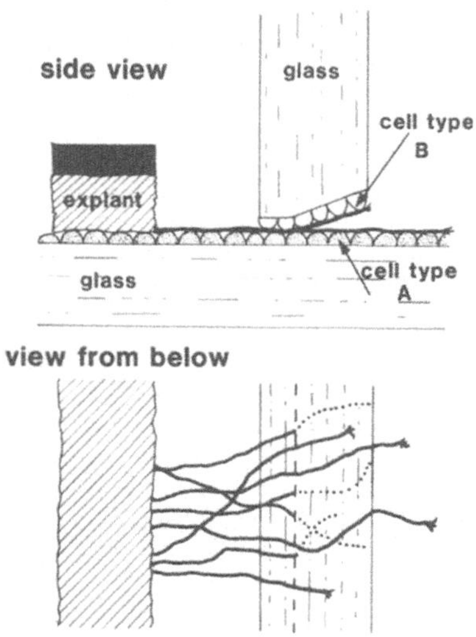

Fig. 2. Experimental design. A retinal explant is placed on a cell monolayer of cell type A and incubated in culture medium. Outgrowing axons reach a second monolayer of cell type B. There they have to decide along which monolayer they prefer to grow. Microscopic observation from below allows the determination of the number of axons staying on monolayer A and of the axons which have a preference for monolayer B. The latter grow out of the focal plane of layer A (*dotted lines*)

lar monolayers. When the choice between two monolayers derived from retina and from tectum was offered, the anterior axons showed a clear preference for the tectal cells: 90% (1575 of 1744) of the axons chose to grow along tectal monolayers. Posterior axons showed clearly a weaker preference: 59% (958 of 1625) grow along the tectal cell layer. When the choice between two monolayers derived from anterior tectal cells and from posterior tectal cells is offered, the axons reveal their ability to distinguish between cells of different position, although these cells are of the same cell type: 88% of the posterior axons prefer anterior tectal cells, as they also project in vivo to the anterior pole of the tectum. Only 47% of the posterior axons grow along the monolayer of anterior tectal cells. So far, these experiments show that (a) cell type and cell position differences can be detected by growth cones in vitro (questions 1 and 2); (b) growth cones from different retinal areas have different recognizing abilities (question 4).

To investigate question 3, whether the position of cues which are recognized by temporal axons have a graded distribution on the tectum, a series of choice experiments has been performed. Tecta have been divided along the antero-posterior axis into five equal parts and the corresponding monolayers have been prepared. Pairs of monolayers in different combinations have been offered between which the axons must choose. The posterior axons always prefer the more anterior to the more posterior monolayer of such pairs, irrespective of the absolute position from which the tectal cells come, i.e. the more anterior the position of the tectal cell, the more attractive it is for the posterior axons. Thus there is a gradient of positional cues.

When instead of the upper cell monolayer a layer of cell membranes is used qualitatively, similar results are obtained [8]. Comparison of the various experiments shows that all membranes contain positional information, which is detected by growth cones in vitro, e.g. membrane layers of anterior tectal cells are more attractive for temporal axons than for nasal axons and vice versa, posterior tectal cells are more attractive for nasal than for temporal growth cones.

In summary the in vitro experiments reported here not only show that growth cones and cell perikarya can exhibit recognition abilities in vitro which correspond to the in vivo projection but also that there is a positional marker which has a graded distribution along the tectal surface and furthermore that this positional marker is membrane-bound.

References

1. Barbera AJ (1975) Dev Biol 46:167-191
2. Barbera AJ, Marchase RB, Roth S (1973) Proc Natl Acad Sci USA 70:2482-2486
3. Bonhoeffer F, Huf J (1980) Nature (London) 288:162-164
4. Bonhoeffer F, Huf J (1982) EMBO 1:427-431
5. Fraser SE (1980) Dev Biol 79:453-464
6. Gierer A (1981) Biol Cybern 42:69-78
7. Gottlieb JD, Glaser L (1975) Biochem Biophys Res Commun 63:815-821
8. Kern-Veits B (1981) Dipl Arb Univ Hohenheim
9. Sperry RW (1963) Proc Natl Acad Sci USA 50:703-710
10. Trisler GD, Schneider MD, Nirenberg M (1981) Proc Natl Acad Sci USA 78:2145-2149

Muscle Transplantation in Mammals: A Tool to Study Neuromuscular Mutations and Specificity of Innervation

H. Jockusch[1]

The transplantation of minced muscle has been used to study muscle regeneration and neuro-muscular diseases in rodents [1,2]. Experiments can be performed (Table 1), in which the implanted muscle either differs in its genetic constitution from the host or is transplanted ectopically (e.g. heart muscle to leg). The first type of experiment could be used to obtain information on the primary focus of mutations affecting the neuro-muscular system [3], the second, to study the tissue specificity of muscle innervation.

Any conclusions from transplantation and regeneration experiments require that the actual cellular contribution of the donor tissue to the regenerated tissue be demonstrated [4]. To this end we have used elec-

Table 1. Some possible types of experiment using muscle grafts in mammals. Results of experiments (a, c, e) are described

MUSCLE TRANSPLANTS		
Donor		Recipient
Skeletal muscle	→	*Skeletal muscle bed*
a) Mutant	→	Wildtype
b) Wildtype	→	Mutant
c) Species 2	→	Species 1
Heart muscle	→	*Skeletal muscle bed*
d) Species 1	→	Species 1
e) Species 2	→	Species 1
Smooth muscle	→	*Skeletal muscle bed*
f) Species 1	→	Species 1
g) Species 2	→	Species 1

1 Lehrstuhl für Entwicklungsbiologie, W 7, Universität Bielefeld, D-4800 Bielefeld, FRG

33. Colloquium - Mosbach 1982
Biochemistry of Differentiation and Morphogenesis

trophoretic variants (of mouse strains or between species) of the enzyme glucose phosphate isomerase (GPI, EC 5.3.1.9, [5]). It was shown in model experiments that the relative activities of donor and recipient type isozymes are proportional to the relative weight contributions of donor and recipient tissues. Apart from the homodimer GPI molecules (pure donor or pure recipient), donor/recipient heterodimer molecules were expected. These occur only when the enzyme subunits have been synthesized in the same cell, and therefore indicated co-fusion of donor and recipient myoblasts [cf.6]. Nude mice were used as recipients in order to avoid graft rejection [7].

The following results were obtained:

1. Mouse to mouse skeletal muscle grafts. Functional and properly innervated muscle fibers were obtained, as was shown by direct and indirect electrical stimulation and by acetylcholinesterase (EC 3.1.1.7) histochemistry of endplates (Fig. 1). Donor and host contributions varied greatly, yet donor tissue did not exceed 80% in the samples of regenerated muscle (5 to 10 mg fresh weight) extracted. However, the sensitivity of GPI analysis is such that fractions of individual muscle fibers could be analyzed. Hybrid GPI was observed which indicated co-fusion of donor and recipient myoblasts (Fig. 2).
2. Rat to mouse skeletal muscle graft. With muscle from newborn rats regenerates of > 90% donor type were obtained. Hybrid GPI was found in several instances, indicating fusion between rat and mouse myoblasts.
3. Rat heart to mouse skeletal muscle bed. Minced heart muscle from newborn rats formed regenerates which showed visible autorhythmic contractions. Thus minced heart muscle is able to organize itself and to retain its tissue specific activity at an ectopic site. GPI analysis showed that cardiac cells contributed only a fraction of the mass of the regenerated tissue, possibly because of their

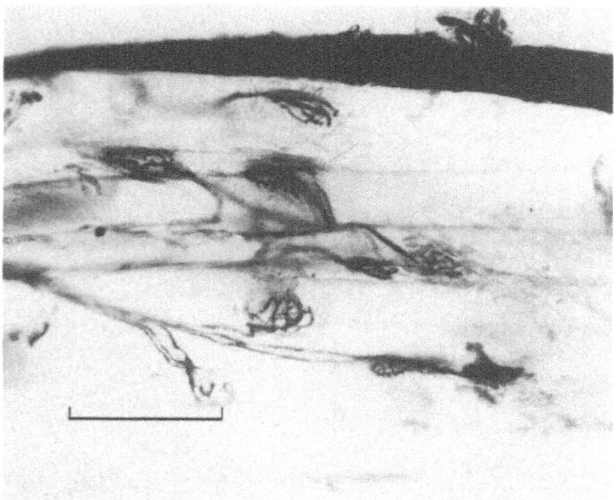

Fig. 1. Histological appearance of regenerated transplanted muscle. Minced muscle from a newborn neuro-muscular mutant (Motor Endplate Disease, genotype med/med) mouse was transplanted to a nude mouse recipient. Incubation period 7 weeks. Vibratom section stained for nerve fibers (silver) and acetylcholinesterase. A thick nerve (*near upper margin*), mature muscle fibers, and nerve fibers terminating in endplates are seen. The area around the branched nerve terminals is positive for acetylcholinesterase as in mature endplates. Scale 100 μm

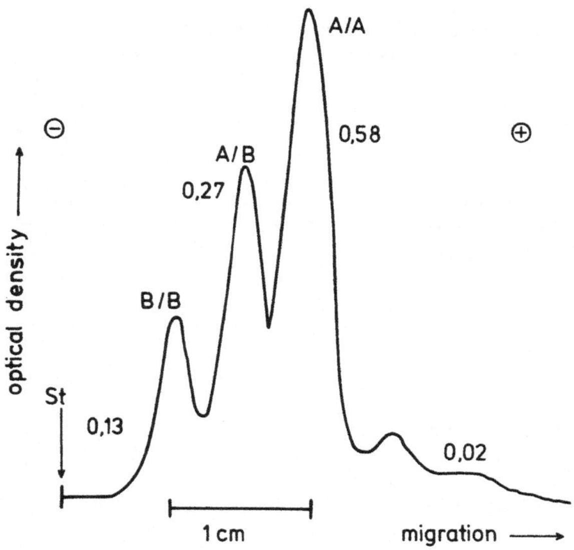

Fig. 2. GPI analysis of a mouse to mouse skeletal muscle graft. Donor: 6-d-old strain AKR (GPI A/A) mouse. Recipient: Adult NMRI nude (GPI B/B) mouse. Incubation period: 6 1/2 weeks. Approximately 1.5 μl of the extract of the regenerated graft was separated by electrophoresis in 1% agarose at pH 7.0 (90 Vh/cm) and developed for GPI activity according to [5] for 15 min. The cellulose nitrate foil containing the trapped formazan dye product was scanned with an LKB "Ultro scan" ruby laser densitometer. *Numbers next to peaks* indicate their relative areas. From these the relative contributions of donor and recipient to the regenerated tissue can be deduced. Note high proportion of hybrid GPI A/B indicating fusion between host and donor myoblasts

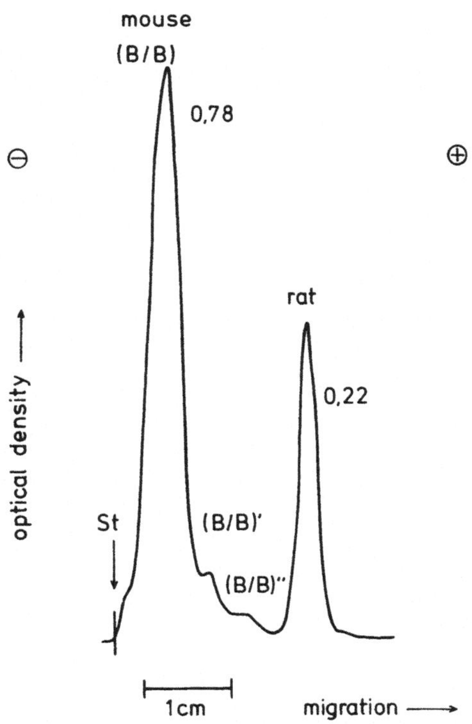

Fig. 3. GPI analysis of a graft of rat heart muscle to a mouse skeletal muscle bed. Incubation period 8 weeks. Other experimental conditions as described in Fig. 2. No hybrid band is found (Minor bands (B/B)' and (B/B)" are also present in pure mouse muscle)

limited potential to divide. As expected, no donor/recipient hybrid GPI was found in these regenerates as heart muscle cells do not fuse (Fig. 3).

The innervation of ectopic heart muscle transplants will be subject to further analysis.

Acknowledgements. I wish to thank G. Mehrke, K.A. Nave, B. Segnitz, J.A. Alexanian, H. Deschner, and E.-M. Füchtbauer for competent and dedicated collaboration, and the Deutsche Forschungsgemeinschaft (Project Jo 84/7) for financial support.

References

1. Carlson BM (1968) J Morphol 125:447-472
2. Zacks SI, Sheff MF (1977) Lab Invest 36:303-309
3. Sidman RC, Cowen JS, Eicher EV (1979) Ann N Y Acad Sci 317:497-505
4. Partridge TA, Sloper JC (1977) J Neurol Sci 33:425-435
5. Peterson AC, Frair PM, Wong GG (1978) Biochem Gen 16:681-690
6. Mintz B, Baker WW (1967) Proc Natl Acad Sci 58:592-598
7. Zalewski A (1978) Exp Neurol 58:261-271

Altered Cell Surface Glycosphingolipids of Murine Lymphocytes Belonging to Different Differentiation Pathways

P. F. Mühlradt, B. Kniep, and R. G. Gruner[1]

Introduction

The very early stages of differentiation of lymphocytes are associated with migrations from primary to secondary lymphoid organs. Thus T lymphocytes, which like all other blood cells originate from the bone marrow, spend some of their early life time in the thymus gland (which is why they are called "T cells"), where they mature and learn to discriminate self from nonself, i.e. own tissue from foreign tissue. After "graduation" they leave the thymus and home to specific areas in lymph nodes and spleen. Other, less sessile lymphocytes, circulate with blood and lymph in the body to watch for foreign substances, invading microorganisms or tumor cells.

Immunological processes, whether occurring in the living organism or in an in vitro cell culture system, involve different populations of cells. These are primarily various subpopulations of lymphocytes, but also include non-lymphoid accessory cells such as macrophages or dendritic cells. Such an interaction of different cells resulting in either antibody synthesis or generation of cytotoxic killer T lymphocytes is - partly science, partly fiction - depicted in Fig. 1. Interactions between various cells take place, some via soluble mediators, the so-called lymphokines, others by cell to cell contacts. Proliferative as well as final differentiation events are involved. Immune cells are therefore fascinating not only to the immunologist, who is primarily interested in their functions, but they are equally fascinating to the cell biologist who wishes to study problems of cell cell cooperation or of cell differentiation.

As biochemists it is our interest to look for cell surface changes during differentiation and activation of lymphoid cells and, if possible, to find cell surface molecules of known chemical structure which are typical for lymphocyte subpopulations with known functions. The less modest, but ultimate goal is of course to eventually correlate such structures with functions in the widest sense, i.e. with the behaviour of these cells in the tissue, their cooperations with other cells, their modulation by soluble factors etc. As we shall see we have still a long way to go.

There are of course the famous alloantisera available to the immunologist. These sera are obtained by cross-immunization of inbred mice with their respective lymphocytes. With these alloantisera several

1 Gesellschaft für Biotechnologische Forschung mbH (GBF), Mascheroder Weg 1, D-3300 Braunschweig-Stöckheim, FRG

33. Colloquium - Mosbach 1982
Biochemistry of Differentiation and Morphogenesis

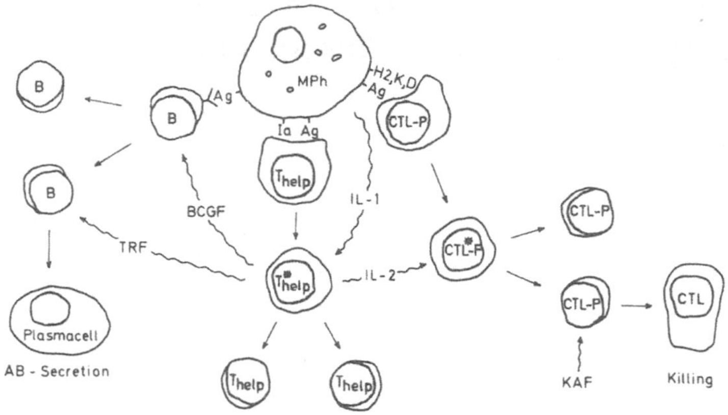

Fig. 1. Interaction of Immune Cells. *MPh* macrophage; *B* B lymphocyte; *Thelp* helper-amplifier T lymphocyte; *CTL-P* cytotoxic T lymphocyte precursor; *CTL* cytotoxic T lymphocyte; *Ag* antigen; *Il-1,2* Interleukin 1 and 2; *IL-2* is identical with T cell growth factor; *BCGF* B cell growth factor; *TRF* T cell replacing factor; *KAF* killer cell activating factor

differentiation antigens such as Thy 1, Lyt 1, 2 and 3 and others were discovered [1,2]. These antigens appear or disappear, as their name implies, during the differentiation of, in this case, T lymphocytes. The alloantisera have made it possible to distinguish between lymphocyte subpopulations, and reaction schemes such as the one in Fig. 1 could not have been elucidated without them. The Lyt 2 and Lyt 3 antigens may in fact be functionally important structures, in that antibodies directed against them can block cytotoxic function [3]. To the biochemist who is interested in chemical structures these differentiation antigens are of less use. Although it is known that they are glycoproteins, the exact peptide-sequences are not known, and also the epitopes which are recognized by the antisera have not yet been determined.

We have been concerned with another class of cell surface components, the glycosphingolipids (GSLs). Among them are the well known A, B, O bloodgroup antigens in man. We wish to review some of our data and include some new results which indicate that in the mouse some GSLs may be regarded as differentiation antigens of lymphocytes. One advantage of these compounds is their relatively simple structure which can be elucidated by conventional methods. This gives us information about defined molecular differences on the cell surfaces of different lymphocyte subpopulations.

Methodological Problems

The obvious main problem which one faces when studying cell surfaces of lymphocyte subpopulations is the situation shown in Fig. 1, i.e. one is dealing with mixtures of cells at various stages of activation and differentiation. There are several ways out of this dilemma, all of which suffer from some shortcomings.

1. One can try to separate cell subpopulations. This is done by differential agglutination with lectins [4,5], by density gradient centrifugations or by elimination of cells with antisera or by other means. The disadvantages may be poor separations, low yields, or insufficient selectivity.

2. One functionally activates lymphocytes in cell culture leaving "innocent bystander cells" in a quiescent state. Addition of suitable radioactive precursors to the culture medium then results in selective uptake and incorporation of radioactivity by only the activated cells [6,7]. Galactose and glucosamine proved to be good precursors for labeling the carbohydrate portion of GSLs [8]. The radioactive GSLs can then be extracted and analysed (Fig. 2). The disadvantage here may be poor selectivity of the activation in that more than one subspecies of lymphocyte may react to a given stimulus.

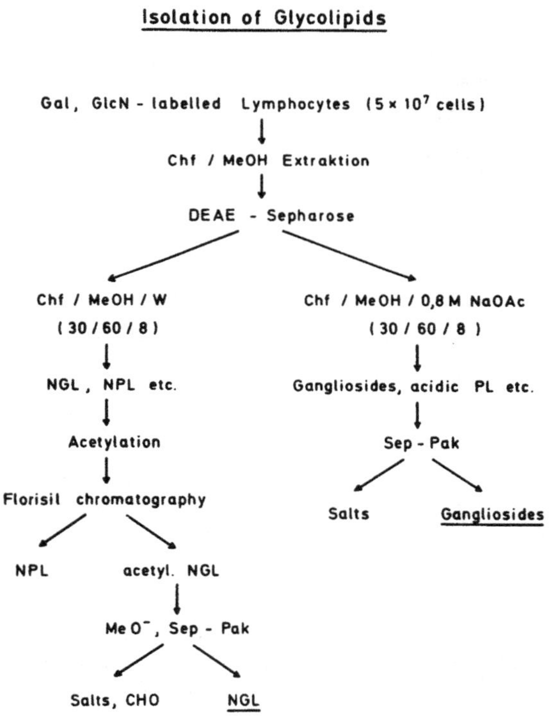

Fig. 2. Flow sheet of the isolation of glycosphingolipids from lymphocytes. *Chf* chloroform; *MeOH* methanol; *W* water; *NaOAc* sodium acetate; *NPL* neutral phospholipids; *NGL* neutral glycosphingolipids; *CHO* carbohydrates; *SEP-PAK* is a C18 reversed phase column

3. One uses cell lines or, better still, cloned cell lines for these studies. As shown in Fig. 1 growth factors exist for T and B lymphocytes which allow clonal expansion of these cells [9,10]. There are also cell lines which have lost the requirement for these factors. The problems one can run into in this system are chromosomal abnormalities or contamination with feeder cells, which can lead to either loss or alteration of cell surface components, or contamination with cell surface components from the feeder cells when such are present.

We have used all three methods either per se or in combination, and rely on those data which are consistently obtained.

For analytical purposes we use either thin layer chromatography, locating radioactive GSLs by autoradiography (Fig. 3), or high performance liquid chromatography [11]. Carbohydrate sequences and anomeric linkages were elucidated by sequential degradation with specific glycosidases [12].

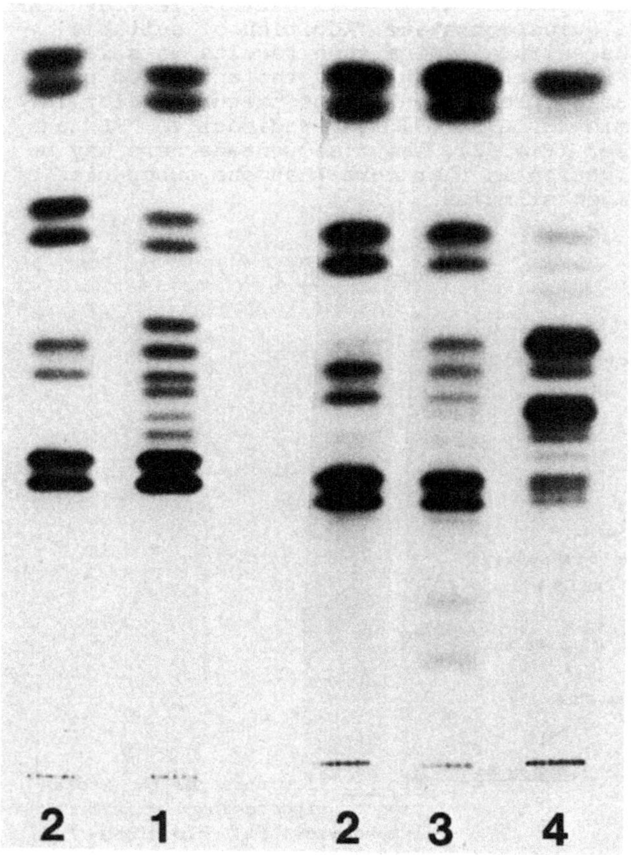

Fig. 3. Autoradiogram of a thin layer chromatogram of radioactively labelled glycosphingolipids extracted from: lymphocytes activated in a primary mixed lymphocyte reaction (*1*) concanavalin A - stimulated lymphocytes (*2*), lipopolysaccharide-stimulated lymphocytes (*3*), and lymphocytes activated in a secondary mixed leucocyte culture (*4*)

Results and Discussion

Globotriaosylceramide (CTH), a Glycolipid on B Blasts

A first glance at the autoradiogram in Fig. 3 shows that lymphocytes, depending on how they were stimulated to divide, synthesize a number of common GSLs, but also some which appear to be specifically present in only certain subpopulations. Care was taken in these experiments by choosing the appropriate stimulants, to a priori stimulate certain subpopulations. For example, it is well established by thymidine incorporation experiments that bacterial lipopolysaccharide brings about primarily B cell proliferation, whereas the lectin concanavalin A is a typical T cell mitogen. Mitogen-induced carbohydrate incorporation is similarly cell population-specific [6]. By comparing lanes 2 and 3 in Fig. 3 one sees that an additional GSL with the migratory properties of a trihexosylceramide appears in lipopolysaccharide-stimulated B cells which is absent from concanavalin A-stimulated T cells. We know from sequencing the carbohydrate portion that this GSL belongs to a special biosynthetic branch of GSLs, the globo series and is identical with CTH [12]. The same GSL is found on three different murine myeloma lines as detected by high performance liquid chromatography (B. Kniep, unpublished). Myelomas are permanently growing cells of the B cell lineage, functionally resembling plasma cells.

Globotetraosylceramide, a Marker for Helper T Cells

However, CTH is not confined to B blast cells but was also detected on lymphocytes which react to an allogenic stimulus in a mixed lymphocyte reaction (Fig. 3, lane 4), [12]. In this case we studied a so-called secondary mixed lymphocyte reaction which is particularly strong. The principle of such a reaction is shown in Fig. 4. It is known that primarily T cells react to such a stimulus. As judged from their GSLs, these T cells are obviously different from the ones reacting to a concanavalin A stimulus (Fig. 3, lane 2). Indeed, concanavalin A reactive cells are known to comprise mostly cytotoxic T lymphocytes [13], whereas in the mixed lymphocyte reaction primarily T cells of the helper type proliferate [14]. These latter cells exhibit another prominent GSL (see lane 4 of Fig. 3) which we have shown to be identical with globoside [12]. Again a comparison of the GSLs from two helper and two cytotoxic T cell lines from different laboratories (T. Hünig [15], A. Schimpl, Würzburg, and F. Fitch [16], Chicago) showed that

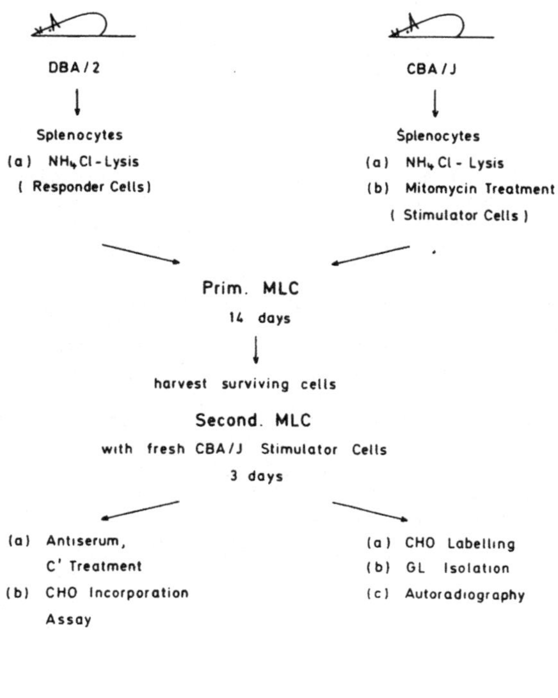

Fig. 4. Flow sheet of the principle of a mixed leucocyte culture (MLC)

globoside was present in the helper cells but absent from the cytotoxic T cells which also react in a mixed lymphocyte reaction. We can therefore quite safely conclude that within the mixture of cells reacting to the alloantigen the helper cells are the ones which carry the globoside. The structures corresponding to the GSL bands seen in Fig. 3 are shown in Fig. 5. They were tentatively assigned to these bands after sequencing the carbohydrate portion of the GSLs with specific glycosidases and by comparison with authentic standards.

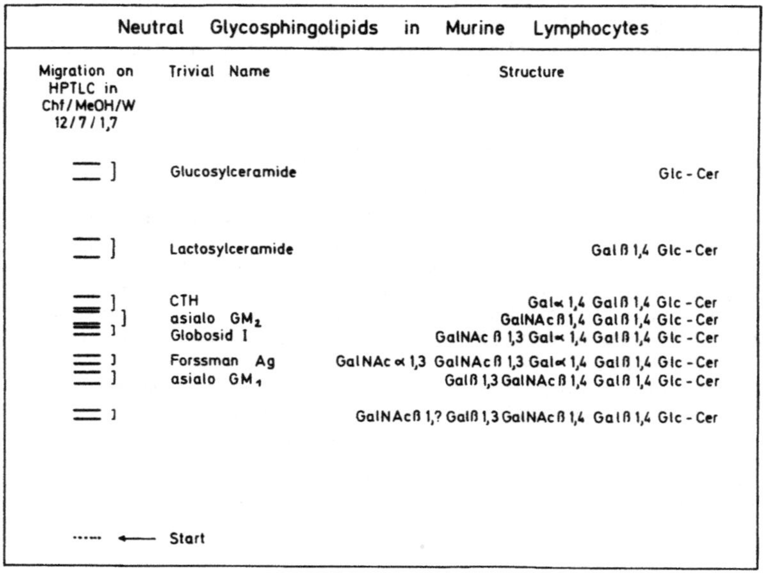

Fig. 5. Schematic drawing of a separation on a thin layer plate of glycosphingolipids found in mouse lymphocytes. The structures are assigned to the bands on the basis of sequential degradation of isolated glycosphingolipids by glycosidases. The linkages at the nonreducing ends of the sugars have not in all cases been confirmed by methylation analysis

Is this globoside molecule accessible to antibodies, could it potentially serve as a serological marker with a known structure? The answer is given in Fig. 6. Treatment with antiglobosíde antibodies, kindly prepared by K. Petzold (Hannover), and complement indeed decreased carbohydrate incorporation into lymphoblasts isolated from a mixed lymphocyte reaction, indicating that globoside is accessible and exposed on the cell surface. The extent of killing was not very impressive in this experiment. There may be two reasons for this: (a) lymphoblasts, as compared to resting lymphocytes, are particularly hard to kill by complement treatment, (b) as mentioned above, only a certain percentage of the proliferating cells are supposedly helper T cells. When the experiment was repeated in a slightly different way, i.e. when antiglobosíde-complement treatment was performed before the mixed lymphocyte culture, inhibition of proliferation was 75%, suggesting that globoside is already present on the alloreactive cells before stimulation and that these cells are easier to kill than when transformed to blasts.

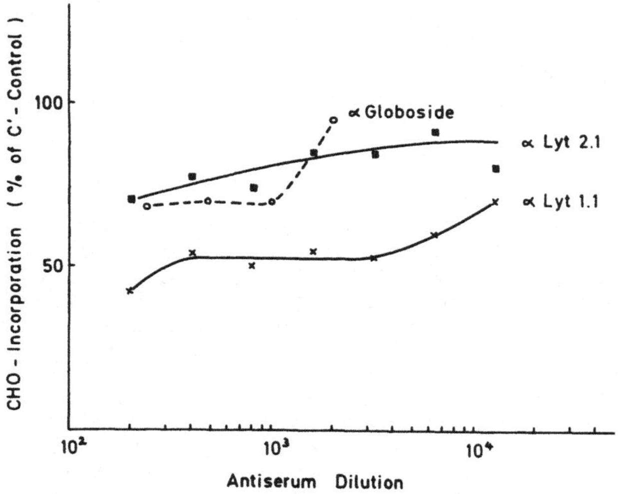

Fig. 6. Carbohydrate incorporation of alloreactive cells after a 3-day-period of a secondary mixed leucocyte culture (see Fig. 4). The cells were treated with various dilutions of antibody and complement. Antibodies were against globoside and the glycoprotein antigens Lyt 1 and Lyt 2

The Problem of Asialo GM_1 as a Marker for Natural Killer Cells

Another interesting aspect has emerged from a comparison of the GSLs of cells participating in a primary and secondary mixed lymphocyte reaction (Fig. 3, lanes 1 and 4, respectively). The patterns look very dissimilar, and since this was contrary to expectations we got excited about this finding, speculating about different cells reacting in the primary and secondary responses. However, the pattern in lane 4 turned out to be an artifact which had not only fooled us but apparently others as well. When the blast cells from a secondary mixed lymphocyte culture were reisolated and labelled in fresh medium, the GSLs looked as in Fig. 3 lane 1. Our explanation is that, as a secondary mixed lymphocyte culture contains many dead cells, glycosidases are released which degrade the GSLs on the cells as they are formed. This results in partial degradation by β-galactosidase of particularly asialo GM_1 and lactosyl ceramide which appear only as weak double bands in Fig. 3, lane 4. The pattern in Fig. 3, lane 4 gives therefore a correct qualitative but a wrong quantitative picture of the GSLs on alloreactive cells. These findings can now explain why other groups have reported the absence of asialo GM_1 on killer T lymphocytes generated in a secondary mixed lymphocyte reaction [17,18], whereas cloned T killer cells *do* carry asialo GM_1 (our own unpublished findings). This GSL, which according to our results is quite common on lymphoblasts, has been described by others to be present on mainly mature T lymphocytes [19]. This point of whether or not asialo GM_1 may be present on T killer lymphocytes is important to the immunologists who are concerned with the so-called natural killer cell. This cell species can kill tumor cells without prior sensitization. The point was made that natural killer and T killer cells differ in that the former carry asialo GM_1 whereas the latter don't [17,18]. Also the ontogenetic relationship of natural killer cells to either the T cell or the monocyte lineage is still a matter of some controversy [20,21].

Unfortunately we are not able to settle this point one way or the other, because we find asialo GM_1 on T killer cells as well as on monocytes (D.A. Monner, unpublished).

Conclusion

Figure 7 summarizes our data in the form of a biosynthetic pathway of neutral GSLs and gangliosides in murine lymphocytes. Since different lymphocyte subpopulations carry partially different GSLs, we conclude that during differentiation and development of lymphocytes, and probably other blood cells, various glycosyl transferase genes are expressed and shut off again as these cells differentiate. This interesting phenomenon may be quite common as it was also observed in early mouse embryos [22] and in differentiating tissue of the small intestine of the mouse [23]. These changes of cell surface carbohydrates may result in proper allocation of cells in the differentiating tissue. For the lymphocytes in particular such alterations may change their behaviour towards other cells with which they cooperate or their homing to specific areas in the lymphoid organs. This hypothesis recently gained support from homing experiments with thymocytes whose cell surface carbohydrates had been changed by treatment with galactose oxidase [24].

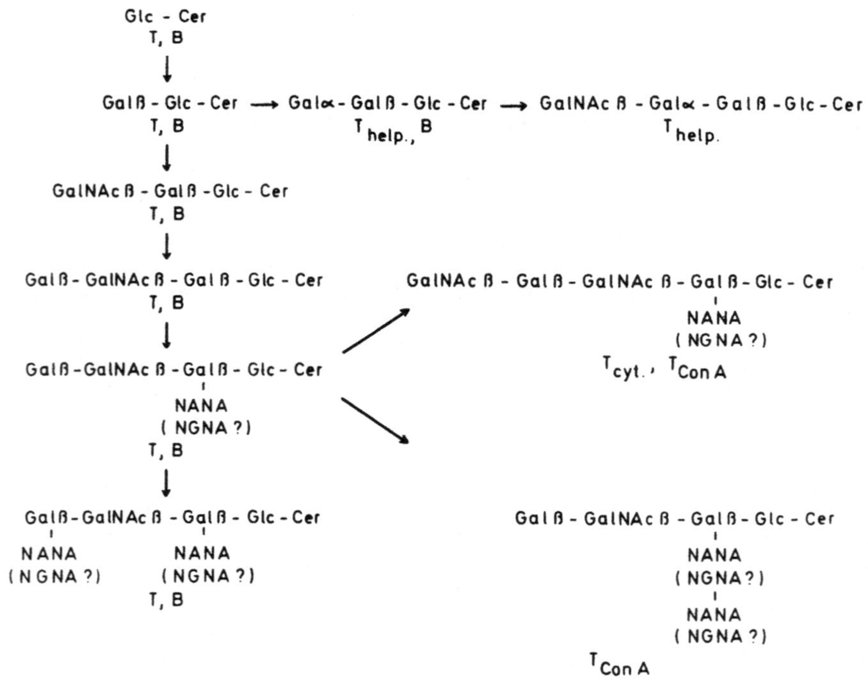

Fig. 7. The structures of neutral glycolipids and gangliosides detected on various lymphocyte subpopulations have been arranged into a biosynthetic pathway. On some cells, e.g. helper T cells, glycosphingolipids are found which belong to the ganglio- and to the globo-series

References

1. Raff MC (1971) Surface antigenic markers for distinguishing T and B lymphocytes in mice. Transplant Rev 6:52
2. Cantor H, Boyse EA (1975) Functional subclasses of T lymphocytes bearing different Ly antigens. J Exp Med 141:1376
3. Nakayana E, Shiku H, Stockert E, Oettken HF, Old LJ (1979) Cytotoxic T cells; Lyt phenotype and blocking of killing activity by Lyt antisera. Proc Natl Acad Sci USA 76:1977-1981
4. Reisner Y, Linker Israeli M, Sharon N (1976) Separation of mouse thymocytes into two subpopulations by the use of peanut agglutinin. Cell Immunol 25:129-134
5. Reisner Y, Ravid A, Sharon N (1976) Use of soybean agglutinin for the separation of mouse B and T lymphocytes. Biochem Biophys Res Commun 72:1585-1591
6. Rosenfelder G, Eijk van RVW, Mühlradt PF (1979) Metabolic carbohydrate labelling of glycolipids from mouse splenocytes. Eur J Biochem 97:229-237
7. Eijk van RVW, Rosenfelder G, Mühlradt PF (1979) Metabolic carbohydrate labelling of glycoproteins from mitogen-stimulated mouse lymphocytes. Eur J Biochem 101:185-193
8. Rosenfelder G, Eijk van RVW, Monner DA, Mühlradt PF (1978) Glycolipids in mouse thymocytes stimulated by concanavalin A. Eur J Biochem 83:571-580
9. Möller G (ed) (1981) In: Immunological Reviews 54 "T cell clones". Stockholm
10. Howard M, Kessler S, Chused T, Paul WE (1981) Long-term culture of normal mouse B lymphocytes. Proc Natl Acad Sci USA 78:5788-5792
11. Ullman MD, Cluer Mc RH (1977) Quantitative analysis of plasma neutral glycosphingolipids by high performance liquid chromatography of their perbenzoylated derivatives. J Lipid Res 18:371-378
12. Gruner KR, Eijk van RVW, Mühlradt PF (1981) Structure elucidation of marker glycolipids of alloantigen-activated murine T lymphocytes. Biochemistry 20:4518-4522
13. Larsson EL, Fischer Lindahl K, Langhorne J, Coutinho A (1981) Quantitative studies on concanavalin A-induced, TCGF-reactive T cells. I. Correlation between proliferation and lectin-dependent cytolytic activity. J Immunol 127:1081-1085
14. Simon MS, Eichmann K (1980) T cell subsets participating in the generation of cytotoxic T cells. Springer Semin Immunopathol 3:39-62
15. Hünig TR, Bevan JM (1982) Antigen recognition by cloned cytotoxic T lymphocytes follows rules predicted by the altered self hypothesis. J Exp Med 155:111-125
16. Glasebrook AL, Fitch FW (1980) Alloreactive cloned T cell lines. J Exp Med 151:876-895
17. Young WW, Hakomori SI, Durdik JM, Henney CS (1980) Identification of ganglio-N-tetraosylceramide as a new cell surface marker for murine natural killer (NK) cells. J Immunol 124:199-201
18. Kasai N, Iwamori M, Nagai Y, Okumura K, Tada T (1980) A glycolipid on the surface of mouse natural killer cells. Eur J Immunol 10:175-180
19. Stein KE, Schwarting GA, Marcus DM (1978) Glycolipid markers of murine lymphocyte subpopulations. J Immunol 120:676
20. Herberman RB, Nunn ME, Holden HT (1978) Low density of thy 1 antigen on mouse effector cells mediating natural cytotoxicity against tumor cells. J Immunol 121:304-309
21. Sun D, Lohmann-Mathes ML (1982) Functionally different subpopulations of mouse macrophages recognized by monoclonal antibodies. Eur J Immunol 12:134-140
22. Willison KR, Karol RA, Suzuki A, Kundu SK, Marcus DM (1982) Neutral glycolipid antigens as developmental markers of mouse teratocarcinoma and early embryos. J Immunol 129:603-609
23. Suzuki A, Yamakawa T (1981) The different distribution of asialo GM_1 and Forssman antigen in the small intestine of mouse demonstrated by immunofluorescent staining. J Biochem 90:1541-1544
24. Madyastha KR, Baker MA, Taub RN (1980) Influence of cell-surface galactosyl determinants on splenic localization of mouse thymocytes. Transplantion 29:252-254

Morphogenesis

Conjugation of Ciliates in Biochemistry of Multicellular Morphogenesis

A. Miyake[1]

Introduction

The surface of ciliates is covered by cilia and other visible structures which are arranged in a pattern characteristic to species, providing a good opportunity to investigate the problem of how the pattern on the cell surface is maintained and reproduced. Therefore, ciliates have been widely used in the study of morphogenesis of individual cells [1-6].

On the other hand, ciliates have seldom been used for the study of "multicellular morphogenesis" in which cells are building-blocks of morphogenesis. The neglect might be due to the fact that ciliate cells are separate from each other during most of the life cycle. However, I think that the unicellularity of ciliates makes their multicellular morphogenetic events particularly accessible to experimental analysis. Because of the predominance of the unicellular life, multicellular morphogenetic events in ciliates appear separately on the unicellular background. We can investigate them one by one, without being disturbed by other multicellular morphogenetic events directly linked to the event in question.

Multicellular morphogenetic events in ciliates are variable. In *Sorogena stoianovitchae*, cells aggregate and form sorocarps (fruiting bodies) as in slime molds [7]. In some ciliates, cells grow and then divide many times without growth, producing a mass of small cells [1,8], as in cleavage of eggs. In sexual reproduction, a fertilized nucleus repeats mitosis within a cell, and the division products move to the periphery of the cell to differentiate into germinal and somatic nuclei depending on their local positions [9], as in the early development of insects. In conjugation, two types of cells temporarily unite at the peristomial region (Fig. 1) and undergo meiosis and other developmental processes of sexual reproduction.

In the following, I will discuss the possible use of ciliate conjugation in the biochemical study of multicellular morphogenesis. The system is simple in the sense that only two cells are involved. Yet here are found many basic elementary processes of multicellular morphogenesis. Three of them are already indicated in the above short description of conjugation: (1) uniting of cells at the particular region of the cell surface, (2) induction of new developmental processes by a contact of two types of cells, and (3) programmed termination of cell union.

---

1 Zoologisches Institut der Universität Münster, Badestraße 9, D-4400 Münster, FRG

33. Colloquium - Mosbach 1982
Biochemistry of Differentiation and Morphogenesis

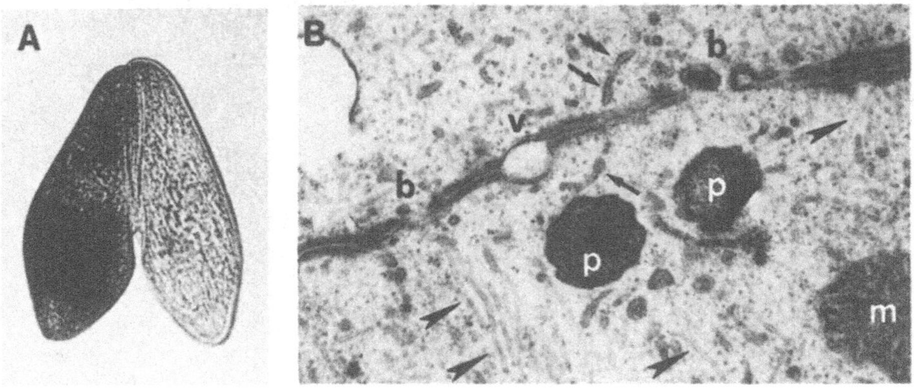

Fig. 1A,B. Conjugant union in *Blepharisma japonicum*. (A) A conjugant pair consisting of a normally pigmented mating type II cell (*left*) and a less-pigmented mating type I cell (*right*). Photograph of living unstained cells. x 170. (B) Cross-section of a conjugant pair (2.5 h old) at the united region. Cell membranes of two conjugants are juxtaposed with a distance of about 20 nm except at the vacuolated place (*v*) and cytoplasmic bridges (*b*). The pair consists of a normally pigmented mating type II cell (*below*) and an albino mating type I cell (*above*). *p* pigment granule, *m* mitochondrion, *arrow* saccule, *arrowhead* PACM microtubule. For materials and methods see [47]. x 30,000. (From [10])

The uniting of cells at a particular region of the cell surface is the basic mechanism by which cells in multicellular organisms make structures such as a hollow sphere of blastula, a tube and a sheet. Ciliates can mimic to some extent such a construction. If doublets (a mutant having a peristomial region on both sides of a cell) of *Blepharisma japonicum* conjugate, cells unite side by side in chains which often close to form rings (Fig. 2). If cells with more attachment regions (e.g., quadruplets) conjugate, tubes or hollow spheres might be formed. Note that this multicellular construction is carried out by simply repeating the same mechanism, namely, the uniting of cells at the particular region of the cell surface. Ciliates use this mechanism only to face each other exactly at the same cell surface during conjugation. One may imagine that nature took such a mechanism to construct more complex structures in other lines of evolution.

The induction of new developmental processes by a close association of different types of cells, often simply called "induction", is a key element in morphogenesis of multicellular organisms, because their embryos increase the spatial and functional complexity through many inductive processes.

The programmed termination of cell union is also essential in morphogenesis of multicellular organisms. During their development, often a group of united cells free themselves, migrate to other places, and engage into a new association of cells following a developmental program. An example is shown in Fig. 3. Although studies on the separation process in conjugation are few [12], the process must be programmed, because conjugants separate at a certain stage (soon after karyogamy).

Fig. 2. Multicellular complexes in *Blepharisma japonicum*. Doublet cells of mating type II were treated by gamone 1 (10^4 U/ml) for 11.5 h at 25 ± 1°C. Photograph of living unstained cells. For materials and methods see [59]. x 38

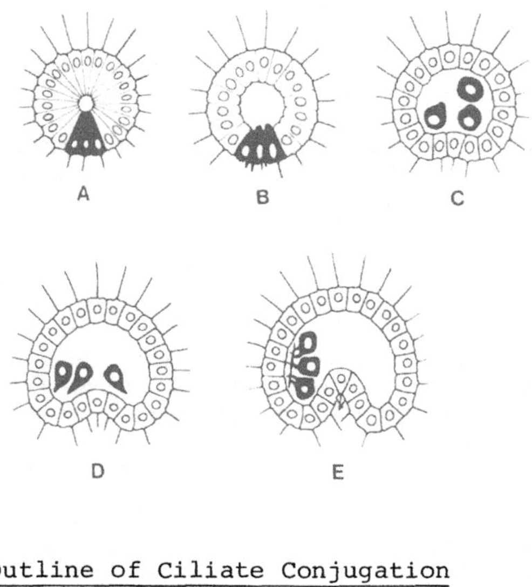

Fig. 3A-E. Formation, dissociation, migration and reassociation of primary mesenchyme cells (*shaded*) in the sea urchin, *Arbacia punctulata*. (A) Early blastula. (B) Late blastula. (C) Primary mesenchyme cells dissociate and enter the blastocoel. (D) Primary mesenchyme cells migrate over the wall of the blastocoel. (E) Reassociation of primary mesenchyme cells to form the skeleton-forming tissue. (From [11])

Outline of Ciliate Conjugation

In conjugation of ciliates two cells unite for fertilization. In this respect ciliate conjugation is similar to conjugation in other unicellular organisms, but it has two unique features: (1) reversibility of cell union, and (2) replacement of the somatic nucleus. In many unicellular organisms, the two conjugating cells (conjugants) fuse totally to form a single cell. However, in ciliates such total conjugation is the rule only in certain groups, e.g., peritrichs. In other ciliates, the two cells in a conjugant pair do not completely fuse. Intercellular bridges are formed between them but each cell retains its identity until the pair separates into individual cells (temporary conjugation) (Fig. 1). Although both, the complete fusion of cells and the persistent cell union, occur in morphogenesis of multicellular organisms, only temporary conjugation is discussed below, because little is known about the biochemistry of total conjugation.

The second characteristic stems from the fact that each ciliate cell has both germinal (micro) and somatic (macro) nuclei. Soon after cells unite in conjugation the germinal nucleus undergoes meiosis and produces two haploid nuclei. One of them is exchanged between conjugants and fertilization follows. Cells then separate. The fertilized nucleus mitotically divides without cytokinesis. The division products differentiate into the germinal nucleus and the somatic nucleus. The old somatic nucleus degenerates and is replaced by the newly developed one. For details of nuclear changes see [4,13].

Ciliate conjugation is regularly induced by bringing together cells of two complementary mating types. Under appropriate conditions they immediately start interacting with each other and, after a few hours of this "preconjugant interaction", they unite into conjugant pairs.

The capacity to undergo preconjugant interaction is controlled by internal and external factors. The main internal factor is the "maturity" which is attained by repeating a certain number of cell divisions after conjugation [14-16]. The main external factor for preconjugant interaction is the deprivation of nutrient.

The mode of preconjugant interaction is the subject of great diversity among ciliates [10,17,18], providing an opportunity for the comparative study of various types of cell-cell interactions. An attempt is being made to apply an unifying view of "gamone-receptor hypothesis" to preconjugant interaction in all ciliates [10].

Preconjugant interaction is "fluid mediated" or "cell mediated" depending on species. In the fluid mediated type, the signal for the interaction is excreted into the medium. In the cell mediated type, the signal is bound to the cell surface and, hence, cells must physically contact with each other to interact (Fig. 4).

However, conjugation sometimes occurs within a clone. This "intraclonal selfing" is common among the cultures established from single cells newly collected from nature, although they are often eliminated during the effort to obtain clones stable for mating type. In several cases, intraclonal selfing was found to be due to the appearance of complementary mating types within a clone [10], but the possibility of conjugation not requiring the presence of two types of cells still looms large. Such conjugation may occur if each cell produces both the conjugation signal and its receptor at the same time, as postulated by Miyake and Bleyman [19]. It may also occur, if the conjugation mechanism of each cell is turned on by an intracellular signal which is autonomously produced by the cell. The fact that simple ionic changes in the external medium can induce conjugation between cells of the same mating type in *Paramecium* [20] indicates that conjugation does not always require the presence of two types of cells.

In multicellular organisms some cells appear to unite not requiring a specific signal from other type of cells. For example, blastomeres in mammals, which are spherical and only loosely stick together, suddenly

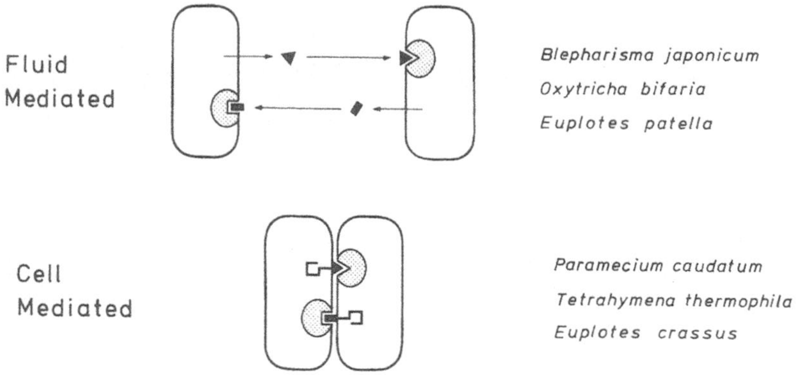

Fig. 4. Schematic view of the two types of preconjugant cell-cell interaction in ciliates based on the gamone-receptor hypothesis. Gamones (*triangles, rectangles*) are excreted into the medium (fluid mediated type) or anchored to the cell (cell mediated type). *Dotted area*: gamone receptor on the cell surface. Ciliates which undergo respective type of interaction are listed on the right

unite firmly to each other to form a compact mass. This process of "compaction" occurs as early as at the stage of eight-cell embryo [21], suggesting that these cells gain the capacity to unite autonomously according to their developmental program. Further study on intraclonal selfing in ciliate is, therefore, worthy not only for the study of ciliate conjugation, but also for the understanding of the cell union in multicellular organisms.

Analytical studies on conjugation are being made in *Blepharisma* [22], *Euplotes* [10,23-26], *Oxytricha* [27], *Paramecium* [28], *Tetrahymena* [29-31], and other ciliates [10] , but this article mainly deals with *Blepharisma* in which the biochemical analysis of conjugation is most advanced.

Cell-Cell Interaction of Positive Feedback

In preconjugant interaction of *Blepharisma japonicum*. two complementary mating types I and II stimulate each other for conjugation as shown in Fig. 5. Type I cells autonomously excrete gamone 1 (step 1 in Fig. 5). This gamone specifically acts on type II cells (step 2), transforms them so that they can unite (step 3) and also induces the excretion of gamone 2 (step 4). Gamone 2 acts back onto type I cells (step 5), transforms them so that they can unite (step 6) and also enhances the excretion of gamone 1. Transformed cells unite in conjugant pairs (step 7). The whole process is completed in several hours [32].

Since each gamone induces the other gamone, the reaction chain consisting of steps 1, 2, 4, and 5 in Fig. 5 is a cycle of positive feedback. Even if the interaction is weak at the beginning, it will soon be intensified to the maximum strength and this condition will be maintained until cells conjugate. Positive feedback is, therefore, a powerful mechanism to amplify the initial subtle interaction of cells and force

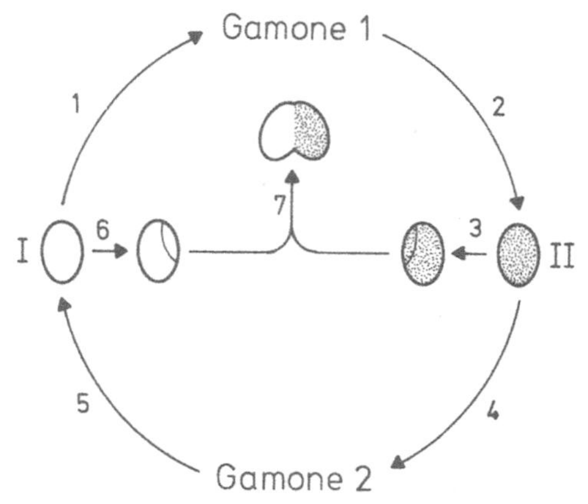

Fig. 5. Preconjugant cell-cell interaction in *Blepharisma japonicum*. Roman numerals, mating types. (Adapted from [32])

them to enter a completely new situation. It might be widely used to control multicellular morphogenesis in which an encounter of different types of cells often triggers a new developmental process.

Positive feedback in preconjugant interaction was first postulated by Heckmann and Siegel [33] to explain the "waiting period" in *Euplotes crassus*. Preconjugant interaction of this ciliate is the cell-mediated type and hence requires direct contact of cells (Fig. 4). When cells of complementary mating types are mixed, there is no visible reaction for about an hour (waiting period). Cells then agglutinate with cilia (mating reaction) until they form conjugant pairs. They assumed that the initial subtle *invisible* mating reaction, which is carried out by colliding of the two types of cells, develops into a stronger *visible* reaction through positive feedback. This hypothesis is strongly supported by later work [26,34]. Positive feedback also occurs in preconjugant interaction of *Oxytricha bifaria* [35].

Preconjugant interaction of *E. crassus* is also interesting as a model of the type of cell-cell interaction in which cells communicate with many distant cells by collisions without jamming the space with excreted signals (as in hormonal systems) or with extensive cytoplasmic projections (as in nervous systems) [26]. This type of cell-cell interaction might be useful in multicellular organisms which have a huge number of interacting cells in a limited space.

Gamone as a Cell-Union-Inducing Substance

Gamone 1 (blepharmone) is a glycoprotein (IEP, 7.5; 20000 D) [36]. It contains 6 sugars (glucosamin 3, mannose 3) and 175 amino acids (Lys 7, His 1, Arg 4, Asp 26, Thr 17, Ser 19, Glu 7, Pro 8, Gly 13, Ala 13, Cys 4, Val 11, Met 6, Ile 8, Leu 12, Tyr 13, Phe 6) not counting tryptophan (not measured because the sample was scanty) [37]. If lyophilized and stored at $< -10^{\circ}C$, blepharmone remains active for years.

Gamone 2 is calcium-3-(2'-formylamino-5'-hydroxybenzoyl) lactate [38] (Fig. 6). It was first named as blepharismin [38], but then renamed as

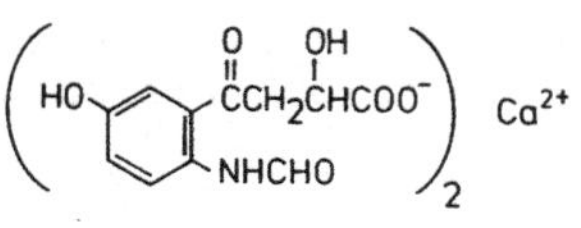

Fig. 6. Blepharismone, gamone 2 of *Blepharisma japonicum*

blepharismone [36]. If dried and stored at < -10°C, blepharismone remains active for years. Blepharismone resembles serotonin chemically (both are tryptophan derivatives) and biologically (both are specific intercellular signals).

The gamone activity is expressed by U, which is the least activity that can induce one or more homotypic pairs (see below) in 500-1000 cells of the complementary mating type suspended in 1 ml of a salt solution [32]. Activities of 1 mg of blepharmone and blepharismone are 1.6×10^7 U and 10^6 U, respectively.

Blepharismone was chemically synthesized starting from 2-amino-5-hydroxyacetophenone via 2-formylamino-5-hydroxyacetophenone by Tokoroyama, Hori and Kubota [39]. Entzeroth and Jaenicke [40] synthesized blepharismone and its derivatives starting from substituted anilin. Among these derivatives, 5'-fluoroblepharismone and 2'-desformamido-blepharismone have the blepharismone activity (about 1/500 and 1/1200 times as active as natural blepharismone). Synthetic blepharismone is about half as active as the natural compound. This result is explained by assuming that the one of the enantiometric forms (L) is more biologically active than the other form (D). This assumption is supported by the finding that L-isomers of blepharismone inhibitors are more effective than D-isomers [22,41,42] (Fig. 7). Figure 7 also shows that some amino acids competitively inhibit the activity of blepharismone and that the inhibition is stronger as their chemical structure is closer to blepharismone. These results suggest that blepharismone acts on type I cells by reversibly binding to the receptor on type I cells.

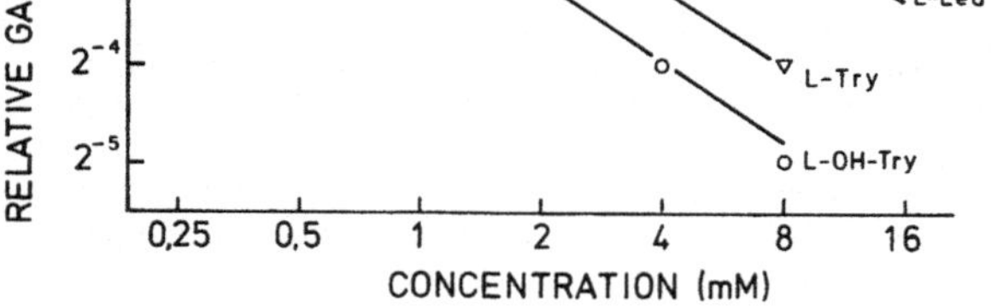

Fig. 7. Inhibition of the blepharismone activity by amino acids which are structurally related to blepharismone. Glycine, L-alanine, L-serine, and L-valine had no inhibiting effect at 0.25-8.00 mM. Blepharismone (0.32 µg/ml) and the amino acid to be tested were added at the same time to mating type I cells of the albino Bangalore strain of *Blepharisma japonicum*. Experiment at 25 ± 1°C. For the preparation of cells see [59]. For the measurement of blepharismone activity see [22]

Among the three known functions of gamones, (1) induction of synthesis and excretion of gamone [22,32,43], (2) chemoattraction (blepharismone only) [44], and (3) induction of cell union [32,43,45,46], only the third one will be dealt with below, because this is most relevant to multicellular morphogenesis.

In the induction of cell union, cells are transformed by the gamone of the other cell type so that they can unite. Once transformed, cells can unite in any combination of cell types. Therefore, in the mixture of two cell types, both heterotypic (type I - type II) and homotypic (type I - type I, type II - type II) pairs are formed. As this implies, if cells of one type are treated by the gamone of the other type, homotypic pairs are induced. Both types of pairs look alike, but meiosis and other nuclear changes of sexual reproduction occurs only in heterotypic pairs [43] (Fig. 8).

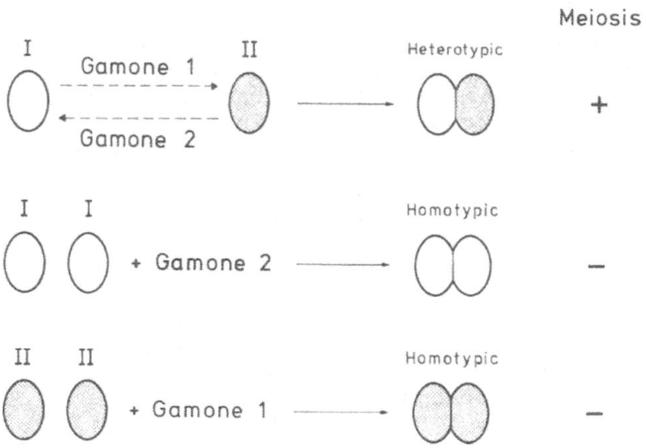

Fig. 8. Inductions of heterotypic (*top*) and homotypic (*middle, bottom*) conjugation in *Blepharisma japonicum*. Roman numerals indicate mating types. Meiosis occurs in heterotypic conjugation (+), but not in homotypic conjugation (-)

The process of the induction of conjugant pairs by gamone is divided in three stages, (1) preunion, (2) ciliary union, and (3) conjugant union [22].

In the preunion stage, which lasts for 1-2 h at 24°C, the sticking together between cells is very slight, if any, except at the late part of this stage and cells do not associate in any spatially fixed way.

In the stage of ciliary union, which also lasts for 1-2 h at 24°C, cells form the face-to-face pair held by the adhesion between adoral zone of membranelle (AZM) of one cell and a row of cilia anterior to the undulating membrane (antUMC) [46]. Since these ciliary structures are arranged at both sides of the peristomial floor, the paired cells are faced to each other with the peristomial floor (Fig. 9). The ciliary union is easily broken up by mechanical stirring, but, if not disturbed, the pair persists until it enters into the next stage of conjugant union. Both heterotypic and homotypic pairs are formed in this way.

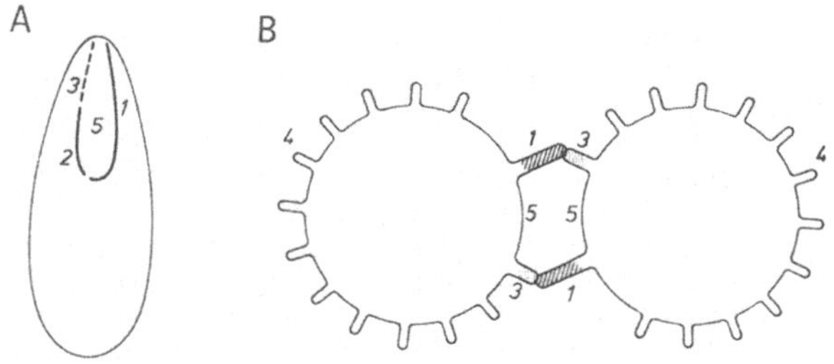

Fig. 9A,B. Diagrams of (A) the peristomial region and (B) a pair of cells attached by ciliary union (cross section) in *Blepharisma japonicum*. *1* Adoral zone of membranelle (AZM), *2* undulating membrane (UM), *3* cilia anterior to undulating membrane (antUMC), *4* somatic cilium, *5* peristomial floor. (A) Adapted from [47]. (B) Adapted from [46]

In the stage of conjugant union, cells unite with the peristomial floor to form conjugant pairs. The peristomial floor lacks cilia and conjugant pairs are held together by a direct contact of cell bodies [45]. The union begins at the anterior end of the peristomial floor and extends posteriorly. Except at the very beginning, cells in a conjugant pair are so firmly united that a surgical operation is usually needed to separate them mechanically. Both heterotypic and homotypic pairs are formed in this way. During the first few hours, these two types of conjugant pairs are morphologically very alike even at the ultrastructural level [47]. The differences between them, e.g., the larger united region, the more extensive cytoplasmic bridges, and earlier disappearance of PACM microtubules (see below) in heterotypic pairs, appear thereafter. Heterotypic pairs persist at least until karyogamy, which takes place about 1 day after the pair formation at 24°C. Homotypic pairs may persist for several days, if enough of the gamone of the other cell type is present.

Gamones thus alter at least three regions of the cell surface in different ways. AZM and antUMC are altered so that they can stick to each other, whereas the peristomial floor is altered so that it can unite with the similarly altered peristomial floor. Although the AZM-antUMC adhesion should facilitate the formation of conjugant pairs by bringing two cells together in a proper arrangement, it does not appear to be essential for the formation of conjugant union [46].

If sufficient gamone (e.g., 10^4 U/ml for gamone 1, 10^2 U/ml for gamone 2) of one type is added to a large number of cells of the other type, virtually all the cells synchronously form homotypic conjugant pairs. Gamone is, therefore, a highly effective cell-union-inducing substance. Since both gamones have been purified and characterized, and gamone 2 has been chemically synthesized, these gamones will provide powerful tools for biochemical analyses of cell union.

Biochemistry and Ultrastructure of Cell Union Formation

Heterotypic and homotypic unions of *Blepharisma japonicum* are both induced by gamones in the same way and they are morphologically indistinguishable until a few hours after conjugant pairs are formed. These results indicate that the two unions are induced by the same mechanism and that the differences between them develop thereafter as the heterotypic union induces the subsequent processes of conjugation [47]. In homotypic pairs, therefore, it is as though conjugation is stopped at the stage of cell union formation.

This situation, so far unique to *Blepharisma*, provides an opportunity to investigate the mechanism of cell union formation, under the condition that it is not complicated by the occurrence of nuclear changes of sexual reproduction. If gamone is used to induce homotypic union, the induction of cell union is separated also from preconjugant interaction. In this way, induction of cell union can be separated from both preceding and subsequent processes and, therefore, "isolated". Otherwise, when a treatment is found to inhibit cell union, it is often difficult to decide whether it does so by directly affecting the mechanism of cell union, or by preventing the occurrence of preconjugant interaction. Similarly, any biochemical changes detected prior to cell union formation might be related to the nuclear change in conjugation and not to cell union. These difficulties, which have plagued the study of ciliate conjugation, are now eliminated.

Miyake and Honda [45] investigated the role of protein synthesis in the induction of homotypic union by gamone in *B. intermedium* (now *B. japonicum* [48]). They first found that the induction is completely inhibited by 10 μg/ml cycloheximide added at the beginning of gamone treatment. Cycloheximide at this concentration rapidly stops most of the ^{14}C-lysine incorporation into 10% TCA insoluble material (Fig. 10B). At lower concentrations, inhibition of the incorporation and inhibition of cell union are closely correlated (Fig. 10A). They concluded that cycloheximide inhibits cell union by inhibiting protein synthesis and, therefore, that protein synthesis is required for the induction of cell union. This conclusion is consistent with most of the work on the role of protein synthesis in ciliate conjugation [26,30,31], (for earlier work see [10]).

They also found that gamone increases amino acid incorporation. The increased incorporation begins about 5 min after the beginning of gamone treatment and continues for about 2 h until the face-to-face ciliary union is formed. The overall separation pattern of labelled proteins of the gamone-treated cells is distinctly different from that of the non-treated cells, if cells are labelled during the first 2 h (but not the following 2 h) of the gamone treatment. They concluded that gamone induces protein synthesis during the pre-union stage and that the accumulation of the induced proteins is a prerequisite to the formation of cell union. The same conclusion has been reached in *Paramecium* using a different method [10,18,49].

Gamone-induced proteins are of various kinds and many of them appear to be synthesized also in nontreated cells in smaller amounts. The results of the density gradient centrifugation and agarose chromatography of cell fractions of the labelled cells indicate that a large portion of the gamone-induced proteins is associated with lipid-containing particles larger than 10^6 daltons. Miyake and Honda [45] assumed that most of the gamone-induced proteins are membrane proteins and that both proteins and lipids participate in the formation of cell union. This assumption is supported by the finding that ficin and lipase (but not trypsin) strongly inhibit the formation of conjugant union in *Paramecium* even if they are applied immediately before the formation of the union [10,49]. A similar effect of lipase was reported also in *Oxytricha* [50].

They found, however, that the accumulation of the induced proteins is not sufficient for the induction of cell union. If 10 μg/ml cyclo-

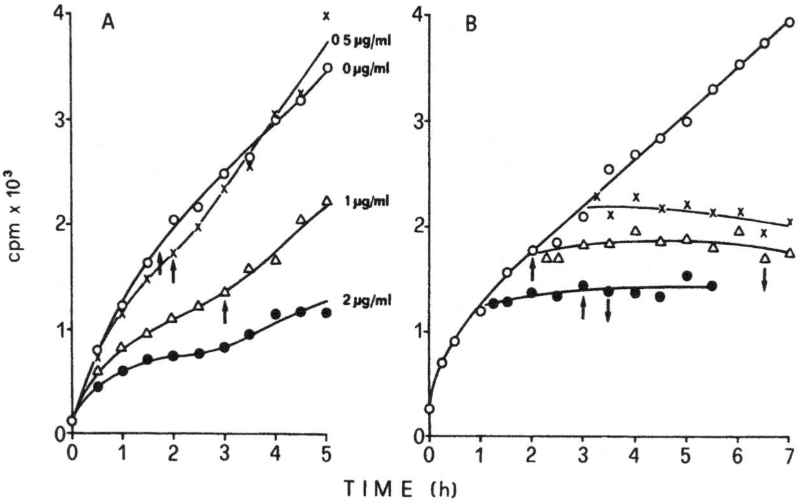

Fig. 10A,B. Effects of cycloheximide on ^{14}C-lysine incorporation and on the formation of cell union in mating type I cells of *Blepharisma japonicum* treated by gamone 2 (1.6×10^2 U/ml). Gamone 2 and ^{14}C-lysine were added at Time 0. (A) Cycloheximide of indicated concentrations were added at Time 0. (B) Cycloheximide (10 µg/ml) was added at 1 h (●——●), 2 h (Δ——Δ) and 3 h (x——x) after the beginning of gamone treatment. No cycloheximide o——o; appearance of pairs ↑; disappearance of pairs ↓. (From [45])

heximide is added shortly after some cells start forming ciliary union, new pairs continue being formed, although most of the protein synthesis is quickly blocked by the treatment (Fig. 10B). Cell union is eventually inhibited, but it is after 4.5 h that all cells separate. On the other hand, if gamone is removed at a comparable time, few pairs are newly formed and all pairs separate within 1/2 h, although the decrease of the incorporated radioactivity is relatively small during this time. These results suggest that, in addition to the accumulation of the gamone-induced proteins, a cycloheximide-insensitive process participates in the induction of cell union.

Work on *P. multimicronucleatum* [49] and *Tetrahymena thermophila* [51-55] suggests that RNA synthesis is needed for the formation of conjugant pairs. In the induction of homotypic union by gamone in *B. japonicum*, actinomycin D (50 µg/ml) added at the beginning of gamone treatment has little effect [41]. However, actinomycin D at this concentration does not affect ^{3}H-uridine incorporation both in gamone-treated and untreated cells [10,42]. Apparently, actinomycin D scarcely enters cells under the condition used. Gamone enhances the ^{3}H-uridine incorporation into 10% TCA insoluble material, if gamone is added to blepharismas suspended in a medium containing ^{3}H-uridine. However, if cells which have been incubated with ^{3}H-uridine are washed and then treated with gamone, no difference is detected in the increase of the incorporation between gamone-treated and untreated cells. Apparently, the enhanced ^{3}H-uridine incorporation in the first experiment is due to an enhanced uptake of ^{3}H-uridine [10]. In *B. japonicum*, therefore, the requirement of RNA synthesis for the induction of conjugant union has not yet been demonstrated.

At the united region of conjugant pairs in *B. japonicum*, cell membranes are juxtaposed with a distance of about 20 nm except at the vacuolated places and cytoplasmic bridges (Fig. 1) [47]. The borderline of the united cells becomes more indented during the first hours of pair formation suggesting an increase of the united surface during this period. In addition, there are two conspicuous structures at this region, (1) PACM (perpendicularly associated with the cell membrane) microtubules, and (2) saccules.

PACM microtubules emanate in bundles from many places on or near the cell surface and usually run at right or slightly obtuse angles, and seldom in parallel, to the cell surface. They are found in both heterotypic and homotypic pairs, but only at the united region of cells exposed to gamone of the other cell type. In heterotypic pairs they start degenerating 3 h after the formation of ciliary union and completely disappear during the next 3 h, whereas in homotypic union they persist for 18 h or longer [47].

Saccules are always present throughout the cytoplasm, but they are more abundant near the united surface in early stages of conjugation [47]. Elliot and Zieg [56] observed a similar increase of saccules in *T. thermophila* (then *T. pyriformis*) and thought that these saccules participate in cell union. Their assumption was criticized by Flickinger and Murray [52] based mainly on the fact that the saccule increase is scarcely inhibited by actinomycin D, which inhibits conjugation. Note, however, that the role of RNA synthesis in the formation of conjugant union in ciliates has not yet been fully established as discussed above.

If blepharismas are treated by gamone under the condition that their peristomial floor can not unite with the other cell (e.g. by isolation of cells), the peristomial floor bulges out, apparently due to the increase of cell surface, and both the formation of PACM microtubules and the accumulation of saccules are observed at the presumptive site of cell union [22,57] (Fig. 11).

These results suggest that PACM microtubules and saccules are devices for conjugant union. Miyake and Honda [45] pointed out that their $>10^6$ m.w. particles containing much of the gamone-induced proteins, might be those saccules and their fragments; and that the cycloheximide-insensitive process which they detected in the mechanism of induction of cell union, might be the transportation process of these saccules to the site of cell union by PACM microtubules which are also induced by gamone.

Based on all these results, a hypothesis on the mechanism of induction of cell union by gamone is constructed, somewhat extending those presented by Miyake and Honda [45] and Miyake [10,22]. During the preunion stage, gamone binds to its receptor, which is postulated to be on the peristomial floor of its target cell (Fig. 12A), and induces protein synthesis. Although most of the induced proteins are membrane proteins which are also synthesized in smaller amount in the cells not treated by gamone, gamone also induces new specific proteins. Some of these specific proteins provide the complementarity between AZM and antUMC so that they can stick to each other. This induces pairs united by the ciliary union. Meanwhile, the gamone-receptor binding induces PACM microtubules emanating from the cluster of gamone-bound receptors (Fig. 12B). Most of the induced proteins are packed in saccules, which are transported to the peristomial floor by PACM microtubules and incorporated into the cell membrane increasing the cell surface at this region. Some of the specific proteins, which are

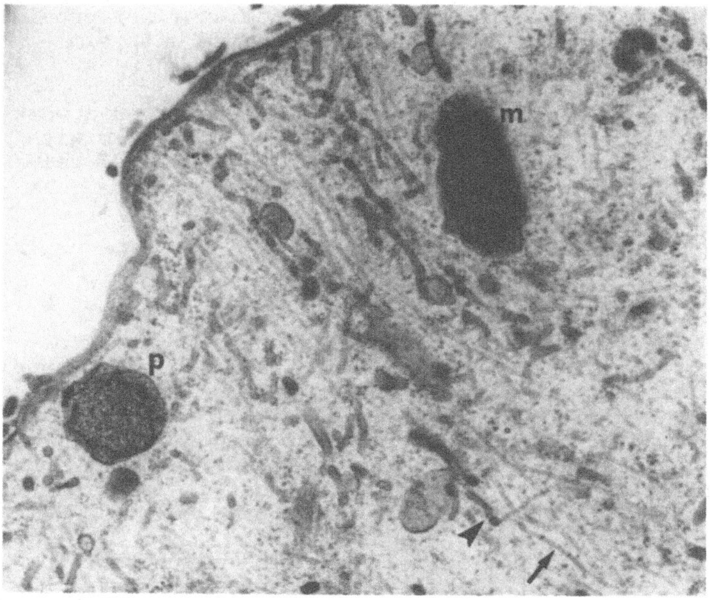

Fig. 11. Cross-section at the presumptive site of cell union (anterior part of the peristomial floor) of a doublet mating type II cell of *Blepharisma japonicum* treated by gamone 1 (10^4 U/ml) for 7 h at 25 $\pm 1^oC$. The cell is at the terminus of a homotypic chain of four doublet cells, thus having one free peristomial floor shown in this figure. *Arrow* PACM microtubule, *arrowhead* saccule, *p* pigment granule, *m* mitochondrion. For methods see [47]. x 21,000. (From [18])

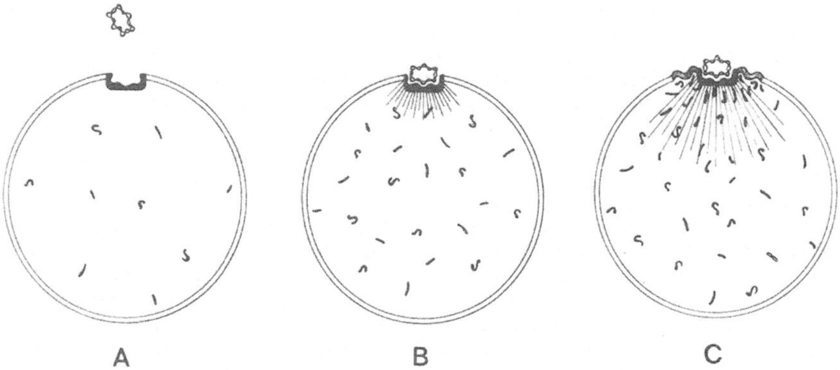

Fig. 12A-C. Diagrammatic illustration of a hypothesis on the mechanism of gamone-induced cell union. (A) Approaching of gamone molecules to their receptors (one of each is shown). (B) Gamone-receptor binding induces, *1* formation of PACM microtubules, and *2* increase of saccules. (C) The cell surface near the gamone-receptor binding is altered so that they can unite with a similarly altered surface. Receptors (*shaded*) are postulated to be in clusters at many places at the presumptive site of cell union. *Double circles* cell, *straight lines* PACM microtubules, *short cords* saccules

packed in saccules, are now on or outside the peristomial floor providing this region of the cell surface the capacity to unite with similarly altered surface (Fig. 12C).

Although the hypothesis is highly speculative, it will serve as a working hypothesis to build a more complete picture of the mechanism with which a chemical signal is translated into a specific pattern of cellular arrangement.

Induction by Heterotypic Cell Contact

In *Blepharisma japonicum*, heterotypic cell union induces a series of developmental processes of sexual reproduction. The first nuclear change is the swelling of the germinal nucleus about 2.5 h after the formation of ciliary union. The nucleus then enters meiosis I and completes meiosis II in about 15 h at 24°C passing through 16 stages (I - XVI) described in [58]. Since meiosis is the most conspicuous event throughout this period, the induction by heterotypic cell contact is often regarded as meiotic activation. This practice is adopted below.

The problem of how long the two types of cells must be in contact for meiotic activation was studied by surgically separating heterotypically-united cells with a fine glass needle at various times after the onset of ciliary union [59,60]. If the occurrence of meiosis was ascertained later in a separated cell, the cell was considered already activated at the time of operation. In this way it was found that cells are activated at 1-2 h after the formation of ciliary union, well before any morphological changes are detected in the nucleus. Therefore, the search for the mechanism of meiotic activation can be narrowed down to the first 1-2 h of heterotypic cell-cell contact.

Meiotic activation is usually all or none. However, if heterotypically united cells are separated shortly before activation, the germinal nucleus in some cells swells, but remains at this stage failing to undergo meiosis [61,62]. The finding of this "partial activation" indicates that meiotic activation consists of at least two steps.

The role of protein synthesis in meiotic activation was investigated by applying the method of surgical operation mentioned above to cycloheximide-treated pairs [60]. If heterotypic pairs are continuously exposed to 10 µg/ml cycloheximide from the stage of ciliary union, no cells are activated for at least 5 h. If such pairs are washed and incubated in an inhibitor-free medium, cells are activated with a delay corresponding to the duration of the cycloheximide treatment. If a 60 min cycloheximide treatment begins at 30 min after the pair formation, the delay of activation is nearly the same as that occurring when the treatment is applied at the time of pair formation. These results not only indicate that protein synthesis is required for meiotic activation, but also suggest that meiotic activation is based on the accumulation of the synthesized protein. It is assumed, therefore, that heterotypic cell contact induces and maintains the synthesis of a protein or proteins, whose accumulation to a certain threshold is required for meiotic activation [60].

Note the similarity of this conclusion to the one for cell union formation. In both, the accumulation of a factor to a certain threshold is considered to turn on a new developmental process. That is, a signal for a new developmental process is translated into and stored as an accumulable factor which works on a concentration basis. Since what is counted here is the total amount of the signal rather than the way with which the signal is given, the occurrence of the developmental process in question is relatively invulnerable to the fluctuation of the underlying processes. Such a mechanism might function as a stabilizer in complex developmental systems.

Meiotic activation does not occur at all in homotypic pairs which are morphologically indistinguishable from heterotypic pairs during the first few hours. Homotypic pairs, therefore, provide a good control in the study of molecular changes which occur in heterotypic pairs during meiotic activation. Although such a biochemical analysis is still to be carried out, the functional difference between the two types of cell union have been exploited in the study of meiotic activation as follows.

If doublet cells of one mating type are treated with the gamone of the other type, they unite side by side to form multicellular complexes consisting of a single type of cells (Fig. 2). As in homotypic pairs, meiotic activation does not occur. If, however, a cell of the other mating type unites at one end of a chain before it closes into a ring (Fig. 13), meiosis occurs in the heterotypically united cell and propagates through the chain [63].

Using this system, Miyake, Maffei and Nobili [59] examined how meiotic activation propagates through the chain. They prepared homotypic chains of various lengths and induced a heterotypic union at one end of each chain. After various times, they surgically separated all cells in the

Fig. 13. A heterotypic chain consisting of 7 red doublet cells of mating type II and one albino singlet cell of mating type I (at the right end of the chain) in *Blepharisma japonicum*. At the *lower right*, a part of another similar chain containing 4 doublet cells is seen. A homotypic chain of 7 type II cells was induced by a 5.5 h treatment with gamone 1 (10^4 U/ml). A type I cell was then induced to unite with this chain. Photograph of living unstained cells taken at 20 h after the type I cell united to the chain. Experiment at 25 ± 1°C. For materials and methods see [59]. x 110

chain. If the occurrence of meiosis was later ascertained in a separated cell, the cell was considered already activated at the time of operation (Fig. 14). In this way, they found that the propagation of meiotic activation is slower in longer chains and that the propagation slows down as it proceeds in the chain. The propagation seldom reaches the 6th doublet cell counted from the heterotypically-united end of a chain. In a long chain, such as shown in Fig. 13, a gradient of responding cells is established. Several cells closest to the heterotypic union undergo meiosis; next comes a partially activated cell; the rest of the cells in the chain remain inactivated. These results indicate that the signal for meiotic activation is something that is "diluted" along the chain, providing evidence that the signal is a substance that is transferable through united cells and requires a certain concentration to work.

That meiotic message can be transferred between united cells was also shown by inducing homotypic conjugation between activated and non-activated cells [61,64,65]. If an activated cell in a heterotypic pair is separated from the partner and forced to unite with a cell of the same mating type which has never united before, the latter undergoes meiosis although the two cells are homotypically uniting. Meiosis oc-

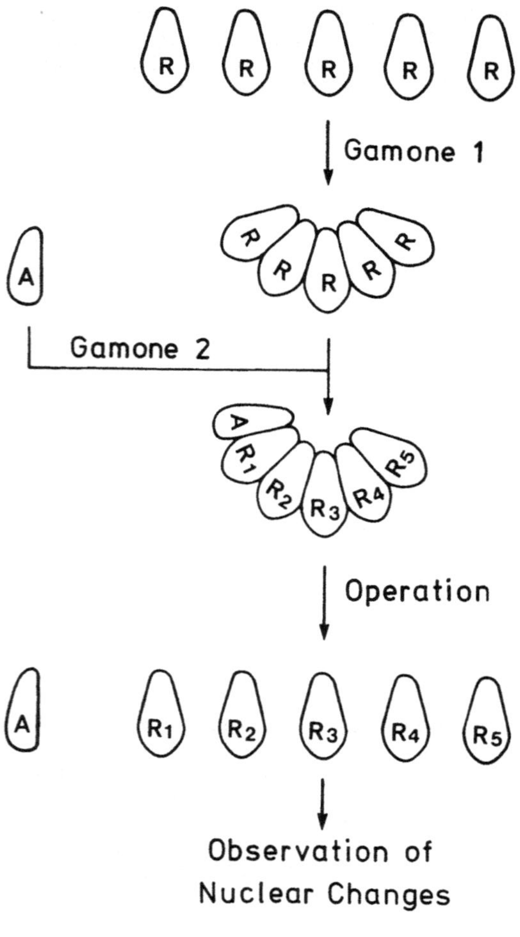

Fig. 14. Experimental design for the study of the cell-cell propagation of meiotic activation in *Blepharisma japonicum*. *A* Albino singlet cell of mating type I, *R* red doublet cell of mating type II. (From [18])

curs in the *induced* cells even if it is separated from the *inducing* cell after several hours of this homotypic association. An activated cell, therefore, can activate the cell with which it homotypically unites.

In such homotypic pairs, the *induced* cell begins meiosis a few hours later than the *inducing* cell. However, meiosis in the *induced* cell progresses faster than usual and the delay disappears by stage XI, indicating that the progress of meiosis in the *induced* cell is accelerated by the *inducing* cell. This suggests that, in addition to the factor for the initiation of meiosis, there is a factor (or factors) which controls the progress of meiosis. This assumption is consistent with the finding that cycloheximide blocks the progress of meiosis in this ciliate at stages I, II, III, IV, XI, and XII, indicating that the progress of meiosis through each of these stages requires at least one protein synthesized at or shortly before the respective stage [66,67].

Based on all these results, a hypothesis on the mechanism of meiotic activation is constructed (Fig. 15). Heterotypic cell contact induces and maintains the production of "master factor" through protein synthesis. This factor "charges" two intracellular components, X and Y, of which X requires a lower concentration of the master factor to be charged. Once charged, X produces, through protein synthesis, "meiosis factor 1" which induces the swelling of the germinal nucleus. When charged, Y produces meiosis factors 2, 3, 4, ---, n in a sequence through protein synthesis. Each of these factors induces the progress of meiosis at a particular step. In this scheme, both X and Y must be charged for the occurrence of meiosis. If only X is charged by the limitation of the available amount of the master factor, meiosis stops at the stage of swelling (stage II), resulting in partial activation.

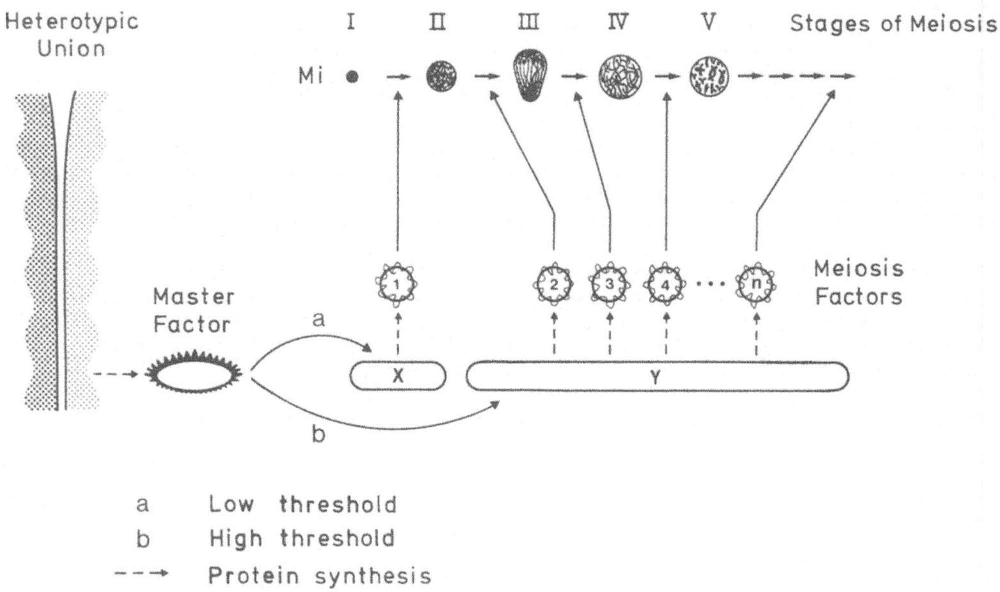

Fig. 15. A hypothesis on the mechanism of meiotic activation in *Blepharisma japonicum*. *Mi* Germinal nucleus (micronucleus). *X* and *Y* hypothetical cellular components. Stages of meiosis *II*, *III*, *IV* and *V* roughly correspond to leptotene, zygotene, pachytene and diplotene, respectively

Although the hypothesis is highly speculative, it will serve as a working hypothesis to build a more complete picture of the mechanism by which a contact between two types of cells induces a series of developmental processes.

Concluding Remarks

Conjugation of ciliates begins with the interaction between two types of cells. Within a few hours cells gain the capacity to unite at the peristomial region and tightly unite to form conjugant pairs. This bicellular condition lasts for several hours in *Paramecium tetraurelia* and for about one day in *Blepharisma japonicum*. Cells then separate to return to the unicellular condition. The system is simple, because there are only two cells, which come together, unite for a while, and then separate.

Conjugation can be induced, regularly and synchronously, in a large number of cells by mixing cells of complementary mating types, or by treating one type of cells with gamone of the other type (in *Blepharisma*), or by a chemical treatment (in *Paramecium*). Because of such a high controllability and the phenomenal simplicity, ciliate conjugation is amenable to biochemical analysis, as shown in the examples described above.

The uniting of cells induces meiosis and other developmental processes of sexual reproduction. Through these processes, the old somatic nucleus is replaced by the product of mitotic divisions of the fertilized nucleus. This phenomenon resembles the replacement of old somatic cells (individuals of older generation) by mitotic products of the fertilized cell (individuals of younger generation) in multicellular organisms. This might be the reason why so many basic elementary processes of multicellular development are found in ciliate conjugation. In this article, only two of them have been dealt with in detail; (1) uniting of cells at the particular region of the cell surface, and (2) induction of new developmental processes by a contact between two types of cells. Other examples which have been touched upon are programmed termination of cell union, cell-cell interaction of various types, chemoattraction, meiosis, fertilization and nuclear differentiation.

It is the combination of these three features, (1) the phenomenal simplicity, (2) the high controllability, and (3) the concentration of many basic developmental processes, that makes conjugation of ciliates a promising system for biochemical analyses of developmental phenomena including multicellular morphogenesis.

Acknowledgements. This work was supported by the Deutsche Forschungsgemeinschaft.

References

1. Lwoff A (1950) Problems of morphogenesis in ciliates. John Wiley, New York
2. Sonneborn TM (1963) In: Allen JM (ed) The nature of biological diversity. McGraw-Hill, New York , pp 165-221
3. Tartar V (1967) In: Chen TT (ed) Research in protozoology, vol I. Pergamon Press, Oxford, pp 1-116
4. Nanney DL (1980) Experimental ciliatology. John Wiley, New York
5. Aufderheide, K, Frankel J, Williams N (1980) Microbiol Rev 44:252-302

6. Frankel J, Nelsen EM (1981) Philos Trans R Soc London Ser B 295:525-538
7. Olive LS, Blanton RL (1980) J Protozool 27:293-299
8. Grell KG (1973) Protozoology. Springer, Berlin Heidelberg New York
9. Sonneborn TM (1954) Caryologia Suppl 307-325
10. Miyake A (1981) In: Levandowsky M, Hutner SH (eds) Biochemistry and physiology of protozoa, vol IV. Academic Press, London New York, pp 125-198
11. Gibbins JR, Tilney LG, Porter KR (1969) J Cell Biol 41:201-226
12. Nobili R (1967) Monit Zool Ital 1:73-89
13. Raikov IB (1972) In: Chen TT (ed) Research in protozoology, vol IV. Pergamon Press, Oxford, pp 147-289
14. Bleyman LK (1971) In: Cameron IL, Padilla GM, Zimmermann AM (eds) Developmental aspects of the cell cycle. Academic Press, London New York, pp 67-91
15. Takagi Y (1970) Jpn J Genet 45:11-21
16. Nanney DL (1974) Mech Ag Dev 3:81-105
17. Nanney DL (1977) In: Reissig JL (ed) Microbial interactions, vol III. Chapman & Hall, London, pp 353-397
18. Miyake A (1978) Curr Top Dev Biol 12:37-82
19. Miyake A, Bleymann LK (1976) Genet Res 27:267-275
20. Miyake A (1968) J Exp Zool 167:359-380
21. Ducibella T, Ukena T, Karnovsky M, Anderson E (1977) J Cell Biol 74:153-167
22. Miyake A (1981) In: O'Day DH, Horgen PA (eds) Sexual interactions in eukaryotic microbes. Academic Press, London New York, pp 95-129
23. Luporini P, Dallai R (1981) J Cell Sci 51:1-13
24. Luporini P, Brachi P, Esposito F (1979) J Cell Sci 39:201-213
25. Lueken WW, Breer H, Hartkemeyer M (1981) J Protozool 28:414-417
26. Dini F, Miyake A (1982) J Exp Zool 220:153-162
27. Ricci N (1981) In: O'Day DH, Horgen PA (eds) Sexual interactions in eukaryotic microbes. Academic Press, London New York, pp 319-350
28. Hiwatashi K (1981) In: O'Day DH, Horgen PA (eds) Sexual interactions in eukaryotic microbes. Academic Press, London New York, pp 351-378
29. Goodenough UW (1980) In: Gooday GW, Lloyd D, Trinei APJ (eds) The eukaryotic microbial cell. Cambridge Univ Press, London, pp 301-328
30. Garfinkel MD, Wolfe J (1981) Exp Cell Res 133:317-324
31. Ron A, Suhr-Jessen PB (1981) Exp Cell Res 133:325-330
32. Miyake A, Beyer J (1973) Exp Cell Res 76:15-24
33. Heckmann K, Siegel RW (1964) Exp Cell Res 36:688-691
34. Miyake A, Nobili R (1974) J Protozool 21:584-587
35. Esposito F, Ricci N, Nobili R (1976) J Exp Zool 197:275-282
36. Miyake A, Beyer J (1974) Science 185:621-623
37. Braun V, Miyake A (1975) FEBS Lett 53:131-134
38. Kubota T, Tokoroyama T, Tsukuda Y, Koyama H, Miyake A (1973) Science 179: 400-402
39. Tokoroyama T, Hori S, Kubota T (1973) Proc Jpn Acad 49:461-463
40. Entzeroth M, Jaenicke L (1981) Z Naturforsch 36c:180-182
41. Beyer J, Miyake A (1973) In: Progr Protozool, 4th Int Congr Protozool, Clermont-Ferrand. Abstr p 42
42. Miyake A (unpublished)
43. Miyake A (1968) Proc Jpn Acad 44:837-841
44. Honda H, Miyake A (1975) Nature (London) 257:678-680
45. Miyake A, Honda H (1976) Exp Cell Res 100:31-40
46. Honda H, Miyake A (1976) Dev Biol 52:221-230
47. Bedini C, Lanfranchi A, Nobili R, Miyake A (1978) J Cell Sci 32:31-43
48. Hirshfield HI, Isquith IR, Dilorenzo AM (1973) In: Giese AC (ed) Blepharisma. Stanford Univ Press, Stanford, Calif, pp 304-332
49. Miyake A (1969) Jpn J Genet 44 Suppl 388-395
50. Ricci N, Esposito F, Nobili R (1975) J Exp Zool 192:343-348
51. Tyler L, Wolfe J (1972) J Protozool 19 Suppl 42-43
52. Flickinger CJ, Murray RL (1974) Cell Tissue Res 153:357-364
53. Allewell NM, Oles J, Wolfe J (1976) Exp Cell Res 97:394-405
54. Ron A, Horovitz O (1977) Experientia 33:1146-1149
55. Ofer L, Mercazi M, Loyter A (1977) Biol Cell 29:23-30

56. Elliott AM, Zieg RG (1968) J Cell Biol 36:391-398
57. Miyake A, Nobili R, Lanfranchi A, Bedini C (cited in [10,18,22])
58. Miyake A, Heckmann K, Görtz H-D (1979) Protistologica 15:473-486
59. Miyake A, Maffei M, Nobili R (1977) Exp Cell Res 108:245-251
60. Miyake A, Tulli M, Nobili R (1979) Exp Cell Res 120:87-93
61. Miyake A (1981) Zool Mag 90:629
62. Miyake A, Takagi Y, Heckmann K (in preparation)
63. Miyake A (1975) Science 189:53-55
64. Santangelo G, Nobili R (1981) J Exp Zool 218:121-126
65. Miyake A, Heckmann K (in preparation)
66. Friedl E, Miyake A, Heckmann K. In: Progr Protozool 6th Int Congr Protozool, Warsaw. Abstr p 103
67. Friedl E, Miyake A, Heckmann K (in preparation)

Molecules Involved in Morphogenesis in the Multicellular Stage of *Dictyostelium discoideum*

K. L. Williams[1]

Morphogenesis involves the formation of an organism by proportioning of cells in a reproducible way to make a structure of recognisable shape. According to this definition, morphogenesis is limited to multicellular organisms and it is largely, although not uniquely (e.g. Myxomycetes, Zusman 1980), a phenomenon seen in eukaryotes rather than prokaryotes. Here I summarise recent studies on morphogenesis in one of the simplest multicellular eukaryotes, the cellular slime mould *Dictyostelium discoideum*. *D. discoideum* is so simple that during the feeding and multiplication parts of the life cycle the amoebae, which are about 10 μ in diameter, are solitary. The amoebae grow readily by eating bacteria (e.g. *Escherichia coli, Klebsiella aerogenes*) or on a partially or fully defined medium (Sussman and Sussman 1967, Franke and Kessin 1977). *D. discoideum* is normally a haploid with seven small chromosomes, each with approximately twice the DNA content of *E. coli*, although diploid strains are easily maintained (Brody and Williams 1974). There are sexual (see Newell 1978) and asexual multicellular stages but only the asexual cycle is discussed here. The best studied developmental phase is the transition between the unicellular and multicellular stages, which involves aggregation of starving amoebae in response to self generated cyclic AMP gradients (Loomis 1975, Gerisch 1982). Depending on the strain used and the conditions, between about ten cells and one hundred thousand cells form a single, polarised, aggregate which has a definite anterior tip (organiser) region. The aggregate, composed of individual cells (it is not a plasmodium), forms a slug which moves along gradients of light or heat inside a self-made slime sheath that is left behind as a trail. This slug stage is most amenable to study as the cell types are differentiated in a linear array (Fig. 1). Slug cells are easily disaggregated by vortexing and groups of cells can be readily transplanted from one slug to another (Durston 1976, Smith and Williams 1980, MacWilliams 1982). Depending on the genotype and environmental conditions, the migration stage is brief or prolonged (Newell et al. 1969, Smith and Williams 1980), but finally the asexual fruiting body, with relatively size invariant proportions of spore, stalk and basal disc cells, is formed (Stenhouse and Williams 1977). The organism can complete the asexual life cycle in the haploid or diploid state, and the genetic system used is independent of this cycle, so morphogenetic mutants are easily studied (Williams and Newell 1976). Two points which simplify morphogenetic studies are (1) that the organism obtains its cells by aggregation of like cells, rather than by a series of hierarchical cell divisions and (2) the fact that only three cell types exist in the fruiting body.

1 Max-Planck-Institut für Biochemie, D-8033 Martinsried b. München, FRG

33. Colloquium - Mosbach 1982
Biochemistry of Differentiation and Morphogenesis

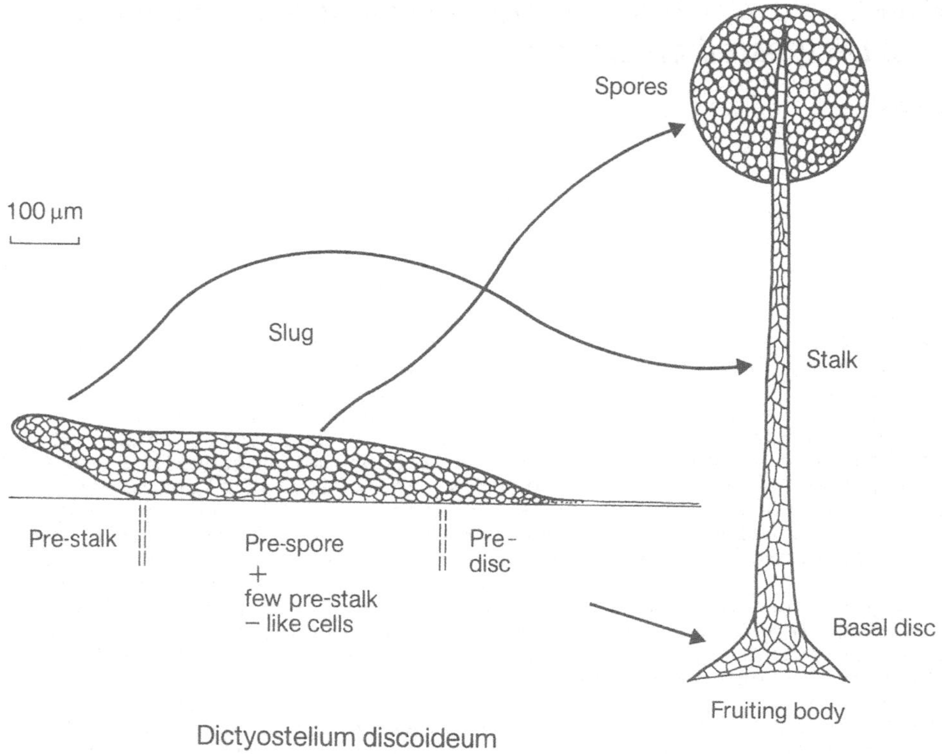

Fig. 1. Relationship between the cells in the migrating slug and asexual fruiting body of *D. discoideum*. Average-sized slugs and fruiting bodies are about 1 mm long and contain $\sim 10^4$ cells. Extremely small slugs and fruiting bodies (a few hundred cells) and large slugs and fruiting bodies (more than 10^5 cells) have similar but not identical proportions of spore: stalk: basal disc cells (Stenhouse and Williams 1977). An "average" fruiting body has about 20% stalk cells, 2% basal disc cells and 78% spores (Stenhouse and Williams 1981). The fate of the prestalk-like cells in the prespore region is not known (Sternfeld and David 1981a)

Biochemical studies on morphogenesis often centre on diffusible morphogens (Gierer and Meinhardt 1972, Maden 1982, Schaller 1982), cell surface molecules (Friedlander 1979, 1980), or extracellular structures (Manasek 1976, Chiquet et al. 1981). In this report recent experiments on *D. discoideum* are described which have started to illuminate the role of these biochemical processes during its morphogenesis. Two other promising approaches using mutants and evolutionary studies are briefly considered.

Diffusible Morphogens

It is believed that morphogenesis is mediated by gradients of diffusible molecules (Child 1941, Wolpert 1971, Gierer and Meinhardt 1972, MacWilliams and Bonner 1979). The difficulty of finding these elusive molecules has led to disillusionment. Not only are they expected to be present in tiny amounts, but in some cases the gradient may be pulsatile rather than static, as has been elegantly demonstrated for the aggregation of *D. discoideum* amoebae to 3'5' cAMP (Gerisch 1982). Furthermore, to understand morphogenesis requires more than identification of the morphogen; for example morphogenesis in plants is still poorly understood despite the discovery of auxin many years ago (Rubery 1981). Subsequently one must discover the details of morphogen action, which cells make it, which cells have receptors, how the binding of morphogen is translated into morphogenetic information (e.g. move to a certain place, differentiate into cell type A), how the sensing system to provide size regulation operates. In contrast to the early ideas of a simple gradient (e.g. Wolpert 1971), it seems not only possible but likely that there are numerous components to the system (Meinhardt 1982). For example *D. discoideum* may have three (interconnected?) components: systems for polarity specification, proportion regulation and the transition between different developmental stages (Table 1). This is not to imply that the situation is hopeless, but it is important to try to come to terms with the complexity of the system.

What follows is highly speculative but I shall outline some likely candidate molecules in *D. discoideum* morphogenesis and then show how it may be possible using mutants to construct pathways of information flow. Since almost nothing is known about the regulation of the basal disc cells (Sussman and Schindler 1978, Stenhouse and Williams 1977, 1981) or stalk-like cells in the prespore region (Sternfeld and David 1981a) these cell types will not be considered (Fig. 1).

As is true in *Hydra* (Schaller 1982), morphogens seem to play an important role in the morphogenesis of *D. discoideum*. In *D. discoideum* at least four self-produced low molecular weight substances are involved in morphogenesis. These are 3'5' cyclic AMP, cAMP (Durston and Vork 1979, Matsukuma and Durston 1979), ammonia (Schindler and Sussman 1977), Differentiation-Inducing Factor, DIF (Gross et al. 1981) and Slug Turning Factor, STF (Fisher et al. 1981; DIF and STF are not yet chemically characterized. There are probably several other substances involved in morphogenesis, in particular compounds involved in the transition between the migration and fruiting modes of development (Sussman et al. 1978). Is this too many? Almost certainly not, particularly if different molecules are involved with different aspects of morphogenesis or if there is redundancy so that several molecules are doing basically the same thing.

Three (cAMP, ammonia, DIF) of the four putative morphogens have been discussed elsewhere recently (Gross et al. 1981), so their possible involvement in morphogenesis will be only briefly considered. However an attempt will be made to determine on which part(s) of the morphogenetic sequence they act: polarity specification, proportion regulation and transition between stages (Table 1). DIF and ammonia may be involved in a two component system controlling proportioning of prestalk and prespore cells (Gross et al. 1981). DIF, a low molecular weight (1000-2000), hydrophobic compound may be an activator of prestalk cell formation, while ammonia may be an inhibitor of prestalk cell formation (Gross et al. 1981). Ammonia may also be involved in maintaining the migration stage or inhibiting the transition between the migration and culmination stages of development (Schindler and Sussman 1977). Cyclic AMP is probably required for differentiation of both stalk and spore cells so it is probably not involved in cell proportioning per se (reviewed in Gross et al. 1981). However, it has been proposed that the tip (organiser region) is the source of cAMP waves to which other cells in the slug move (Rubin and Robertson 1975, Rubin 1976, Durston and Vork 1979). Hence cAMP may be an activator in the specification of polarity. STF may have an antagonistic role

Table 1. Morphogens in *Dictyostelium discoideum*

	Aspect of morphogenesis affected[d]		
	Polarity	Proportions	Transitions
NH_3	Tip[a] inhibitor?	Stalk[b] inhibitor	Slugger[a]
cAMP	Tip[b] activator	-	-
STF	Tip[c] inhibitor	Not known	Slugger[c]
DIF	Not known	Stalk[b] activator	Not known

[a]Sussman and Schindler (1978), [b]Gross et al. (1981), [c]this volume.

[d]Morphogenesis is divided into three components: *polarity*, the establishment and maintenance of the tip (organizer); *proportions*, the establishment of the correct proportions of prestalk, prespore (and predisc) cells; *transitions*, control of the change between migration and culmination modes of development

to cAMP as will be explained shortly. If this is correct, morphogenesis in *D. discoideum* involves at least two activator/inhibitor systems, one (cAMP/STF) to specify polarity, the other (DIF/ammonia) to specify the proportions of the stalk and spore cells (Meinhardt 1982).

STF - a Low Molecular Weight Factor Affecting Slug Orientation

STF was discovered as a low molecular weight (< 500 daltons) compound which at high uniform concentrations disoriented phototaxis and thermotaxis by slugs of *D. discoideum*. Slugs were repelled by a gradient of STF when migrating in the dark and STF release was shown to be altered in the light. Since STF is the only known molecule to which slugs orient, slug behaviour (phototaxis and thermotaxis) was explained in terms of modulation of STF release (Fisher et al. 1981).

STF, which is released by developing aggregates, can be extracted from aqueous solution with pentane and hence it is hydrophobic. STF activity (i.e. compound(s) causing disorientation of slugs) has also been found in yeast extract and 40°-60°C distillate of crude petroleum. The hydrophobic nature of STF suggests it may be similar to DIF, although in preliminary collaborative experiments with Dr. R. Kay (ICRF Mill Hill London U.K.) it was found that STF did not copurify with DIF, nor did a petroleum distillate extract with high STF activity have any DIF activity. Tentatively it can be concluded that STF and DIF are different molecules, although they may be closely related.

A Role for STF in Morphogenesis?

a) Is STF a tip inhibitor? or "Do slugs turn by respecifying the position of their tips?" Although it has been claimed that slug turning is mediated by a differential change in speed at the tip (Poff and Loomis 1973), this is not correct (Smith et al. 1982), nor does STF itself induce a change in speed (Smith et al. 1982). Slugs orient chemotactically away from STF (Fisher et al. 1981) so the simplest explanation for a turn is direct chemotaxis. Alternatively (or in addition) it seems possible that STF may lead to tip respecification. It seems likely that slug movement involves chemotaxis to cAMP (Durston and Vork 1979, Sternfeld and David 1981a) and that the tip is the pacemaker and source of cAMP waves (Rubin 1976). Since we believe that STF may feed into the same sensory transduction pathway as cAMP (see Fig. 2), it is possible that STF antagonizes the pulsing of cAMP. This would result in those cells away from the high STF concentration taking over as pacemakers due to their higher cAMP signalling rate.

b) A role for STF in proportion regulation? We know that light and temperature changes influence the proportions of spore and stalk cells (see below). STF is a common intermediate in the phototaxis and thermotaxis pathways and there is some preliminary evidence that the spore:stalk ratio is altered in phototaxis mutant HU120 (Williams, unpublished). Conversely the proportion mutant HS3 of MacWilliams (1982) is defective in phototaxis (Williams unpublished). These findings make it clear that behavioural (phototaxis, thermotaxis) and proportioning pathways share common components although it is not yet certain that STF itself has a role in proportion regulation (Fig. 2).

c) STF inhibits the transition between migration and culmination. STF is a "slugger" compound in that it prolongs migration of strains that normally construct fruiting bodies after a short period of migration (Fisher, Mahurin and Williams, unpublished). Both ammonia and STF cause prolonged migration, and "slugger" mutants are known which have alterations in ammonia sensitivity (strain KY3, Newell and Ross 1982) and STF levels (strain NP84, Fisher and Williams, unpublished). Whether this is coincidental or the effect occurs via a common pathway, is not yet clear.

Pathways of Information Flow in Morphogenesis

The roles of the four diffusible morphogens proposed here (Table 1) are highly speculative and other workers have different views. For example Sussman and Schindler (1978) have proposed that ammonia, cAMP and the slime sheath together control all three aspects of morphogenesis considered here. The important point is to establish a framework within which these differing hypotheses can be evaluated. This is best attempted using a genetic approach and we have commenced this task (Fig. 2). There are now several partially characterized mutations which were isolated on the basis of altered phototaxis or morphogenesis. These mutations make it clear that morphogenesis shares common components with slug behaviour (Fig. 2). Furthermore there are a number of environmental factors which can be inserted into the scheme shown in Fig. 2, that have been until now unconnected observations.

These factors affect all three aspects of morphogenesis considered here. Oxygen has been implicated in polarity specification (Sternfeld and David 1981b). Light (Bonner and Slifkin 1949, Stenhouse and Williams

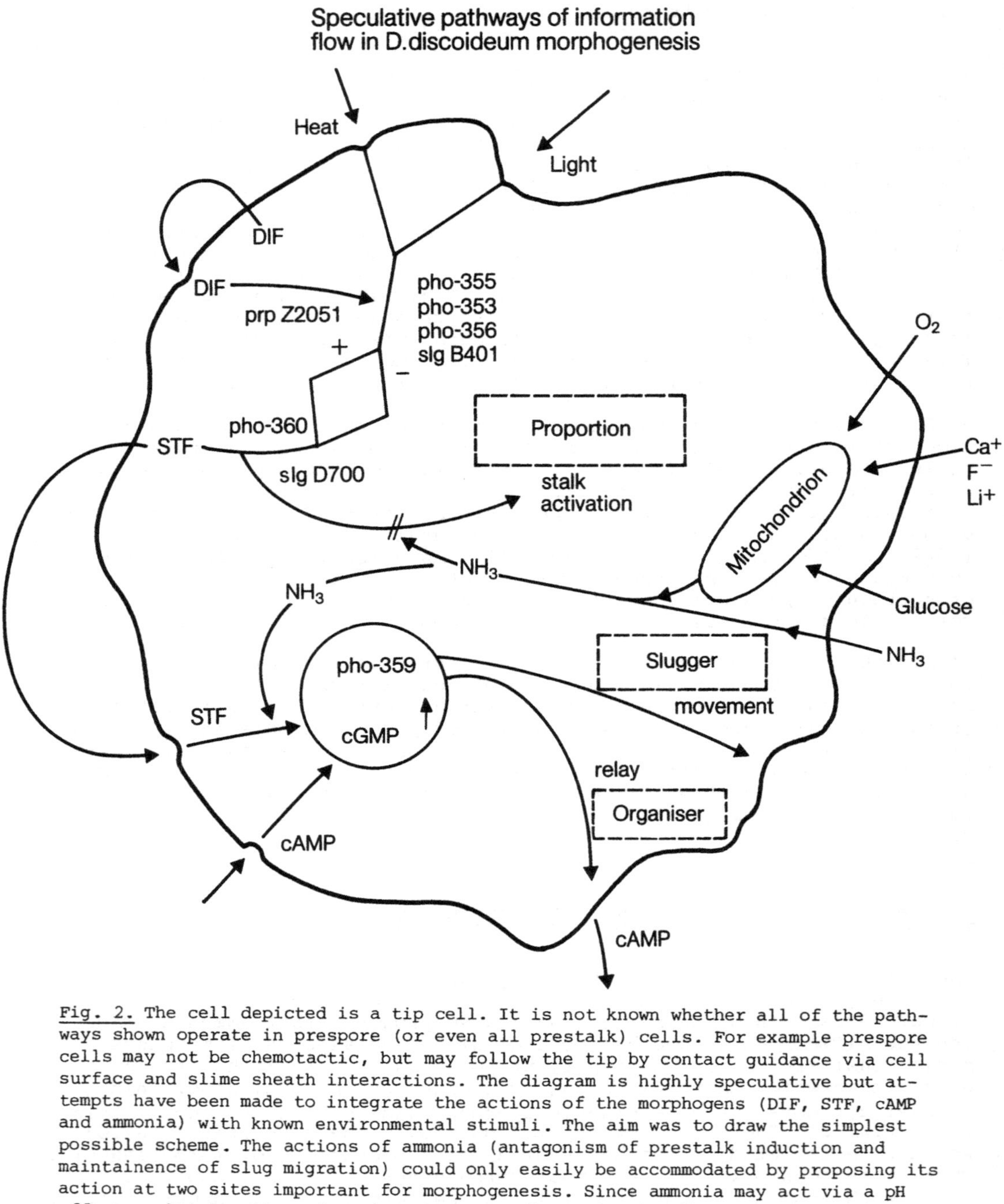

Fig. 2. The cell depicted is a tip cell. It is not known whether all of the pathways shown operate in prespore (or even all prestalk) cells. For example prespore cells may not be chemotactic, but may follow the tip by contact guidance via cell surface and slime sheath interactions. The diagram is highly speculative but attempts have been made to integrate the actions of the morphogens (DIF, STF, cAMP and ammonia) with known environmental stimuli. The aim was to draw the simplest possible scheme. The actions of ammonia (antagonism of prestalk induction and maintainence of slug migration) could only easily be accommodated by proposing its action at two sites important for morphogenesis. Since ammonia may act via a pH effect, multiple sites of action seem more plausible than for molecules such as STF, DIF or cAMP which probably act via specific receptors. Key features are the proposed convergence of behavioural pathways (phototaxis, thermotaxis) with the proportioning pathway. Full details of the behavioural pathway, which involves an adaptor controlling sign reversals depending on signal strength, will be described elsewhere (Fisher and Williams, in prep.). Signals of opposite sign from the putative STF receptor and cAMP receptor are proposed to converge, possibly at the level

1977), heat (Bonner and Slifkin 1949, Farnsworth 1975, Stenhouse and Williams, unpublished), cations (Li^+, Ca^{++}) and anions (F^-) (Maeda 1970, Ishida 1982) and nutritional status (Garrod and Ashworth 1972, Forman and Garrod 1977) all alter the proportions of spore and stalk cells, while factors affecting transitions between developmental stages include ionic strength, overhead light, humidity (Newell et al. 1969) and a low molecular weight anionic compound that is partially mimicked by propionate (Sussman et al. 1978).

Genetic Analysis

With the exception of studies on *Drosophila* (Yund and Germeraad 1980) and the yeast mating type locus (Herskowitz et al. 1980) genetic analysis has not made a major impact on our understanding of morphogenesis. A general problem is that mutations affecting developmental processes often lead to sterility and this makes genetic analysis difficult. Moreover when one deals with a complex organism, making many tissues, it becomes impossibly difficult to analyse morphogenetic mutations which are usually pleiotrophic.

Genetic analysis of *D. discoideum* uses the parasexual cycle involving fusion of two haploids to form a diploid and subsequent haploidisation. Two points are very important. First the pathway to asexual fruiting body formation is entirely irrelevant to the genetic system (i.e. even non-aggregating mutants are easily crossed), and second, diploids formed from two haploid strains can be passaged through the entire asexual development cycle so that complementation tests are easily conducted. This and the ease of making isogenic diploids (Welker and Williams 1980) has led to not only complementation tests but also experiments in which ploidy specific effects have been analysed (Spiegel and Cox 1980, Stenhouse and Williams, 1981). The genetic system is becoming well established with over 100 loci placed on the chromosomes (Loomis 1980, Williams and Welker, in prep.) and approximately 50 loci ordered with respect to each other (Welker and Williams 1982a). The availability of a transformation system in *D. discoideum* would greatly facilitate studies at the molecular level. We have recently identified a plasmid which almost certainly codes for heavy metal resistance in *D. discoideum* and which we are using in transformation studies (Metz, Ward, Welker, Williams, unpublished). It should not be long before fine structure

---

of cGMP modulation (Gerisch 1982). Subsequently signals lead to a movement response (slugger?) and signal relay (organiser?). The steps between convergence of thermotactic and phototactic pathways and the sign reversal adaptor are defined by phototactic (and thermotactic) mutations *pho*-353, *pho*-355, *pho*-356 (Fisher and Williams 1981, 1982). The proportioning mutation *prpZ*2051 (MacWilliams 1982) and 'slugger' mutation *slgB*401 (Newell and Ross 1982) have also been localised in this region by us (Dohrmann and Williams unpublished). The mutation *pho*-360 defines the region between the sign reversal adaptor and STF production on the basis of its phenotype with respect to phototaxis and thermotaxis (Fisher and Williams, in prep.). The "slugger" mutation *slgD*700 (Newell and Ross 1982) also lies in this region (Dohrmann and Williams, unpublished). The mutation *pho*-359 is placed after convergence of signals from the putative STF receptor and cAMP receptor as its phenotype is impairment of phototaxis and thermotaxis in slugs and amoebal chemotaxis to cAMP.

The channeling of O_2 (Sternfeld and David 1981b), ionic effects (Maeda 1970, Ishida 1982) and nutritional effects (Garrod and Ashworth 1972) via mitochondrial metabolism into ammonia regulation is mainly to stimulate argument

studies using cloned developmental genes are possible, and such techniques in combination with classical genetic analysis should lead to the elucidation of morphogenetic pathways. A recent example of successful fusion of genetics and biochemistry in the study of a complex trait is the analysis of sensory transduction in bacteria (*E. coli* and *S. typhimurium* Koshland 1979), and our approach to sensory behaviour and morphogenesis in *D. discoideum* is similar (Fig. 2).

The Cell Surface and Development

It is widely believed (e.g. McMahon 1973) that the cell surface is involved somehow in morphogenesis. There is a hint that cell contact is involved in polarity specification from studies on microtubule inhibitors colchicine (O'Day and Durston 1978), CIPC (White et al. 1981, Welker 1982), thiabendazole, cambendazole (Welker and Williams 1982b) and coumarin (Welker and Williams 1982b), all of which induce multiple tips in *D. discoideum* slugs. These compounds cause rounding up of cells and may therefore exert their effects by disrupting surface receptors (Welker and Williams 1982b). Cytochalasin B (a microfilament inhibitor) causes rounding of vegetative cells (Jones et al. 1977) but it is yet to be tested on slugs.

Gross et al. (1981) have argued that cell contact is not required for cell proportioning in *D. discoideum*, since they can induce either stalk cell or spore cell differentiation under certain conditions in solitary cells with only added diffusible factors. If this finding is also true in normal morphogenesis, cell surface involvement in proportion regulation is restricted to the receptors for the diffusible morphogens. There is indirect evidence for this proposal (Stenhouse and Williams 1981).

The cells must be correctly positioned and be sufficiently co-ordinated with each other to undergo the movements involved in morphogenesis, for example, the change in shape from slug to mature fruiting body. Cell contact molecules with similar roles to those characterized in aggregating amoebae (Müller and Gerisch 1978) are found on the surface of slug cells (Parish et al. 1978, Geltosky et al. 1979, Ochiai et al. 1982). Little is known about the distribution of these putative cell contact molecules in the slug but at least one, a cell surface glycoprotein gp150, is not unique to a particular cell type (Geltosky et al. 1979). However, gp150 may be involved more with prespore than prestalk cell adhesion (Lam et al. 1981). If so, cell surface differences could play a role in cell sorting out in the slug.

There are likely to be cell type specific surface molecules. The challenge is to discover and characterize these molecules. A search has been made, using dissected slugs, to find cell type specific plasma membrane differences and five such membrane molecules which bind wheat germ agglutinin have been described (West and McMahon 1979). Three are apparently prestalk cell specific and two are apparently prespore cell specific. However the resolution of this technique does not allow a confident statement that these molecules are truly cell type specific.

In principle, an immunological approach to the question of cell type specificity is much more powerful. The recent development of methods for obtaining hybridomas producing monoclonal antibodies (Milstein and Lennox 1980) should revolutionise these studies. Monoclonal antibodies have been used very effectively, particularly in the study of subpopulations of lymphocytes and their relationship to other cell types (Williams 1980). Of particular importance may be the finding that the major cell surface antigens of lymphocytes are not ubiquitous antigens, but antigens of restricted specificity (Williams 1980). This raises the possibility that each differentiated cell type carries at its surface a very clear message as to its type. The slug of *D. discoideum* provides a simple system in which this hypothesis can be tested in a developing multicellular organism. We have started to isolate and characterize monoclonal antibodies which recognize the cell surface of slug cells. As well as providing markers to specific cell types, these antibodies should tell us how many different cell types there are in the slug and their relationship to each other e.g. are the prestalk cells (at the front of the slug) and predisc cells (at the rear of the slug) similar, identical or unlike each other? We might also discover what an organiser region (the tip in the case of *D. discoideum*) means in cellular and molecular terms.

Although our studies are only just beginning it is clear from the hybridomas isolated and partially characterized, that two monoclonal antibodies which recognize single cell surface antigens on the basis of immunoblotting are not uniformly distributed (Krefft Voet and Williams, unpublished). One is specific to prespore cells (Fig. 3) and the second is found on the surface of spores. Antibodies from several other hybridomas recognise multiple antigens on the surface of both vegetative amoebae and slug cells, but the patterns of recognition are not necessarily the same in both classes of cells. These results provide a hint that differentiating cells may indeed have rather specialized surfaces.

The extent to which these monoclonal antibodies define cell surface molecules involved in sorting out between prestalk and prespore cells (Garrod and Nicol 1981) and cells of different slime mould species (Garrod et al. 1978, Gregg et al. 1982) remains to be established. The role of sorting out of different cell types in patterning and morphogenesis has been discussed elsewhere (Sternfeld and David 1981a, Durston and Vork 1979, Gross et al. 1981). Cell sorting is probably important in the maintenance and possibly formation of the pattern of prestalk cells at the front and prespore cells at the rear of the slug (Fig. 3). Whether proportion regulation in *D. discoideum* is positional or non-positional is still a hotly debated question (Tasaka and Takeuchi 1981, Gregg and Davis 1982, Meinhardt 1982).

Extracellular Structure - What Does the Slime Sheath Do?

As we have already observed, slugs have a definite polarity - under normal circumstances the tip of the slug controls morphogenetic movements and orientation in response to stimuli. This may be because only the cells at the tip have turning receptors - in terms of the above model, STF receptors. However, there is another feature of the slug, the slime sheath, which may be relevant. Until recently almost nothing was known of the chemistry of this external 'skeleton' apart from the fact that it contains a cellulosic backbone (Hohl and Jehli 1973, Freeze and Loomis 1977) embedded in a proteinaceous matrix (Hohl and Jehli 1973, Smith and Williams 1979). Longitudinal differences in sheath properties are known; at the tip it is more fluid (Francis 1962; Wolfe, Smith, and Williams, unpublished) and the front of the slug has a thinner sheath than the back (Francis 1962, Farnsworth and Loomis

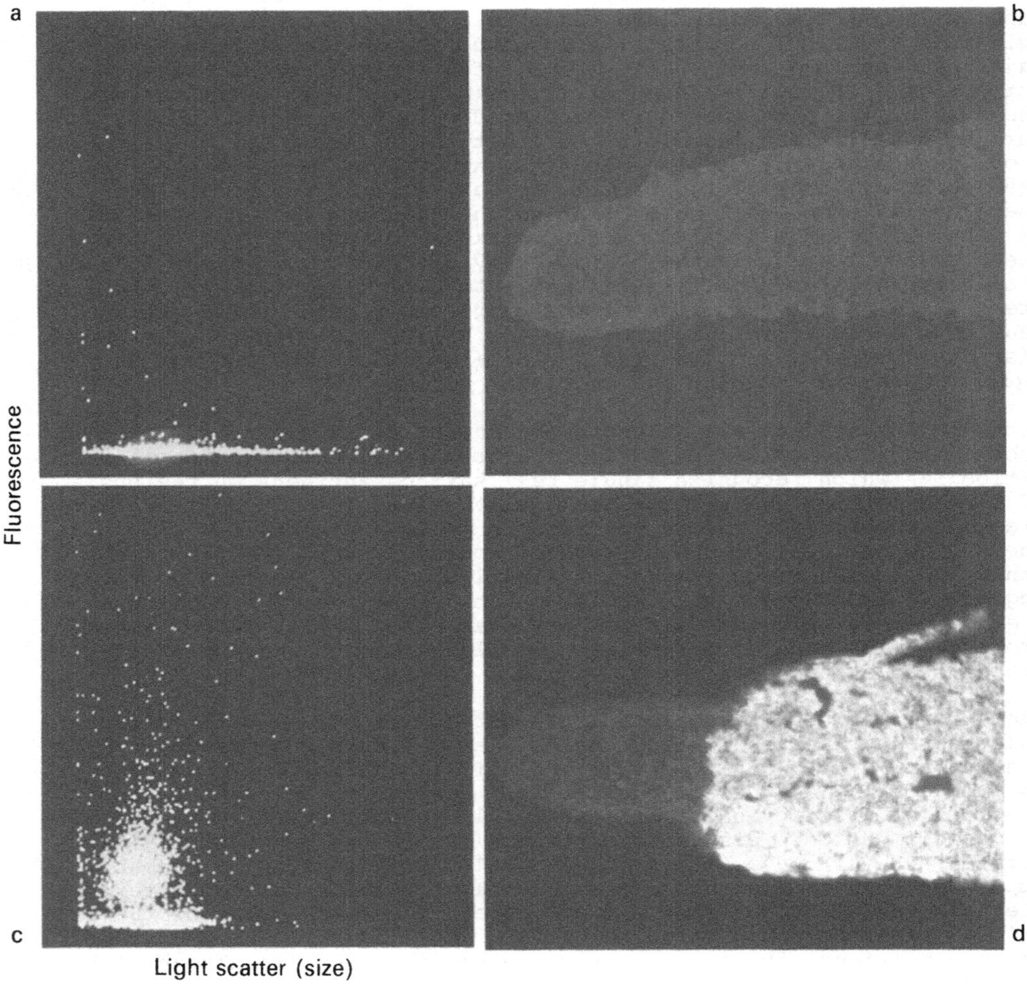

Fig. 3a-d. FACS IV profiles (a,c) and indirect fluorescence (FITC) staining of paraformaldehyde fixed 5 µm frozen sections of *D. discoideum* (strain NP73) slugs (b,d). Control (a,b) culture supernatant of MUD7, an IgG producing hybridoma whose antibodies do not recognize *D. discoideum* cells, and prespore specific antibody (c,d), culture supernatant of hybridoma MUD1. With the FACS IV profiles the control (a) shows a single population with less than 2% of cells labelled nonspecifically. The prespore antiserum (c) shows two populations, one of which is unlabelled while the other (labelled) population contains about 60% of cells. The tip (prestalk region) of the slug is on the left in the frozen sections, and it is clear that only the prespore cells are labelled with monoclonal antibody MUDI (d). The slugs are about 1 mm long

1975). A simple explanation is that all cells in the slug can in principle respond to external stimuli, but only at the tip is the sheath sufficiently flexible for the cells to disturb it. Thus the sheath may be an additional constraint, although the arguments (above) concerning

STF respecification of tip position suggest that the cells at the tip are different from those further back.

Some theories invoke differential permeability of the sheath (Sussman and Schindler 1978) or sheath modification (Ashworth 1971) as key elements in morphogenesis. To investigate these questions one needs to know about the chemistry of slime sheath. We have recently characterized the major proteins (all of which are glycoproteins) of slime sheath (Grant and Williams, unpublished) and using monoclonal antibodies we are attempting to test theories which involve sheath inhomogeneity. A more likely role for the slime sheath is that it provides a substratum for the amoebae in the slug to move along (Smith et al. 1982).

Evolutionary Perspective

In *D. discoideum* we have been able to examine different aspects of morphogenesis separately (diffusible signals, cell surface, structural components) and try to think of how they interrelate (e.g. Fig. 2). An examination of related organisms may help probe the origins of the components of morphogenesis in order to clarify how *D. discoideum* arrived at its present complexity. There are four levels of organization observable in slime moulds (Olive 1975; Fig. 4). Most simple is *Protostelium mycophaga* which makes a tiny fruiting body comprising a single spore and an acellular stalk; next comes *Acytostelium leptosomum* which forms a multicellular fruiting body which comprises only spores and an acellular stalk (i.e. there is no cell proportioning mechanism operating). *D. discoideum* follows with its simple multicellular fruiting body and three cell types. Finally *Polysphondylium* species make a more complex segmented pattern which is one of the simplest examples of a repeated pattern as seen in body segmentation. So far, there has been little study of the simpler slime moulds but such organisms may provide clues about the evolution of morphogenesis. It is of interest that all four species have a slime sheath which Olive (1975) speculates was originally used to prevent desiccation. If *D. discoideum* uses the sheath for additional purposes (Schindler and Sussman 1977, Ashworth 1971), study of the comparative chemistry of the slime sheath of these organisms may give some interesting clues about its role.

On the other hand there is also much to be learned from study of *D. discoideum* in its unicellular vegetative mode. It is possible that all four putative morphogens (STF, cAMP, ammonia, DIF) also have roles during aggregation. For example Lonski (1976) has observed an effect of ammonia on the aggregation territory size of a *D. discoideum* mutant; cAMP is the chemotactic agent for amoebae; Gross et al. (1981) have speculated that DIF may be involved in the determination of aggregate size; negative chemotaxis (due to STF?) has been observed in *D. discoideum* amoebae (Kakebeeke et al. 1979). We now suspect that STF is a primitive molecule which was used to sense the environment even before the *D. discoideum* parent learned to be social. Perhaps an early event in the development of the amoebae was to set up a transduction chain involved with sensing food and environment. Such a food sensor is known (folic acid which is secreted by bacteria, Pan et al. 1972), while STF was (and is) used to seek a comfortable environment via light and heat sensors. Recent studies on amoebal phototaxis (Hong et al. 1981) with slug phototaxis mutants suggest that vegetative amoebae also orient via the sensory transduction chain involving STF (Fig. 2; Häder, Williams and Fisher, in prep.). STF and folate receptors may well have preceded receptors for the better known aggregative molecules

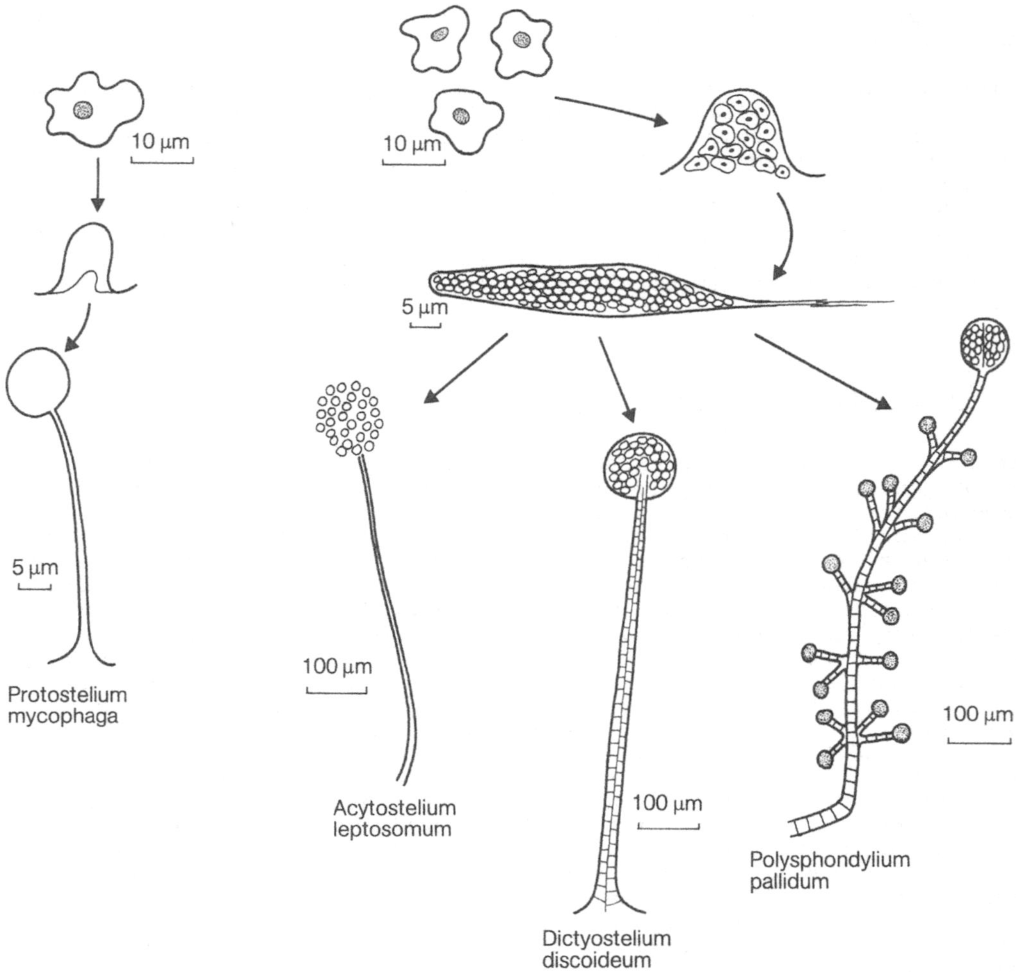

Fig. 4. Slime mould species which may be useful for comparative studies of morphogenesis. In *Protostelium mycophaga* a single amoeba constructs a fruiting body with a tiny acellular stalk and solitary spore. Amoebae of *Acytostelium leptosomum* aggregate and construct a multicellular fruiting body, but all cells form spores and the stalk is acellular. Note that the slime sheath which encases the aggregate disintegrates in the mature fruiting body (Olive 1975). *Dictyostelium discoideum* and *Polysphondylium pallidum* both construct fruiting bodies in which cells are specialised to be stalk or spores (or disc cells in *D. discoideum*). *P. pallidum* also constructs whorls of fruiting bodies along the main stalk

(cAMP in *D. discoideum*). Cells of all *Dictyostelium* and *Polysphondylium* species tested are sensitive to folic acid. Strangely enough and perhaps consistent with a more recent origin, the molecules used as chemoattractants for aggregation are quite varied between different species. In addition to cAMP (*D. discoideum*), three other chemoattractants are used within the genus *Dictyostelium* (*D. lacteum*, Mato et al. 1977; *D. minutum*, Kakebeeke et al. 1978; *D. mexicanum*, Cavender et al. 1981), while *Polysphondylium* species also use a different chemoattractant (Wurster et al. 1976).

Outlook for the Future

A major problem concerning morphogenesis is its complexity. The main components of *D. discoideum* morphogenesis have probably been identified (small morphogens, cell surface, extracellular structures) although their precise roles are unknown. More puzzling is how it all fits together. In principle it is possible to discover major aspects by studies on individual components of the system, but just as in the elucidation of a metabolic pathway finally the scheme has to be put together. *D. discoideum* is almost unique since it is an organism in which all of the major components of a morphogenetic system are accessible for experimental analysis (except those involved with hierarchical divisions). It has the added bonuses of the possibility of genetic and evolutionary approaches. The broad framework is becoming clear. The next few years should be exciting as we begin to fill in the picture.

Acknowledgements. I thank my collaborators Ulrike Dohrmann, Paul Fisher, Warwick Grant, Marianne Krefft, Ron Mahurin, Birgit Metz, Ludwig Voet and Dennis Welker for helpful comments and permission to cite unpublished material. I also thank Anneliese Kühnlein for typing the manuscript.

References

Ashworth JM (1971) Cell development in the cellular slime mould *Dictyostelium discoideum*. Symp Soc Exp Biol 25:27-47

Bonner JT, Slifkin MK (1949) A study of the control of differentiation: The proportions of stalk and spore cells in the slime mold, *Dictyostelium discoideum*. Am J Bot 36:727-734

Brody T, Williams KL (1974) Cytological analysis of the parasexual cycle in *Dictyostelium discoideum*. J Gen Microbiol 82:371-383

Cavender JC, Worley AC, Raper KB (1981) The yellow pigmented Dictyostelia. Am J Bot 68:373-382

Child CM (1941) Patterns and problems of development. Univ Press, Chicago

Chiquet M, Eppenberger HM, Turner DC (1981) Muscle morphogenesis: evidence for an organizing function of exogenous fibronectin. Dev Biol 88:220-235

Durston AJ (1976) Tip formation is regulated by an inhibitory gradient in the *Dictyostelium discoideum* slug. Nature (London) 263:126-129

Durston AJ, Vork F (1979) A cinematographical study of the development of vitally stained *Dictyostelium discoideum*. J Cell Sci 36:261-279

Farnsworth PA (1975) Proportionality in the pattern of differentiation of the cellular slime mould *Dictyostelium discoideum* and the time of its determination. J Embryol Exp Morphol 33:869-877

Farnsworth PA, Loomis WF (1975) A gradient in the thickness of the surface sheath in pseudoplasmodia of *Dictyostelium discoideum*. Dev Biol 46:349-357

Fisher PR, Williams KL (1981) Bidirectional phototaxis by *Dictyostelium discoideum* slugs. FEMS Microbiol Lett 12:87-89

Fisher PR, Williams KL (1982) Thermotactic behaviour of *Dictyostelium discoideum* slug phototaxis mutants. J Gen Microbiol 128:965-971

Fisher PR, Smith E, Williams KL (1981) An extracellular chemical signal controlling phototactic behaviour by *D. discoideum* slugs. Cell 23:799-807

Forman D, Garrod DR (1977) Pattern formation in *Dictyostelium discoideum* I. Development of prespore cells and its relationship to the pattern of the fruiting body. J Embryol Exp Morphol 40:215-228

Francis DW (1962) Movement of pseudoplasmodia of *Dictyostelium discoideum*. PhD Thesis, Univ Wisconsin, Madison, Wisconsin

Franke J, Kessin R (1977) A defined minimal medium for axenic strains of *Dictyostelium discoideum*. Proc Natl Acad Sci USA 74:2157-2161

Freeze H, Loomis WF (1977) Isolation and characterization of a component of the surface sheath of *Dictyostelium discoideum*. J Biol Chem 252:820-824

Friedlander M (ed) (1979) Immunological approaches to embryonic development and differentiation. Curr Top Dev Biol 13 Part 1

Friedlander M (ed) (1980) Immunological approaches to embryonic development and differentiation. Curr Top Dev Biol 14 Part 2

Garrod DR, Ashworth JM (1972) Effect of growth conditions on development of the cellular slime mould *Dictyostelium discoideum*. J Embryol Exp Morphol 28:463-479

Garrod DR, Nicol A (1981) Cell behaviour and molecular mechanisms of cell-cell adhesion. Biol Rev 56:199-242

Garrod DR, Swan AP, Nicol A, Forman D (1978) Cellular recognition in slime mould development. Soc Exp Biol Symp 32:173-202

Geltosky JE, Weseman J, Bakke A, Lerner RA (1979) Identification of a cell surface glycoprotein involved in cell aggregation in *D. discoideum*. Cell 18:391-398

Gerisch G (1982) Chemotaxis in *Dictyostelium*. Annu Rev Physiol 44:535-552

Gierer A, Meinhardt H (1972) A model of biological pattern formation. Kybernetik 12:30-39

Gregg JH, Davis RW (1982) Dynamics of cell redifferentiation in *Dictyostelium*. Differentiation 21:200

Gregg JH, Krefft M, Haas-Kraus A, Williams KL (1982) Antigenic differences detected between prespore cells of *Dictyostelium discoideum* and *Dictyostelium mucoroides* using monoclonal antibodies. Exp Cell Res 142:229-233

Gross JD, Town CD, Brookman JJ, Jermyn KA, Peacey MJ, Kay RR (1981) Cell patterning in *Dictyostelium*. Philos Trans R Soc London Ser B 295:497-508

Herskowitz I, Blair L, Forbes D, Hicks J, Kassir Y, Kushner P, Rine J, Sprague G, Strathern J (1980) Control of cell type in the yeast *Saccharomyces cerevisiae* and a hypothesis for development in higher eukaryotes. In: Leighton T, Loomis WF (eds) The molecular genetics of development. Academic Press, London New York, pp 79-118

Hohl HR, Jehli J (1973) The presence of cellulose microfibrils in the proteinaceous slime track of *Dictyostelium discoideum*. Arch Mikrobiol 92:179-187

Hong CB, Häder MA, Häder DP, Poff KL (1981) Phototaxis in *Dictyostelium discoideum* amoebae. Photochem Photobiol 33:373-377

Ishida S (1982) A mutant of *Dictyostelium discoideum* with alternative pathways of differentiation depending on culture conditions. J Gen Microbiol 128:411-414

Jones GE, Pacy J, Jermyn K, Stirling J (1977) A requirement for filopodia in the adhesion of pre-aggregative cells of *Dictyostelium discoideum*. Exp Cell Res 107:451-455

Kakebeeke PIJ, Mato JM, Konijn TM (1978) Purification and preliminary characterization of an aggregation-sensitive chemoattractant of *Dictyostelium minutum*. J Bacteriol 133:403-405

Kakebeeke PIJ, DeWit RJW, Konijn TM (1979) Negative chemotaxis in *Dictyostelium* and *Polysphondylium*. Exp Cell Res 124:429-433

Kessin RH (1981) Conservatism in slime mold development. Cell 27:241-243

Koshland DE (1979) A model regulatory system: bacterial chemotaxis. Physiol Rev 59: 811-862

Lam TY, Pickering G, Geltosky J, Siu CH (1981) Differential cell cohesiveness expressed by prespore and prestalk cells of *Dictyostelium discoideum*. Differentiation 20:22-28

Lonski J (1976) The effect of ammonia on fruiting body size and microcyst formation in the cellular slime molds. Dev Biol 51:158-165

Loomis WF (1975) *Dictyostelium discoideum*: A developmental system. Academic Press, London New York

Loomis WF (1980) Genetic analysis of development in *Dictyostelium*. In: Leighton T, Loomis WF (eds) The molecular genetics of development. Academic Press, London New York, pp 179-212

MacWilliams HK (1982) Transplantation experiments and pattern mutants in cellular slime mold slugs. In: Developmental order: its origin and regulation. Alan R. Liss, New York, pp 463-483

MacWilliams HK, Bonner JT (1979) The prestalk-prespore pattern in cellular slime molds. Differentiation 14:1-22

Maden M (1982) Vitamin A and pattern formation in the regenerating limb. Nature (London) 295:672-675

Maeda Y (1970) Influence of ionic conditions on cell differentiation and morphogenesis of the cellular slime molds. Dev Growth Differ 12:217-227

Manasek FJ (1976) The extracellular matrix: A dynamic component of the developing embryo. Curr Top Dev Biol 10:35-102

Mato JM, Haastert Van PJM, Krens FA, Konijn TM (1977) An acrasin-like attractant from yeast extract specific for *Dictyostelium lacteum*. Dev Biol 57:450-453

Matsukuma S, Durston AJ (1979) Chemotactic cell sorting in *Dictyostelium discoideum*. J Embryol Exp Morphol 50:243-251

McMahon D (1973) A cell-contact model for cellular position determination in development. Proc Natl Acad Sci USA 70:2396-2400

Meinhardt H (1982) Biological pattern formation. Academic Press, London New York

Milstein C, Lennox E (1980) The use of monoclonal antibody techniques in the study of developing cell surface. Curr Top Dev Biol 14:1-32

Müller K, Gerisch G (1978) A specific glycoprotein as the target site of adhesion blocking Fab in aggregating *Dictyostelium* cells. Nature (London) 274:445-449

Newell PC (1978) Genetics of the cellular slime moulds. Annu Rev Genet 12:69-93

Newell PC, Ross FM (1982) Genetic analysis of the slug stage of *Dictyostelium discoideum*. J Gen Microbiol 128:1639-1652

Newell PC, Telser A, Sussman M (1969) Alternative developmental pathways determined by environmental conditions in the cellular slime mold *Dictyostelium discoideum*. J Bacteriol 100:763-768

Ochiai H, Schwarz H, Merkl R, Wagle G, Gerisch G (1982) Stage specific antigens reacting with monoclonal antibodies against contact site A, a cell-surface glycoprotein of *Dictyostelium discoideum*. Cell Differ 11:1-13

O'Day DH, Durston AJ (1978) Colchicine induces multiple axis formation and stalk cell differentiation in *Dictyostelium discoideum*. J Embryol Exp Morphol 47: 195-206

Olive LS (1975) The Mycetozoans. Academic Press, London New York

Pan P, Hall EM, Bonner JT (1972) Folic acid as a second chemotactic substance in the cellular slime moulds. Nature (London) 237:181-182

Parish RW, Schmidlin S, Parish CR (1978) Detection of developmentally controlled plasma membrane antigens of *Dictyostelium discoideum* cells in SDS - polyacrylamide gels. FEBS Lett 95:366-370

Poff KL, Loomis WF (1973) Control of phototactic migration in *Dictyostelium discoideum*. Exp Cell Res 83:236-240

Rubery PH (1981) Auxin receptors. Annu Rev Plant Physiol 32:569-596

Rubin J (1976) The signal from fruiting body and conus tips of *Dictyostelium discoideum*. J Embryol Exp Morphol 36:261-271

Rubin J, Robertson A (1975) The tip of *Dictyostelium discoideum* pseudoplasmodium as an organiser. J Embryol Exp Morphol 33:227-241

Schaller CH (1982) Morphogenetic substances and their role in pattern formation in *Hydra*. 33 Mosbacher Kolloquium

Schindler J, Sussman M (1977) Ammonia determines the choice of morphogenetic pahtways in *Dictyostelium discoideum*. J Mol Biol 116:161-169

Smith E, Williams KL (1979) Preparation of slime sheath from *Dictyostelium discoideum*. FEMS Microbiol Lett 6:119-122

Smith E, Williams KL (1980) Evidence for tip control of the 'slug/fruit' switch in slugs of *Dictyostelium discoideum*. J Embryol Exp Morphol 57:233-240

Smith E, Fisher PR, Grant WN, Williams KL (1982) Sensory behavior in *Dictyostelium discoideum* slugs: phototaxis and thermotaxis are not mediated by a change in slug speed. J Cell Sci 54:329-339

Spiegel FW, Cox EC (1980) A one-dimensional pattern in the cellular slime mold *Polysphondylium pallidum*. Nature (London) 286:806-807

Stenhouse FO, Williams KL (1977) Patterning in *Dictyostelium discoideum*: the proportions of the three differentiated cell types (spore, stalk and basal disk) in the fruiting body. Dev Biol 59:140-152

Stenhouse FO, Williams KL (1981) Investigation of cell patterning in the asexual fruiting body of *Dictyostelium discoideum* using haploid and isogenic diploid strains. Differentiation 18:1-9

Sternfeld J, David CN (1981a) Cell sorting during pattern formation in *Dictyostelium*. Differentiation 20:10-21

Sternfeld J, David CN (1981b) Oxygen gradients cause pattern orientation in *Dictyostelium* cell clumps. J Cell Sci 50:9-18

Sussman M, Schindler J (1978) A possible mechanism of morphogenetic regulation in *Dictyostelium discoideum*. Differentiation 10:1-5

Sussman M, Sussman RR (1967) Cultivation of *Dictyostelium discoideum* in axenic medium. Biochem Biophys Res Commun 29:53-55

Sussman M, Schindler J, Kim H (1978) "Sluggers" a new class of morphogenetic mutants of *D. discoideum*. Exp Cell Res 116:217-227

Tasaka M, Takeuchi I (1981) Role of cell sorting in pattern formation in *Dictyostelium discoideum*. Differentiation 18:191-196

Welker DL (1982) Coumarin and antimicrotubule agents as probes of microtubule function in *Dictyostelium discoideum*. In: Cappuccinelli P, Morris NR (eds) Microtubules in microorganisms. Marcel Dekker, New York, pp 99-108

Welker DL, Williams KL (1980) Mitotic arrest and chromosome doubling using thiabendazole, cambendazole, nocodazole and benlate in the slime mould *Dictyostelium discoideum*. J Gen Microbiol 116:397-407

Welker DL, Williams KL (1982a) A genetic map of *Dictyostelium discoideum* based on mitotic recombination. Genetics (in press)

Welker DL, Williams KL (1982b) Genetic analysis and phenotypic characterization of effects on the cytoskeleton of coumarin-sensitivity mutations in *Dictyostelium discoideum*. J Gen Microbiol 128:1329-1343

West CM, McMahon D (1979) The axial distribution of plasma membrane molecules in pseudoplasmodia of the cellular slime mold *Dictyostelium discoideum*. Exp Cell Res 124:393-401

White E, Scandella D, Katz ER (1981) Inhibition by CIPC of mitosis and development in *Dictyostelium discoideum* and the isolation of CIPC-resistant mutants. Dev Genet 2:99-111

Williams AF (1980) Cell-surface antigens of lymphocytes: markers and molecules. Biochem Soc Symp 45:27-50

Williams KL, Newell PC (1976) A genetic study of aggregation in the cellular slime mould *Dictyostelium discoideum* using complementation analysis. Genetics 82: 287-307

Wolpert L (1971) Positional information and pattern formation. Curr Top Dev Biol 6:183-222

Wurster B, Pan P, Tyan GG, Bonner JT (1976) Preliminary characterization of the acrasin of the cellular slime mold *Polysphondylium violaceum*. Proc Natl Acad Sci USA 73:795-799

Yund MA, Germeraad S (1980) Drosophila development. In: Leighton T, Loomis WF (eds) The molecular genetics of development. Academic Press, London New York, pp 237-360

Zusman DR (1980) Genetic approaches to the study of development in the myxobacteria. In: Leighton T, Loomis WF (eds) The molecular genetics of development. Academic Press, London New York, pp 41-78

Hormonal Control of Regeneration in *Hydra*

H. C. Schaller and H. Bodenmüller[1]

Summary

We use *Hydra* as a model system to understand how pattern formation, cellular growth, and differentiation are controlled at the molecular level. We have found that four substances influence head and foot formation in *Hydra*: an activator and an inhibitor of head formation and an activator and an inhibitor of foot formation. The two inhibitors are small molecules with molecular masses below 500 daltons, they have an overall positive charge, and they do not contain peptide bonds. The two activators have an overall negative charge, both are peptides, the foot activator has a molecular mass between 500 and 1000 and the head activator of 1124. The head activator consists of 11 amino acids and has the sequence pGlu-Pro-Pro-Gly-Gly-Ser-Lys-Val-Ile-Leu-Phe. All four substances act at concentrations below 10^{-8} molar and their action is specific. At such low concentrations the head factors only influence head formation and the foot factors foot formation. Within the animal, all four substances occur as gradients, the head factors with a maximum in the head, the foot factors with a maximum in the foot region. The release from the structure-bound form is specific, and it determines the polarity of the tissue. Thus release of head factors is necessary to induce head regeneration, and that of foot factors to induce foot regeneration. In normal hydra all four substances are produced by and stored in nerve cells. We try to understand how these substances act and interact to create the spatial and temporal pattern of growth and differentiation typical for hydra.

Hydra is a small freshwater coelenterate about 1 cm long. It has a head (hypostome and tentacles) on one end and a sticky foot (peduncle and basal disk) at the other end of a cylindrical body column (Fig. 1). Under normal circumstances *Hydra* reproduces asexually by budding with a doubling time of 3 days. The animal also has a sexual cycle, that is, mutants can be and have been produced [1]. *Hydra* has a simple structure: It consists of only two cell layers, ecto- and endoderm, made up basically of epithelio-muscular cells between which gland cells, nerve cells, nematocytes (cells to capture prey), and interstitial cells are interspersed. Epithelial cells, gland cells, and interstitial cells are dividing cell types, nerves and nematocytes are differentiation products derived from the interstitial stem cell. In a sexual animal, interstitial cells give riese also to oocytes and sperm. The advantage of *Hydra* is that it has few cell types, that the differen-

1 Max-Planck-Institut für Medizinische Forschung, Abteilung für Biophysik und Universität Heidelberg, Zoologisches Institut, D-6900 Heidelberg, FRG

33. Colloquium - Mosbach 1982
Biochemistry of Differentiation and Morphogenesis

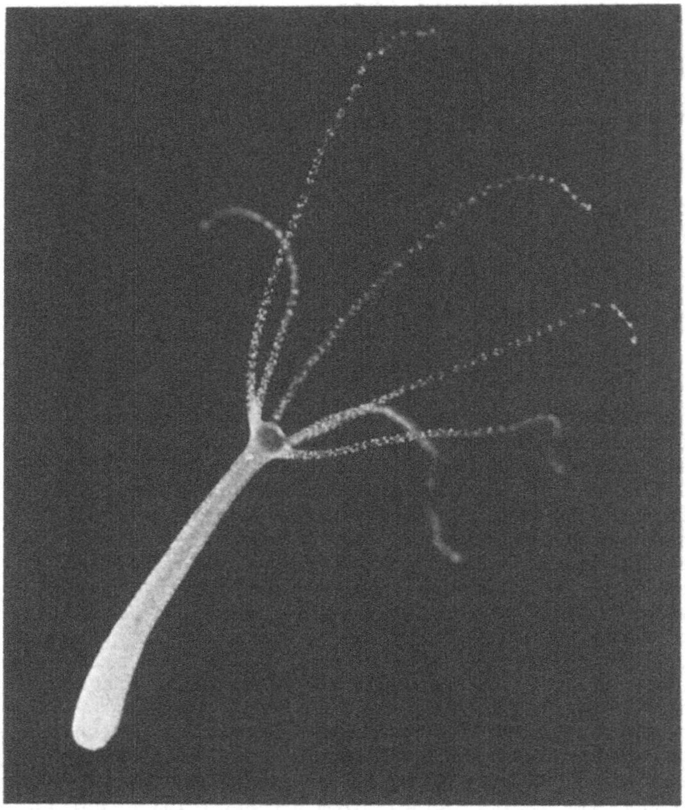

Fig. 1. *Hydra attenuata*: Animal before it produces its first bud

tiation pathways are short (Fig. 2), and that it is very easily amenable to experimental manipulations such as regeneration, tissue grafting and even reaggregation experiments.

Major cell types in hydra

Epithelial cells		20 000
Gland cells		5 000
Interstitial cells		15 000
Nerve cells		5 000
Nematocytes		55 000
(Sexual cells)		
	Total	100 000

Fig. 2. Major cell types in *Hydra*. *Arrows* indicate numbers of cell divisions between stem cell and final differentiation

Within the animal kingdom *Hydra*, or coelenterates in general, are the first to possess something like a nervous system. *Hydra* has no special nervous organ like a ganglion or a brain and there are no nerve-supporting cells like glia. The nerve cells are organised as a network with a higher density in the head and in the foot region. According to morphological criteria the nerve cells are normal, with a small cell body and long processes (Fig. 3), and at the ultra-structural level with synapses, vesicles, neurosecretory granules, neurofilaments etc. The nerve cells may be primitive in so far as they seem to be multifunctional by having sensory, motoneuron and transmittory functions simultaneously [2] with some specialisation in certain regions [3].

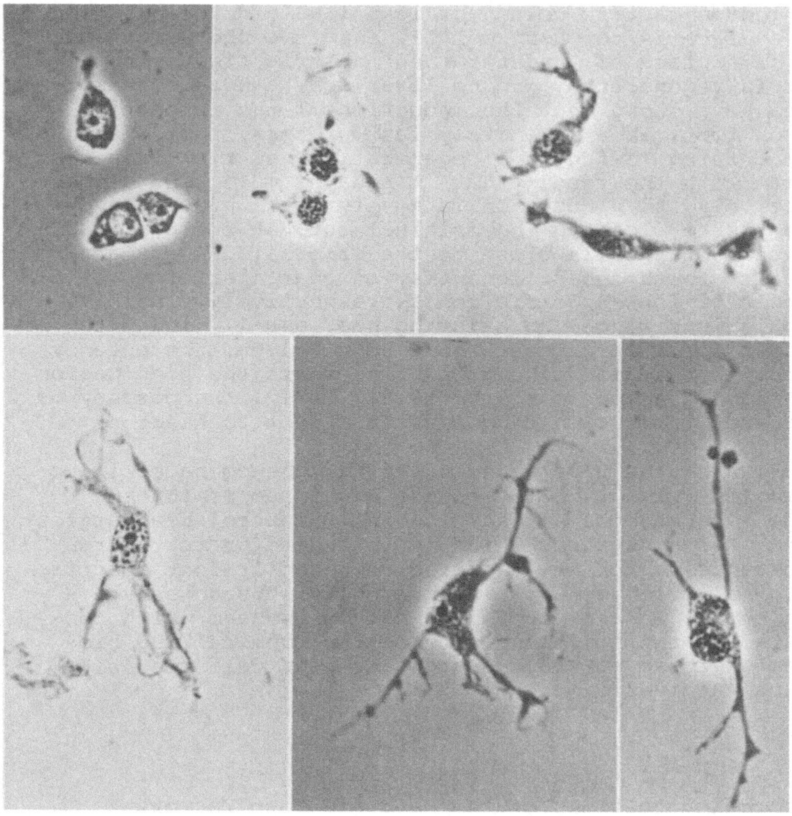

Fig. 3. Interstitial precursor cells, developing nerve cells and mature nerve cells. A nerve cell body has a diameter of approximately 10 μm

The nervous system in *Hydra* serves two functions. It is necessary for the coordination of fast processes like body movement, and it controls slow processes like cellular growth and determination. So far not much is known how and with what transmitter substances the fast neuronal communication system works. Preliminary experiments suggest that non-peptidergic transmitters such as acetylcholine [4] or the catecholamines are not present in nerve cells in *Hydra*. Recently, C.J.P. Grimmlikhuijzen found that antibodies against mammalian and other neuropeptides such as cholecystokinin, substance P, neurotensin, bombesin, and FMRF-amide cross-react with molecules in *Hydra* and that the nerve cells which contain them show a characteristic distribution within *Hydra* [5]. The exact structure and possible function of these peptides as transmitters or neuroregulators remains to be elucidated.

We know more about the second function of the nervous system in *Hydra*, namely the control of pattern formation, cellular growth and differentiation by hormone-like substances. To understand pattern formation we want to know why a certain structure is induced at a definite position and time during development, and why it subsequently inhibits the formation of an identical such structure nearby. In *Hydra* only the tissue at the ends is terminally differentiated, the tissue of the body column is undetermined. Depending on the stimulus, it can become head, foot, bud, remain gastric column or less likely develop into a gonade. Thus, if a *Hydra* is cut horizontally into two pieces of equal length, from the same piece of "undetermined" gastric tissue in the upper half a foot is regenerated, in the lower half a head. Transplantation experiments have shown that the induction of the respective structures, head or foot, is a relatively fast process, that within a few hours after cutting (3-6 h) the regenerating gastric tissue has changed its program with the result that it behaves more like a foot or a head, respectively [6]. Induction is a very local process and restricted to the tissue close to the original cut. At the cellular level this means that only cells close to the original cut surface are changed in their growth and differentiation characteristics, i.e. become head-specific or foot-specific cells, respectively. Inhibition on the other hand is long range: an existing head usually inhibits that another head is formed along the whole body column. The potency to induce or inhibit a head is highest in or close to the head region decreasing in a graded fashion to the foot. Similarly, the potency of foot induction or foot inhibition decreases from foot to head.

Our contribution to a better understanding of the phenomena of induction and inhibition in *Hydra* or to the choice mechanism is that we could show that four substances act as signals to control head or foot formation in *Hydra*, two substances which inhibit head or foot formation, and two substances which are required for head or foot specific determinations (for a review see [7]). Since these substances are products of nerve cells, in *Hydra* it is the nervous system which controls morphogenesis. In the following some of the chemical and biological properties of these substances are discussed with special emphasis on the head activator.

Isolation and Chemical Properties of the Head and Foot Factors from *Hydra*

The effect of substances which influence head or foot formation in *Hydra* can most easily be measured as acceleration or inhibition of head or foot regeneration [8]. An example of such an assay is given in Fig. 4, where it is shown that increasing concentrations of foot inhibitor lead to an increasing retardation or inhibition of foot regeneration. The response is dose-dependent, and allows the quantitation of the respective factor per animal or in a fraction during the purification procedure [8].

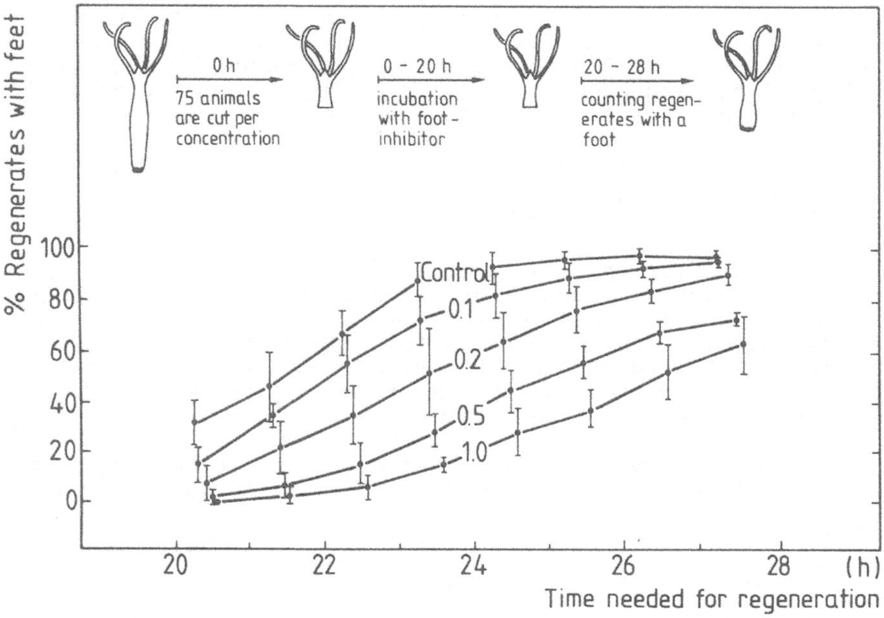

Fig. 4. Assay for the foot inhibitor. The numbers in the figure indicate the concentration of the crude extract expressed in $O.D._{280}$/ml

By means of such assays we have purified the four substances. the head activator to homogeneity (10^9-fold enrichment), the other three at least 10^5-fold. The two activators are destroyed by proteases and acid hydrolysis (6 N HCl, 110°, 24 h), the two inhibitors survive such a treatment. The two activators under neutral conditions absorb to anion exchangers, i.e. have an overall negative charge, the two inhibitors have an overall positive charge. The head activator has a molecular mass of 1124, the foot activator between 500 and 1000. The two inhibitors have molecular masses around or below 500. The inhibitors are very hydrophilic molecules, the activators are more hydrophobic. Thus, the head activator at high concentrations (10^{-3} M) is insoluble in water, soluble in organic solvents, and it has a very high surface adhesivity as demonstrated by its tendency to adhere to glass, plastic, other molecules and even to itself forming aggregates of undefined multiplicity. We believe that the striking parallelism in chemical properties of the inhibitors with the activators must have biological relevance. In particular, the difference in hydrophobicity or surface adhesivity resulting in completely different diffusion rates provides a plausible explanation for the biological finding that the action of the inhibitors is long-range, whereas that of the activators is short-range.

From the degree of purity and the molecular weight we can calculate that all four substances act at very low concentrations. According to the present state of purity the two foot factors and the head inhibitor act at concentrations below 10^{-8} molar, the head activator at 10^{-13} molar. This also means that a *Hydra* needs very little of these substances and explains our difficulties in obtaining in pure form workable quantities from *Hydra* for a chemical analysis. For example, we were able to isolate from 3×10^6 *Hydra* (3 kg), accumulated over several years, 0,5 µg of the pure head activator. Fortunately, we discovered that other coelenterates, in particular sea anemones, which are 10^4-fold larger and abundant in the ocean, contain the same set of substances as *Hydra* in similar concentrations. Meanwhile we have processed 200 kg of the sea anemone, *Anthopleura elegantissima*, and used it so far for the chemical analysis of the head activator [9].

Sequence Analysis of the Head Activator

The isolation procedure for the head activator, the purification steps and the respective yields are summarised in Table 1. After an extensive extraction with methanol the head activator was purified by ion exchange and molecular sieve chromatography followed by reverse phase low and high pressure liquid chromatography. From 200 kg of sea anemones we extracted 20 nanomoles (20 µg) and from 3 kg of *Hydra* 500 picomoles (0.5 µg) of the pure head activator [9].

The amounts of head activator available were still minute and required use and improvement of micromethods. To determine the amino acid com-

Table 1. Purification procedure for the head activator

Step	Column material	Eluent	Purification (x-fold)	Yield (%)
1.1	Sephadex G-10 (6 l)	5 mM ammoniumbicarbonate (ABC)	10^4	90
1.1	DEAE A-25 (batchwise)	pH gradient		
1.2	G-10 (250 ml)	distilled H_2O	10^4	70-80
2.1	Biogel P-2 (660 ml)	5 mM ABC		
2.2	G-10 (25 ml)	0.1 M NaCl, 0.01 M Tris-HCl, pH 7.6		
2.3	P-2 (40 ml)	0.1 M NaCl, 0.01 M Tris-HCl, pH 7.6	5×10^5	60-70
3.1	LiChrosorb RP-8 (500 µl, low pressure)	Step gradient of methanol in 5 mM ABC		
3.2	RP-8 (500 µl, low pressure)	Step gradient of acetonitrile in 0.1% trifluoracetic acid (TFA)	10^7	40-50
4.1	HPLC, RP-8 (250 x 4 mm)	50% methanol in 5 mM ABC (isocratic)		
4.2	HPLC, RP-8	30% acetonitrile in 0.1% TFA (isocratic)		
4.3	HPLC, RP-8	Gradient of 20-80% methanol in 5 mM ABC		
4.4	HPLC, RP-8	Gradient of 20-40% acetonitrile in 0.1% TFA	10^9	10-20

position of the head activator the purified peptide was hydrolysed, dansylated and the dansylated amino acids were separated for a qualitative analysis on micro thinlayer plates and for quantitation on HPLC [9]. We found that the head activator contained the following amino acids, Gly(2), Glu(1), Ile(1), Leu(1), Lys(1), Phe(1), Pro(2), Ser(1), Val(1).

The sequence analysis was complicated by the fact that the head activator had no free amino terminus, that its amino end was blocked by pyroglutamic acid. Therefore, most of the sequence was obtained by enzymatic degradation methods and, in part, by a chemical degradation of the biologically inactive Glu-peptide which contained glutamic acid instead of pyroglutamic acid as amino terminal amino acid [9]. In Table 2 the different sequencing steps and their results are outlined. The head activator from *Anthopleura* was found to have the sequence

pGlu-Pro-Pro-Gly-Gly-Ser-Lys-Val-Ile-Leu-Phe.

Table 2. Sequence analysis of the head activator

Method	Amino acid sequence
Edman degradation	Glu - Pro - Pro - Gly - Gly -
Carboxypeptidase Y	- Lys - Val - Ile - Leu - Phe
AEP[a]	- Gly -
AEP and end group	- Gly - Ser -
AEP and Edman	Glu - Pro - Pro - Gly -
AEP and aminopeptidase M	- Gly - Ser - Lys - Val - ILe - Leu - Phe

[a]AEP (*Astacus* endopeptidase) is a low molecular weight endopeptidase isolated from *Astacus fluviatilis* [10]

We also isolated the head-activator from *Hydra*, from rat intestine and from bovine and human hypothalami. We found that the peptides from all these different sources are identical [11]. The function of the head-activator peptide in mammals is unknown. Meanwhile we have chemically synthesised the head activator and could show that there is both chemically and biologically no difference between the synthetic and the native peptides [12].

Characterisation as Neurohormones

In normal animals all four substances seem to be produced by nerve cells. One line of evidence is indirect. All four substances are present in *Hydra* as gradients, the head factors with a maximum in the head, the foot factors with a maximum in the foot (Fig. 5). Since only the nerve cells and no other cell type correlates with such a distribution [13], one could argue that they must be contained in nerve cells. The more convincing evidence came from the finding that on

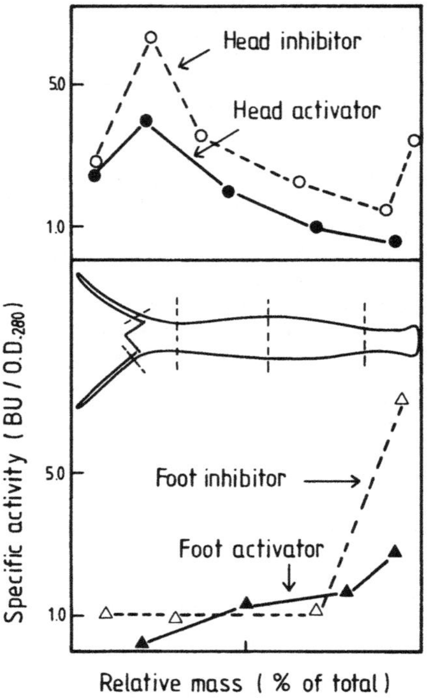

Fig. 5. Distribution of morphogens in different body regions of *Hydra*

glycerol gradients where cells can be separated, the four morphogens co-purify with the nerve cells and with no other cell type [14-17]. Meanwhile we could show that antibodies raised against synthetic head activator (purchased from BACHEM, Switzerland) react immunocytochemically with nerve cells in cut sections of *Hydra* . These nerve cells are abundant in the head region and are located predominantly in the ectoderm (unpublished results).

Within the nerve cells all four substances seem to be bound to intracellular structures. This was shown most elegantly for the head activator by its isolation in or copurification with membraneous vesicles resembling in all criteria (density: 1.09, S-value: 800, average diameter : 1200 Å) neurosecretory granules [14]. All four factors thus occur in two states, an inactive, structure-bound state and an active free state. Under normal conditions very little (less than 1%) seems to be in the free state. Evidence for this comes from the finding that all four substances act if added to the medium at concentrations one thousandfold lower than what is present in the animal itself. The advantage of such a system is that the release can be regulated, that larger amounts can be set free during special events such as during regeneration or after feeding.

Biological Effects on Morphogenesis, Cellular Growth and Differentiation

As mentioned above all four factors act at very low concentrations (below 10^{-8} molar), and they act specifically, which means that at such low concentrations the head factors only influence head and the foot factors foot formation [8]. The release of all four factors is strictly regulated and it is specific. For the induction of head regeneration release of head factors is required and that of foot factors to induce foot regeneration. The release of the head factors is controlled by the head inhibitor. It functions as a release inhibiting factor for its own release and for that of the head activator as well ([18] and W. Kemmner, unpublished results), whereas the head activator has no effect neither on its own release nor on that of the head inhibitor. For head induction thus the absence of head inhibitor is one prerequisite, presence or release of head factors another. Based on the chemical finding that the head inhibitor is able to diffuse fast, whereas the head activator due to its surface adhesivity is locally restricted close to the cut, an area is created with a high concentration of free head factors and dominance of head activator over head inhibitor allowing head-specific inductions to occur. Further away from the cut head inhibitor predominates leading to a restriction of the activation to the region adjacent to the cut. Preliminary experiments suggest that for foot induction also absence of foot inhibitor is needed to allow foot-specific release. We assume that such a specific release is necessary to ensure the reprogramming of the gastric tissue to head- or foot-specific growth and differentiations, respectively.

At the cellular level the head activator has two measureable effects: It stimulates cells to divide, i.e. acts as a growth hormone and it affects head-specific cellular differentiations [19-21]. The head inhibitor has antagonistic effects [22]. So far nothing is known about the action of the foot factors at the cellular level.

Head activator and head inhibitor thus interact or compete with each other at several levels. For the control of release only the head inhibitor is important. For the control of production both factors play a role. This control is indirect because not the production as such is affected, but the determination and differentiation of the producing nerve cells. An interstitial precursor cell only enters the nerve-cell pathway if in a critical phase of its cell cycle (early S phase) head activator is present (Fig. 6a). Once determined the prospective nerve cell is again subject to the controlling influence of head acti-

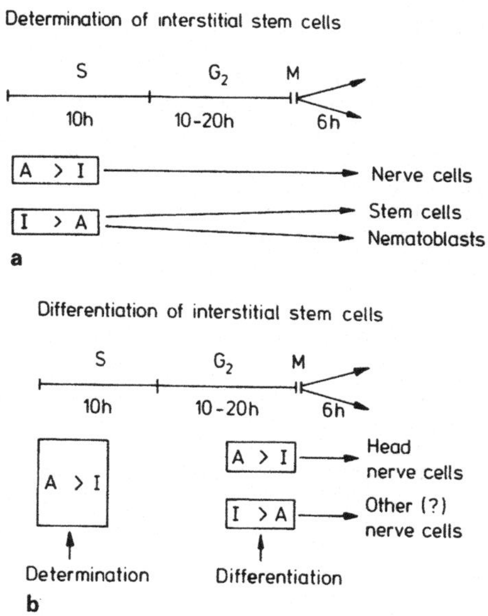

Fig. 6a,b. Influence of the head factors on (a) the determination and (b) the differentiation of interstitial stem cells to nerve cells. *S* S phase, G_2 G_2 phase, *M* mitosis, *A* head activator, *I* head inhibitor

vator and head inhibitor. It can undergo its final mitosis and differentiation to a head-specific nerve cell only if head activator is present and if more head activator is present than head inhibitor (Fig. 6b). The new head-specific nerve cells produce both head activator and head inhibitor thus reinforcing or amplifying the headspecific influence. Such a feedback mechanism may ensure that wherever head activator predominates over head inhibitor, head-like structures will be formed and maintained. As a third type of interaction there is the influence of the head factors on target cells other than interstitial stem cells such as epithelial cells, gland cells and nematocytes. In how far the two foot factors also interact or compete for the same targets is unclear at present.

The second function of the nervous system in *Hydra* is thus the control of cellular growth and differentiation resulting in pattern formation. This control is achieved by means of hormone-like substances such as our inhibitors and activators of head and foot formation. These substances act as slow carriers of intercellular communication. This slow transfer of information is not a strictly neuronal property. We have found that in the absence of nerve cells, epithelial cells take over the hormone-producing function of nerve cells [23]. This was not unexpected, since in an embryo the slow flow of intercellular communication must also exist before the nervous system becomes established. It also suggest that the slow intercellular communication by means of hormone-like substances is the more ancient, or primitive, type of information transfer between cells and that the fast conduction system requiring nerve cells and neurotransmitters may be the more modern version of intercellular communication.

Acknowledgements. This work was supported by the Deutsche Forschungsgemeinschaft. H.C. Schaller is a recipient of a Heisenberg fellowship.

References

1. Sugiyama T, Fujisawa T (1977) Dev Growth Differ 19:187-200
2. Westfall JA (1973) Ultrastruct Res 42:268-282
3. Kinnamon JC, Westfall JA (1981) J Morphol 168:321-329
4. Erzen J, Brzin M (1978) Comp Biochem Physiol 59c:39-43
5. Grimmelikhuijzen CJP, Dockray DJ, Schot LPC (1982) Histochemistry
6. Berking S (1979) Wilhelm Roux's Arch 186:189-210
7. Grimmelikhuijzen CJP, Schaller HC (1979) Trends Biochem Sci 4:265-267
8. Schaller HC, Schmidt T, Grimmelikhuijzen CJP (1979) Wilhelm Roux's Arch 186: 139-149
9. Schaller HC, Bodenmüller H (1981) PNAS 78:7000-7004
10. Zwilling R, Dörsam H, Torff H-J, Rödl J (1981) FEBS Lett 127:75-78
11. Bodenmüller H, Schaller HC (1981) Nature (London) 293:579-580
12. Birr C, Zachmann B, Bodenmüller H, Schaller HC (1981) FEBS Lett 131:317-321
13. Bode H, Berking S, David CN, Gierer A, Schaller HC, Trenkner E (1973) Wilhelm Roux's Arch 171:269-285
14. Schaller HC, Gierer A (1973) J Embryol Exp Morphol 29:39-52
15. Berking S (1977) Wilhelm Roux's Arch 181:215-255
16. Grimmelikhuijzen CJP (1979) Cell Differ 8:267-273
17. Schmidt T, Schaller HC (1980) Wilhelm Roux's Arch 188:133-139
18. Kemmner W, Schaller HC (1981) Wilhelm Roux's Arch 190:191-196
19. Schaller HC (1978) In: Papaconstantinov J, Rutter WJ (eds) Molecular control of proliferation and differentiation. Academic Press, London New York, pp 231-241
20. Schaller HC (1976) Cell Differ 5:1-11
21. Schaller HC (1976) Cell Differ 5:13-20
22. Berking S (1979) J Cell Sci 40:193-205
23. Schaller HC, Rau T, Bode H (1980) Nature (London) 283:589-591

Inflammatory Mediators and Wound Hormones: Chemical Signals for Differentiation and Morphogenesis in Tissue Regeneration and Healing

J. H. Wissler[1]

Introduction: The Secretory Leukocytic Tissue and the Basic Chemical Mechanisms for Cellular Reactions and Communication in Tissue Regeneration and Morphogenesis Operated by Effector Molecules of the Inflammatory and Healing Processes

Physically, chemically and biologically caused injuries provoke inflammatory reactions. Examples for exogenous and endogeneous inducers for inflammatory processes are immune (anaphylactic), atherosclerotic and (heart muscle) ischemic injuries. They participate in reproductive cycles and survival, outgrowth or regression and destruction of tissues, grafts and tumors [1-8]. The inflammatory process is a physiological helper reaction of the natural defense system of an organism to its tissues and cells for maintenance of their morphogenic patterns, injury repair and control of their social and regeneration behaviour [5-7]. It is activated if cellular auto-repair and homoestatic regulatory mechanisms fail to prevent non-physiological changes and are unable to maintain the structural and functional integrity in cells and tissues. The inflammatory process (healing reaction) aims at restoration of life-compatible morphogenic patterns and maintenance of the structural and functional integrity in tissue. This may be achieved either by complete pattern renewal. Or, as impaired reaction, by formation of a scar [1-6], if an impaired regeneration capability of the remaining viable, highly differentiated or matured cells of a tissue adjacent to the reaction site of injury does not allow full restoration of the structural and functional integrity. Therefore, the inflammatory and healing reactions are integral parts. Like reproductive cycles, they are an example for a tissue-regeneration process.

A complex array of interactive biological signaling devices operated by effector substances as chemical signals of the inflammatory and healing reactions govern the organization of the tissue regeneration process [1-9]. The blood components (plasma, leukocytes, thrombocytes, erythrocytes) as sources of these effector substances have a prominent role in the chemical activation and regulation of the inflammatory and wound-healing processes. Humoral and cellular (cytokine) effectors exist. By regulated, limited, bond-specific proteolysis of distinct components of plasma proteins (e.g. of the kinin, complement and coagulaprotein system), humoral effector polypeptides are formed upon contact reactions with intruded exogenous, foreign matter (e.g. antigens

1 Arbeitsgruppe Biochemie, Max-Planck-Institut für Physiologische und Klinische Forschung, D-6350 Bad Nauheim, FRG

33. Colloquium - Mosbach 1982
Biochemistry of Differentiation and Morphogenesis

or allergens); or with exposed unmasked endogenous molecules (e.g. cardiolipin, complex glycosphingolipids, proteinases) which escaped their biological compartments or biochemical regulation [1-6,10-20,23]. Cellular (cytokine) effectors are formed by cells participating in the tissue-regeneration process, foremost by white blood cells which accumulate at the reaction site of tissue injury. Accordingly, cytokines produced by the different types of activated leukocytes (granulocytes, monocytes, lymphocytes) are termed leukokines, monokines or lymphokines [24].

These effector substances may transmit their information by autocrine and paracrine mechanisms (mediators of inflammation) [6,7,25]. Or they may operate by endocrine (hormonal) information transmission mechanisms [25] as wound hormones [6,7,25]. Effector substances represent activating or inhibiting chemical signals operating chemical mechanisms for cellular reactions and communication in tissue regeneration and morphogenesis processes. Chemopoiesis, chemorecruitment, chemokinesis, chemotaxis, chemotropism and chemostasis of cells are basic categories of the effector-operated chemical mechanisms for cellular reactions and communication in tissue regeneration and morphogenesis processes, as exemplified in reproductive cycles and wound healing. The definitions of these mechanisms are compiled in Table 1. In Fig. 1, the meanings of chemically operated cell-biological reactions and communication mechanisms are illustrated: The morphogenesis of new life-compatible patterns and functions of tissues at injured sites is intimately associated with the formation of a "secretory leukocytic tissue" in terms of organized leukocyte accumulation, recruitment and reproduction (by proliferation, differentiation, maturation and cell storage).

Nature, Biotechnical Preparation and Properties of Inflammatory Mediators and Wound Hormones

A rough estimate suggests that about 10,000 different substances might exist which participate in and effect the complex interplay of these interactive mechanisms for cellular reactions and communication in tissue regeneration and morphogenetic processes. Most of such substances are still unknown: Leber [26] discovered the inflammatory mediator action associated with leukocyte accumulation in 1888. The possible existence of inflammatory mediators of defined structure was postulated in 1927 by Sir Thomas Lewis [27] on account of the "triple response" [4,27] which histamine mimicked. Today, about 100 defined substances are known as inflammatory mediators. Again about the same number of different mediator activities as associated with a certain biological assay system have been evaluated. As far as is known, mediators of inflammation are organic molecules of non-protein as well as of polypeptide structure. Dold [28] and, especially, Menkin [1,2] devised the concept of polypeptides being structural elements of biologically specific mediators of inflammation. The peptide nature of inflammatory mediators was then first definitely recognized as structural equivalent of kinin action [29] which had been detected in 1937 by Werle and associates [30]. With the classical anaphylatoxin, the cocytotaxin and the serum-derived leukorecruitin, for the first time, humoral mediator proteins of distinguishable activity could be concomitantly separated and obtained molecularly homogeneous in biologically specific, active form and in crystaline state in my laboratory [5,6,12-20]. Thus Menkin's basic concepts [1,2] became substantiated on a chemical basis.

Table 1. Basic chemical mechanisms operative in cellular reactions and communication for morphogenesis of tissue patterns by reproduction and healing processes: Autocrine, paracrine and endocrine signal transmission. Compiled subcategories are not claimed to be complete

---

1. *Chemopoiesis of cells and organisms:*
Chemical control of growth at random direction, determination and storage

Definition: Chemopoiesis is a reaction by which growth at random direction, determination and storage of cells and organisms are determined by chemicals in their environment

1.1. Hyperplastic and hypertrophic growth
1.2. Differentiation (switching on / off of un- / expressed genes)
1.3. Maturation (modulation of phenotype profile)
1.4. Storage and compartmentation (of biologically quiescent cells)

2. *Chemorecruitment of cells and organisms:*
Chemical control of mobilization from storage pools

Definition: Chemorecruitment is a reaction by which the mobilization of (biologically quiescent) cells and organisms from their storage pools into sites of functional readiness is determined by chemicals in their environment

2.1. Mobilization from primary (poietic) storage pools
2.2. Mobilization from secondary (marginal) storage sites

3. *Chemokinesis of cells and organisms:*
Chemical control of random locomotion

Definition: Chemokinesis is a reaction by which the motility of cells and organisms migrating at random is determined by chemicals in their environment

3.1. Orthokinesis (speed and frequency of locomotion)
3.2. Klinokinesis (frequency of turning)
3.3. Haptokinesis (adhesion and aggregation)
3.4. Endocytosis
3.5. Mechanokinesis

4. *Chemotaxis of cells and organisms:*
Chemical control of the direction of locomotion by recognition of dissipative (metastable) physical and chemical structures

Definition: Chemotaxis is a reaction by which the direction of locomotion of cells and organisms is determined by chemicals in their environment

4.1. Chemotaxis (locomotion along solute gradients of informative molecules)
4.2. Haptotaxis (locomotion along surface gradients of informative molecules)
4.3. Mechanotaxis

5. *Chemotropism of cells and organisms:*
Chemical control of the direction of (hyperplastic and hypertrophic) growth by recognition of dissipative (metastable) physical and chemical structures

Definition: Chemotropism is a reaction by which the direction of growth of cells and organisms is determined by chemicals in their environment

5.1. Chemotropism (growth along solute gradients of informative molecules)
5.2. Haptotropism (growth along surface gradients of informative molecules)
5.3. Mechanotropism

6. *Chemostasis of cells and organisms:*
Chemical control of homeostatic regulatory mechanisms for

Definition: Chemostasis is a reaction by which homeostatic regulatory mechanisms of cells and organisms are determined by chemicals in their environment

6.1. Maintenance and auto repair of cells and organisms
6.2. Maintenance of microenvironment of cells and organisms
6.3. Maintenance of system of organization of cells and organisms

---

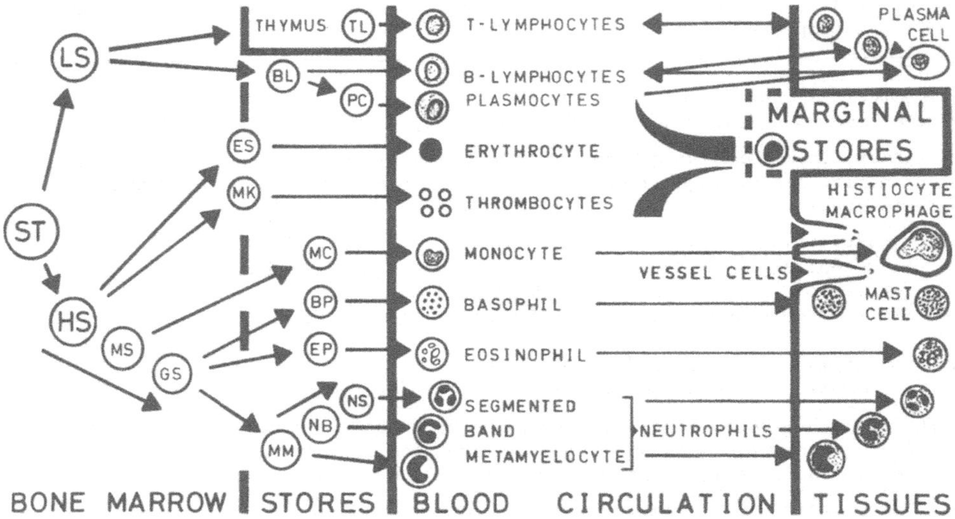

Fig. 1. Schematic presentation of the course of life of blood cells from their reproduction in the bone marrow (in the myeloid phase of hematopoiesis) to morphogenesis of a *"secretory leukocytic tissue"* at injured tissue sites. This secretory leukocytic tissue operates by different (autocrine, paracrine and endocrine) transmission and processing modes of the specific informations intrinsic to chemical signals of the effector network the various distinct types of chemical mechanisms (Table 1) for cellular reactions and communication to which regeneration of life-compatible morphogenic structures at injured tissue sites are subjected. The basic categories of chemical mechanisms for cellular reactions and communication (Table 1) may not be considered as restricted to occur only in a distinct tissue or a topochemical area, as visualized by possible recruitment reactions directed to poietic or marginal storage pools [5-8,15-19,41]. Abbrevations in use are (alphabetically): *BL* B-lymphocyte, *BP* basophil, *EP* eosinophil, *ES* erythroid stem cell, *GS* granulocyte stem cell, *HS* hematopoietic stem cell, *LS* lymphoid stem cell, *MC* monocyte, *MK* megakaryocyte, *MM* metamyelocyte, *MS* myelopoietic stem cell, *NB* neutrophilic band, *NS* neutrophilic segmented leukocyte, *PC* plasma cell, *ST* undifferentiated, pluripotent bone marrow stem cell, *TL* T-lymphocyte

All those known effector substances operate the basic chemical mechanisms for cellular reactions and communication by (pseudo-) autocrine and paracrine information transmission [6,7,25]. Whether or not there are existing effectors which convey information by endocrine (hormonal) mechanisms, could only recently be clarified [6,7,20,31-35]. Nevertheless, historically, the terms "wound hormones", "necrohormones", "tissue repair hormones", and as overhead names, "autacoids" and "cybernins", have been coined alternatively [25,36-38]. They were als in use for inflammatory mediators. Figure 2 shows the relations between the different modes of information transmission mechanisms. Thus, to qualify an effector molecule as "wound hormone", an inflammatory mediator, or exactly spoken, an effector should display its signal specifically by an endocrine transmission mechanisms. This would require leucocytes - or another cell type participating in inflammation and healing *at the reaction site* - to secrete a substance to be isolated and defined which has no phlogistic or other activity at its extravasal production site, the wound. But for display of its activity, it should be transported in the blood circulation to target cells remote from this production site in inflammation and healing. Although some effects

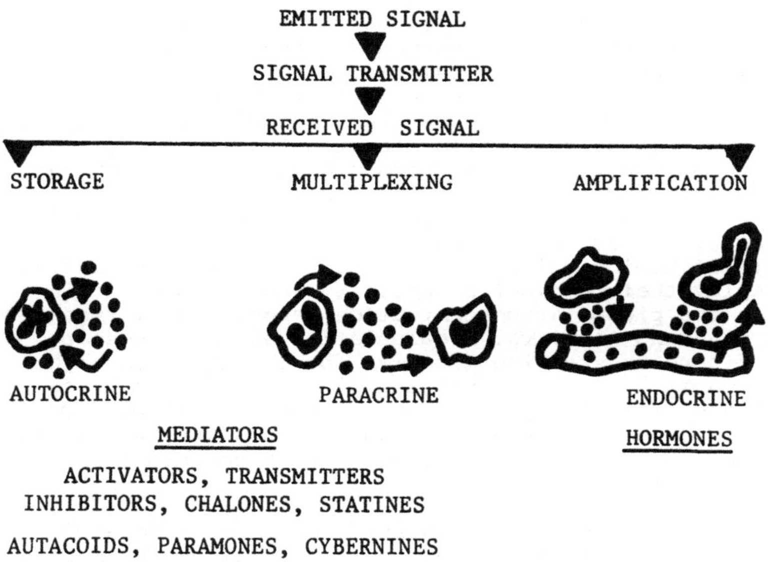

Fig. 2. Chemical signaling devices in cellular communication: Autocrine, paracrine and endocrine information transmission mechanisms. Note that transmission of a chemical message in the intercellular space may occur by different physical structures of messenger molecules; i.e. in form of their concentration equilibrium (no gradient), or in form of a dissipative diffusion non-equilibrium of the solute (boundaries, concentration gradients). They represent (positional) information terms additional and superimposed to the chemical information residing in the solute structure itself. Signal transmission as shown in a simplified, schematic from in any case is a multistep process: The formation, secretion, diffusion, reception and transmembrane transmission reactions of one single messenger in line from emission of its signal to membrane reception, storage, multiplexing, amplification and translation into biological functions, in principle, are each separate, but interactive signaling devices. Each of them must be subject to complex (positive and negative) feedback regulatory loops. A more elaborated discussion of the terms autocrine, paracrine and endocrine mechanisms has been published by Karlson [25]. In this context, the term "mediator" is used for effector molecules with autocrine and paracrine information transmission mechanisms. Whereas, the term "hormone" ("wound hormone") has been reserved for effector molecules of the inflammatory and healing reactions fulfilling the definitions of hormones, (1) when they transmit their information by endocrine mechanisms, (2) when they are formed by cells in the biological fluids at the reaction site of tissue injury, inflammation and healing; (3) when they have no apparent local activity on adjacent cells, but when their action is directed to cells and biological units remote from this reaction site. Some further overhead names proposed for mediators in general or with specific functions are listed in the figure according to proposals discussed or compiled in references [6,25,36-38,49]

produced by inflammation and healing suggested the existence of wound hormones, none of the cell-derived inflammatory (cytokine) mediators so far isolated or biologically demonstrated, fulfilled the criteria of hormone definitions [25]. The action of all of them is associated with autocrine and paracrine transmission mechanisms. This also applies to the recently published serum protein-derived leukocytosis-inducing protein preparations [39-44]. As far as investigated or specified at all, they have local phlogistic (mediator activity in addition to their leukocytosis-inducing capacity, or are not cell-borne. Hence,

until recently by our work [6-8,19,20,31-35,56] which clearly distinguished mediator effects from wound hormone action, specific endocrine effects of the wound and its cellular constituents could not be defined in terms of an endogenous, cell-secreted chemical substance. Therefore, the term "wound hormone" remained without an object and was also largely left. The reasons for the relatively slow progress in research on polypeptide cytokine effectors are mainly based on the problems encountered with their production and isolation in physical (mg) amounts by efficient and economic processes: Biologically highly specific polypeptide cytokines and, in particular, wound hormones, are trace components of biological fluids and cell culture supernatant solutions. They display their biological activity and are present in biological fluids in the picomolar concentration range, when formed physiologically in vivo and in vitro. A calculation of the necessary amount of cells or tissue as starting material for isolation in highly purified form of physical (mg-) quantities of a polypeptide cytokine effector substance ($M \sim 10^4$, yield $\sim$ 10%) shows that about 100 trillions or 50 kg of cells or tissue are required [6,31,46,56]. Thus, in the case of leukocytes, a minimum of 10,000 l of blood has to be processed for such purpose.

Biotechnical methods have recently been established in my laboratory to isolate for the first time in a high degree of purity in mg-amounts such biologically active, chemical trace signal polypeptides physiologically derived from white blood cells for further functional, clinical and structural investigations. In vivo and in vitro sources for their biotechnical isolation are inflamed tissue sites (is chemically injured, infarcted heart muscle sites), serum and ($\sim$ 1000 ℓ) supernatant solutions of large scale, serum-free culture of ($\sim$ 50 kg $\triangleq 10^{14}$) white blood cells [5-7,12-14,18-20,31-35,45,46,49-58]. A prerequisite was the evaluation of methods devised for batchwise isolation of the different leukocyte types in physiological mixture or as homogeneous population in daily kg-quantities within the short life span of cells [6,31-33,45,46,55-58].

For each of the basic categories of chemical mechanisms for cellular reactions and communication in tissue regeneration and morphogenesis processes (Table 1), distinct effector polypeptides of high biological, reaction and cell specificity can now be shown to exist. They can be obtained highly purified in physical quantities as separate entities. Beside by effectors transmitting information by autocrine and paracrine mechanisms (Fig. 2), leukocytes also may regulate distinct steps in tissue regeneration and morphogenesis processes by endocrine functions (e.g. chemorecrutiment of new leukocytes). This was definitely shown by the existence of true wound hormones [6,19,20,31-35]. Thus, the secretory leukocytic tissue chemically effects the organization of tissue regeneration and morphogenesis by all types of known information transmission mechanisms. In the following, some effector substances operative in chemopoiesis of leukocytes, and chemotropism and morphogenesis of blood vessels will be discussed in more detail.

Chemopoiesis and Chemotropism of Cells: Two Types of Effector-Operated Morphogenesis Reactions

Cytokine Effector Substances in Chemopoiesis of Leukocytes: Models for Studies of Cell Proliferation, Differentiation, Maturation and Storage

Chemopoiesis is a reaction by which hyperplastic and hypertrophic growth at random direction, determination and storage of cells and organisms are determined by chemicals in their environment (Table 1). So far, only effector substances acting on distinct, matured leukocyte types and on some of their progenitor cell types of the bone marrow shall be considered as operative in this mechanism. For this selection of cells out of the variety of cells participating in tissue morphogenesis, Fig. 1 may schematically exemplify several major steps of chemopoietic mechanisms: Hyperplastic and hypertrophic growth, differentiation, maturation and storage of cells (Table 1). Bradshaw and Rubin [59] recently gave an excellent review in relation to the terminology of this subject. In Fig. 1, such steps have been detailed for the development of the hematopoietic system in the bone marrow (left).

Endogenous chemical substances which activate and regulate such processes of cell development, maturation and storage in leukopoiesis, are called leukopoietins [41]. Different types of leukopoietins controlling hematopoiesis as a first step in cell and tissue regeneration process can also be obtained by physicochemical fractionation of the variety of substances formed by leukocytes in the supernatant solutions of biotechnical cultures [6,7,31-35,57,61-64]. They represent activating (mitogen) and inhibitory (chalone) chemical signals for cell proliferation, differentiation and colony formation. Both classes of effector substances prepared from the same biotechnical culture supernatant solutions together verify a (negative) feedback regulatory circuit [6,60], controlling leukocyte proliferation and colony formation [57,61-64]. They are pairs of interfering activities acting directly on the cell surface as one of several possible regulatory principles to which cellular reaction and communication systems may be confined in effector action [6,57,60-64].

Endogenous substances activating the mitotic cycle of leukocyte types capable of proliferation, promote hyperplastic growth and colony formation of bone marrow stem cells of the granulocyte-monocyte line, or of peripheral lymphocytes or macrophages (histiocytes) (Fig. 1). Their physicochemical and biological properties are described in detail elsewhere [46,57,61-64]. On the other hand, endogenous, reaction- and cell-specific, species-nonspecific, non-cytotoxic, reversibly acting, cytostatic effector substances and inhibitors of proliferation and colony formation of bone marrow leukocyte progenitor cells in culture formed by the biotechnically cultured leukocyte types, conform to the biological definition of the term "chalone" [37,49,60-63]. It is used for a class of locally acting reversible inhibitors of cell proliferation [37,60]. It was originally proposed by Schäfer [25,37] to verify negative feedback control of hormone action, a term introduced by Starling in 1906 [25,37,60]. A pattern obtained for separation into cell-specific chalone entities has been published elsewhere (Fig. 3 in [60]) and accomplishes other data on these substances [61-63] so far known.

Novel Effector Substances in Chemotropism of Blood Vessel Cells: Angiogenesis and Neovascularization of Tissue as a Model for Directional Hyperplastic Growth of Cells

Chemotropism is a reaction by which the direction of hypertrophic and hyperplastic growth of cells is determined by chemicals in their environment (Table 2). Pfeffer [65,66] was the first to distinguish this basic mechanism from chemotaxis and chemokinesis of cells [67]. Rosen [68] and Delbrück [69] reviewed some relations of these mechanisms to general cellular recognition phenomena. Directional hypertrophic growth of tissues may be exemplified by nerve growth [70]. Directional hyperplastic growth of (endothelial) cells operated by effector substances of the inflammatory and healing process, may be analyzed as model of tissue regeneration and morphogenesis process.

"Blood vessels go where they are needed" [71]. This statement of Hunter made already over 200 years ago indicates that the homeostasis of the body and its organs and tissues depends on regulatory mechanisms of angiogenesis (lateral and directional growth of blood vessel capillaries). This homeostasis of blood vessel patterns influences tissue repair processes, wound healing, tissue neoformation in embryogenesis and other reproductive cycles as well as survival, out-growth, regression and destruction of tumors, grafts, and of non-vascularized tissues [1-4,6,21,22,49-53,70-87].

Chemotropism of blood vessels is expressed by sprouting of cells present in the proximate environment. This sprouting as a basic ontogenic mechanism and common to all organisms may occur in direction to a substance along its increasing concentration gradient; or away from the

Table 2. Summary of major physicochemical and biological properties of reaction- and cell-specific monocyto-angiotropin. Compiled from [49-53,56]

1. Specific chemotropic activity on blood vessels in vivo:
   - 1.1. Induction of directional growth (chemotropism) of blood vessels along its concentration gradient
   - 1.2. Induction of neovascularization of tissues
   - 1.3. Morphogenesis of vascular patterns in tissues
   - 1.4. During sprouting of vessels, tips of growing capillaries have increased vascular permeability
   - 1.5. Activity threshold is ~2.5 fmol
2. Mitogenic activity on endothelial cells in vitro (activity threshold: ~50 pmol/l)
3. It is substantially free of other biological and toxic effects up to more than 10'000-fold of its threshold
4. Physicochemical properties:
   - 4.1. Protein with a hydrodynamic equivalent of molecular weight 4500
   - 4.2. No quartenary structure in form of physically bound peptide subunits: the native protein consists of only one peptide unit
   - 4.3. It is salt-soluble in ammonium sulfate solutions at concentrations $\geq$ 3.6 mol ammonium sulfate/l

substance along its decreasing gradient. Accordingly, directional growth is called positive and negative chemotropism, respectively [66,68]. As far as known for blood vessels, probably only the first mode (positive chemotropism) is biologically relevant. Angiogenesis is a common feature associated with most inflammatory and tissue regeneration processes provoqued by endogenously and exogenously induced injuries. Numerous factors which themselves are not chemotropically active, but provoque inflammatory reactions in injured tissues, may indirectly induce ingrowth of blood vessels. Such are e.g. silver nitrate, sodium hydroxide, microbial infections (like tuberculosis), immune reactions, etc. Fromer and Klinthworth [72-74] have reviewed the relation of such factors to angiogenesis. Accordingly, soluble substances presumably formed at reaction sites of inflammation, leukocyte accumulation and regenerative growth of tissues have been postulated to act as morphogens for induction and regulation of neoformation of blood vessel systems [3,4,49,50,71-75,77,78]. Several authors have been able to show that soluble extracts of such tissue and the supernatant solutions of (tumor) cell lines in culture can stimulate blood vessel growth [3,70,72-75,78,80-91]. Thus, e.g., tumor angiogenesis factors of so far not well defined nature of molecular weights of about 100,000 [70,78,80,82,83], of 3000 [70], in the molecular weight range of 400-800 [88], and about of 200 [70,89-91], were reported as identified. Of the latter, low molecular weight group of factors identified, one is supposed not to be a prostaglandin or polypeptide [70,89], another one is identified as a prostaglandin (PGE_1) [70, 91]. A third one of this group contains high amounts of zinc and copper ions [70,90]. Therefore, it is assumed by some authors [70,90, 91] that prostaglandins and copper ions exert angiogenic activity in the corneal vascularization assay. The very high molecular weight tumor angiogenesis factor was found to express strong chemotactic activity for mononuclear leukocytes and macrophages [82]. This may point on a relation between inflammation and angiogenic activity displayed by tumors [71,75].

Several authors have pointed on leukocytes and wounds as possible source of angiogenic activity [3,70,75,77,86,87]. Granulocytes, lymphocytes and phagocytosing macrophages appeared to be involved in formation of angiogenic activity of unknown nature and specificity of action [72-75,86,87]. We showed [49] that mitogenic activation of mononuclear leukocytes can provide such activity. According to our proposed terminology [49], specifically acting, endogenous chemical effectors with mitogenic activity for blood vessel (endothelial) cells (chemotropic mitogens) which cause chemotropism of blood vessels along their concentration gradients, are called "angiotropins". Such specific polypeptide effectors can be isolated from supernatant solutions of homogenates of inflamed tissue sites (infarcted heart muscle) and of biotechnical cultures of the different types of leukocytes (Fig. 3) [7,49-53,56]. Granulocytes and mononuclear cells are the sources of such effector substances. In Table 2, some of the properties of monocyto-angiotropin as a most efficient substance in stimulating directional blood vessel growth are summarized. Its action on neovascularization of the cornea of a rabbit is shown in Fig. 4. Monocyto-angiotropin can also induce directional growth of blood vessels in embryonic membranes (Fig. 5) and rabbit skeletal muscle (Fig. 6). Its molecular homogeneity is indicated by several criteria, e.g. by electrophoretic analysis.

As is evident from the data available, monocyto- and granulocyto-angiotropins are polypeptide effectors of high biological specificity which show no cross-reactivity to the activities of the other effector substances so far evaluated [56]. In addition, amongst a set of about

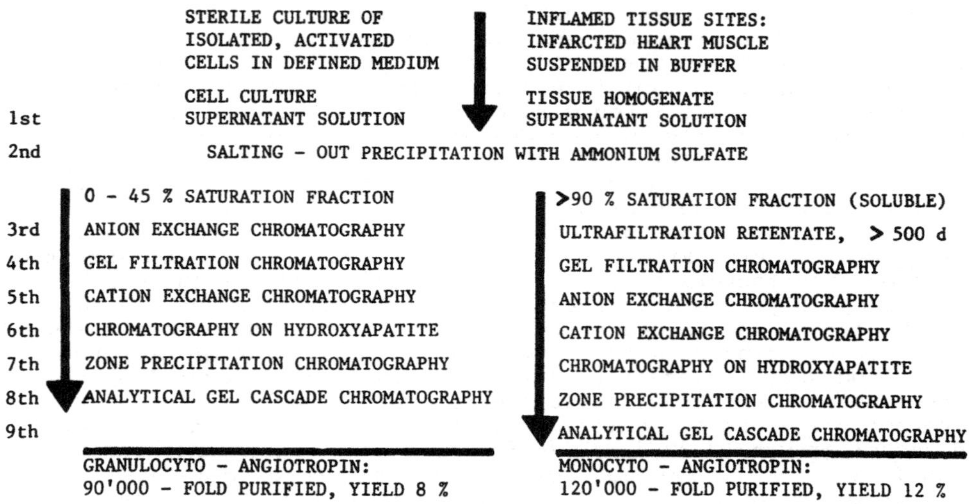

Fig. 3. Strategy for purfication of polypeptide cytokines of leukocytes from cell culture supernatant solutions and inflamed tissue sites at technical scales. The sequences of steps for physicochemical isolation of the cytokine effectors are schematically exemplified for purification of "angiotropins" (Figs. 4-7) produced by different leukocyte types in culture in vitro and in inflamed tissue sites in vivo. Such angiotropins according to our nomenclature [8,9,22,47,49-53] are natural endogenous, chemical mitogens for chemotropism (directional sprouting) of blood vessels. Similar sequences of steps are used for purification of other polypeptide cytokines (e.g. chemopoietins) which are formed concurrently to angiotropins by different leukocyte types under equivalent in vivo and in vitro conditions. Details on used methods are reported elsewhere [49-53,56]

200 substances, including salts, vitamins, hormones, fatty acids and their derivatives (e.g. prostaglandins A_1, A_2, E_1, E_2), biogenic amines (e.g. histamine), biologically active peptides (e.g. bradykinin) and several proteins and enzymes, only the angiotropins express activity in terms of efficient blood vessel growth. Thus, for example, none of the first 93 substances of a cell culture medium [54-56] display angiogenic activity. In summary, it is conceivable that evolution has equipped leukocytes with a set of highly specific polypeptide effector substances for operation of chemical mechanisms in regeneration and morphogenesis of blood vessel patterns. Obviously, leukocytes as carriers of specific endogenous information for tissue regeneration and pattern formation processes have retained the capability to activate physiological mechanisms for blood vessel sprouting and morphogenesis of blood vessel patterns by provision of adequate, topochemically specific signal substances. It represents one prerequisite for tissue regeneration and wound healing: The newly formed growing tissue can be nourished through supply of blood vessels; *and* through blood supply it becomes easily accessible to defense cells (leukocytes) for control of its compatibility.

This is evident from a variety of situations which lead to changes in vascular patterns following local ischemic injury, leukocyte (monocyte) diapedesis and infiltration of the tissue as well as in atherosclerosis and tumor growth [3,4,6,50,71,75-78]. Examples are the formation and obliteration of collateral vessels and arteriovenous communications in muscles [22,71], lung, tongue, ileum, embryonic and reproductive tissue as well as in atherosclerosis [71,75-77,84].

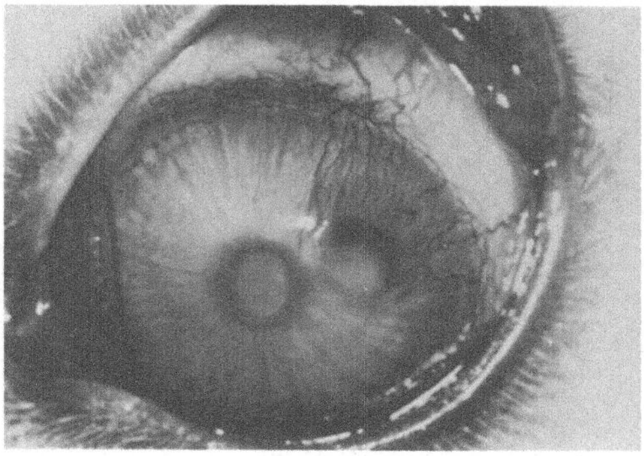

Fig. 4. Biological action of highly purified monocyto-angiotropin (MAT) as one representative of endogenous chemotropism polypeptide signals of leukocytes (monocytes) for communication with blood vessel cells: Chemotropism (directional sprouting) of blood vessels out of the limbal plexus of the corneo-scleral junction at the corneal circumference and neovascularization of a rabbit cornea centripetal in direction to the intracorneal application site of 10 fmol MAT. It can be estimated that a maximum of 10^4 to 10^5 cells are sufficient to produce this biologically active amount of MAT in an in vivo situation. The assay is performed according to our modification of classical methods of intracorneal, topical application of solutions [49,50]. The blue spot (exuded, extravasal Evans blue as a previously intraveniously applied dye marker) in the cornea around the tips of growing vessel sprouts demonstrates an increased vascular permeability of these tips (*"leaky tip"*) in relation to the lateral walls of growing sprouts. This property of sprouts which is lost upon morphogenesis of closed vascular circulation patterns allows blood solutes (e.g. dye bound to plasma proteins) to penetrate easily out of vessels at their tips into the tissue. But the blood cells are retained from passive exudation within the vessel sprout, as are blood proteins on lateral vessel walls. The demonstration of this property of growing vessel sprouts I consider as important for evaluation of molecular mechanisms operative in directional sprouting morphogenesis of blood vessel patterns thereof and in atherosclerosis plaque formation [21,22,49]: Angiotropins are the first mediators known which produce "leaky tips" in vessels. From [56]

Neovascularization and the formation of abnormal vessel patterns in tissues may be induced by only one endogenous polypeptide effector substance (e.g. monocyto-angiotropin). If either endogenously produced by activated leukocytes, or exogenously applied in a highly purified form at the wrong place or at the wrong time, it may fully mimic pathological manifestations of morphogenesis of blood vessel patterns (Fig. 4). Such may occur, e.g. in ocular tissues, in diabetes, tuberculosis, hypersensitivity and other inflammatory reactions, e.g. in response to contact lens carriage, in tumor growth and, in general, following ischemic tissue damage and leukocyte accumulation [9]. It can be calculated that about 10^6 molecules monocyto-angiotropin derived from a maximum of 10 to 10^2 (secreting) cells are sufficient to achieve full vascularization, e.g. of a cornea surface.

Formation, Recognition and Turnover of Morphogenic Patterns of Tissues

Muscle and Embryonic Tissue Neovascularization. Our experimental approach to the problem of formation, recognition and turnover of a morphogenic pattern of a tissue as the final aim of the inflammatory and healing process has been concentrated on the blood vessel system in non-vascularized (cornea) and vascularized tissues (embryonic membrane and

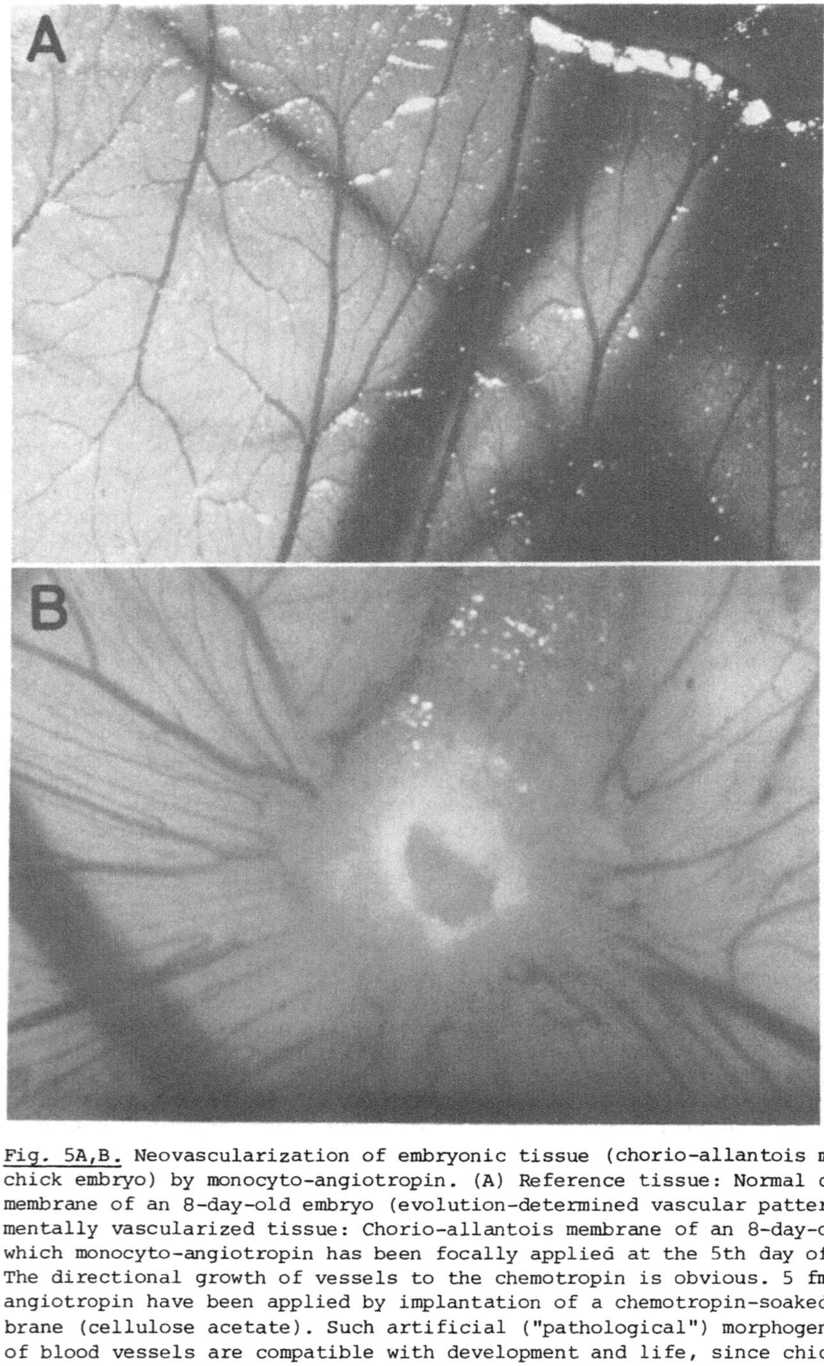

Fig. 5A,B. Neovascularization of embryonic tissue (chorio-allantois membrane of a chick embryo) by monocyto-angiotropin. (A) Reference tissue: Normal chorioallantois membrane of an 8-day-old embryo (evolution-determined vascular pattern). (B) Experimentally vascularized tissue: Chorio-allantois membrane of an 8-day-old embryo to which monocyto-angiotropin has been focally applied at the 5th day of its life. The directional growth of vessels to the chemotropin is obvious. 5 fmol monocyto-angiotropin have been applied by implantation of a chemotropin-soaked filter membrane (cellulose acetate). Such artificial ("pathological") morphogenic patterns of blood vessels are compatible with development and life, since chicken can be bred and borne from such eggs

skeletal muscle) by use of monocyto-angiotropin. The neovascularization by this effector of such different tissues (of different species) is shown in Fig. 4 (rabbit cornea), Fig. 5 (chorio-allantois membrane of a chicken embryo) and Fig. 6 (rabbit skeletal muscle).

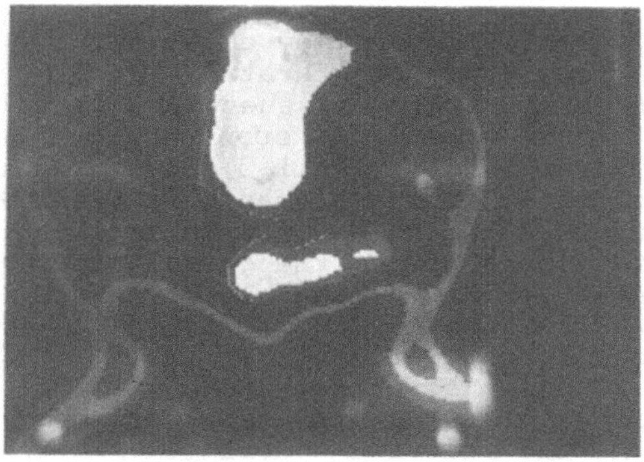

Fig. 6. Neovascularization of skeletal muscle of rabbits by monocyto-angiotropin: Demonstration of (hemodynamic) functions of newly formed blood vessels. 50 fmol monocyto-angiotropin has been focally applied to the skeletal muscle of the right hind leg. After 2 days, a ^{99}technetium scintigram is made according to usual procedures as detailed in references [22,47]. Asymmetric distribution of ^{99}Tc shows that monocyto-angiotropin is able to induce fast formation of functionally efficient, additional blood circulation patterns with active hemodynamics in muscles as compared to the reference (left hind leg): ^{99}Tc marker appears in the circulation of the right muscle as fast and as intense as in the large axial vessels of the body (figure taken 3 s after intravenous application of the marker). The reference muscle in the right hind leg was treated as the left one, but only culture medium composed of the first 93 components as given in references [54-56] was applied

These experiments show that the action of the leukocyte-derived polypeptide effectors for chemotropism of blood vessel cells is not restricted to a certain tissue. In a non-vascularized tissue, such as the cornea, new vessels grow in from adjacent vessel sites. This situation (for a case of "pathological" vessel formation) is similar to ingrowth of vessel cells into embryonic tissue [71] and into tumors [71,77,78,80-83,92]. In already vascularized tissue (membrane and muscle), a completely other morphogenic pattern with additional vessels is formed by this polypeptide effector. A basic question was, whether or not such altered patterns, e.g. as shown for the embryonic membrane (Fig. 5B), are compatible with development and life; in particular, since e.g. the chorio-allantois membrane has important functions for the embryo: At least, it serves as a kind of "fetal lung" for supply of the embryo with oxygen and gas exchange. Experiments to answer this question clearly show: Embryos subjected to such an altered blood vessel pattern in their chorio-allantois membrane early (third to sixth day) in their life, develop normally and show no different behaviour and functions with reference to chicken not treated with polypeptide effectors. Thus, beside the lack of toxicity of the endogenous polypeptide effector for chemotropism of blood vessel cells for a very sensitive embryo, the "pathological" blood vessel patterns formed artificially are life-compatible. Such artificial patterns are maintained in the membrane during the embryonic life until birth. They seem not to be replaced in this (21-day) period by "normal" patterns.

Blood vessel patterns artificially formed in (resting) muscles of the hind leg of non-trained rabbits have been determined to exist for about 20 days upon one single application of a bolus of microliters of diluted angiotropin solution (50 fmol monocyto-angiotropin), as far as detectable in terms of the existence of blood vessels with active flow. In contrast, blood vessels formed in the cornea upon application of one single dose of monocyto- or granulocyto-angiotropin, may exist for three month or longer, until they are completely obliterated. Such results may show that the newly formed blood vessel tissues in various host tissues have functions [22], but a different life span and turnover, when they are not maintained by additionally applied polypeptide effectors. These different turnover rates of formed elements in tissues indicate the different needs of tissue to adapt to environmental changes: Whereas muscle has to adapt fast to such needs, the cornea, as representing a tissue normally resistant to invasion with blood vessels, does not. This also may indicate that formation of a new tissue in a host tissue and its turnover is subject to regulation to other factors, so far not identified in nature and mechanism of action: Morphogenesis of a tissue results from a multifactoral interplay and the predominant factor at a certain time will determine the reaction occuring.

A Possible Role of Leukocytic Effectors in a Novel Mechanism for Atherogenesis. In principle, since the polypeptide effectors are not toxic and the morphogenetic patterns formed by them have functions and a normal turnover and are compatible with the functions of the host tissue, they may well be considered as a future clinical tool to vascularize tissues by endogenous means when necessary: Blood vessel sprouting may be induced by monocyto-angiotropin within 12 h. After 2-3 days already, complete blood vessel circulation pattern begin to form from sprouts [22,56]. Several authors have found differences in structure and function of newly formed vessel sprouts. In particular, they emphasized the different permeabilities of sprout walls when compared to normal vessels [71,75,76,79]. However, which role plays the tip? We have investigated possible mechanisms of this sprout "finding". Figure 4 exemplifies such an experiment concerning the different vascular permeabilities lateral walls have in comparison to tips of sprouts. The higher permeability of the tips as indicated by exudation of intravenously applied marker dye (Evans blue) and proteins show that chemokinesis and chemotaxis mechanisms may be well involved in locomotion and direction finding of different sprouts for formation of a closed circulation: Humoral and cellular endogenous effector substances may be derived or present in blood to make endothelial cell motile and finding direction, as this is the case for leukocytes [5,6,12-17,46,58]. Hence, structural stability and morphogenesis of blood circulation patterns formed from different sprouts may be subject to multiple factors, as also reasoned by others on theoretical grounds [92,93].

On the other hand, this property of growing sprouts ("*leaky tips*") [21] may also play a paramount role in formation of atherosclerotic plaques: Exudation and insudation of components are well known mechanism and characteristics intrinsic to them [76,94,95]. Thus, monocyte adhesion and diapedesis inside vessels shown [76] as early event in atherogenesis, may result in an endogenously formed reversible "lesion" ("*intravasally leaky tip*") by topochemically false formation of monocyto-angiotropin. Interstitially formed, it would result in a vessel sprout growing on surfaces: Therefore, a first step in atherogenesis in this context [21] is considered not to be a "response-to-injury" [96]; in the meaning of this hypothesis, that mechanical, chemical or immunological injury would form a true lesion in the endothelial surface layer of vessels which could lead to clinical sequelae, e.g. aberrated smooth muscle cells, thrombosis and infarction. But an endogenous reversible "lesion" ("intravasally leaky tip") as *cellular reaction* to topochemically false formed monocyto-angiotropin may result alone from metabolic disorders or false nutrition: Pathological manifestation ("plaque") of this "intravasally leaky tip" occurs only upon perfusion perturbation by disordered composition or metabolism of biological fluids and cells [21], as conceptually earlier supposed by Doerr [94]. Thus, e.g., on the one hand, hyperlipidemic plasma does not allow serum albumin to play its role as positive chemokinesin [97,98]; and, thus, to minimize the tendency of leukocytes to adhere to surfaces and to aggregate. In contrast, such cells even may die [98]. On the other hand, as in diabetes, pathological glucose levels may fire direct formation of reducing equivalents in form of NADPH by D-xylose: NADP-oxidoreductase [99] in blood vessel cells, without being subjected to normal metabolic control. Polyol formation and accumulation, as well as increased fatty acid synthesis may directly contribute by cycling of NADPH equivalents to disorders in membrane transport systems and ordered membrane phases in blood vessel cells. Together with other factors involved [98-100], they can bring about wrongly programmed monocyte adhesion and functions at surfaces of endothelial cells: It is a common method to bring monocytes to adhere to surfaces, to emigrate and to accumulate in interstitial cavities by means of some hydrophobic substances (mineral oil, fatty acids, etc.). Such mechanism also can be used to make monocytes producing monocyto-angiotropin in vitro. However, concerning the in vivo situation of focal, intravasal production of monocyto-angiotropin which may produce "leaky tips" in vessels, as chemotropin it is not a wound hormone, but an interstitially acting and regulated mediator (Fig. 2). This emphasizes the importance of the topochemistry of effector formation dictated by cellular behaviour for physiological function of the effector system.

On the other hand, the determination of final "physiological" pattern formation is doubtless also a major function of the host tissue. Thus, e.g. monocyto-angiotropin once formed by a macrophage, induces formation of various vessel sprouts. Only few of them at last make up formed blood vessel circulation patterns. In tissues, they are recognized as additional vessels and not intrinsic to the normal pattern program of the tissue. Later they are obliterated or tolerated only, when needed for tissue function. By our recognition of the enormous chemical repertoire of specific effector substances which the chemical machinery of the monocyte-macrophage-histiocyte phenotype family can produce and control, I may suggest that situations which lead to changes in vascular patterns of tissues under physiological conditions without apparent tissue injury, are mainly influenced by tissue-resident histiocytes.

Acknowledgements. The author gratefully acknowledges the assistance and cooperation of Mr. Helmut Renner.

References

1. Menkin V (ed) (1950) Newer concepts of inflammation. Thomas, Springfield, Illinois
2. Menking V (ed) (1956) Dynamics of inflammation, 2nd ed. Thomas, Springfield, Illinois
3. Glynn LE, Houck JC, Weissmann G (eds) (1979-1981) Handbook of inflammation, vol 1 (Houck JC, ed) Chemical messengers of the inflammatory process, vol 2 (Weissman G, ed) The cell biology of inflammation, vol 3 (Glynn LE, ed) Tissue repair and regeneration. Elsevier, North Holland, Amsterdam
4. Zweifach BW, Grant L, McCluskey RT (eds) (1973, 1974) The inflammatory process, vols 1-3, 2nd ed. Academic Press, New York
5. Wissler JH (1975) In: Urbaschek B, Urbaschek R, Neter E (eds) Proc Immunosymp Wien 1973: Gram-negative bacterial infections and mode of endotoxin actions; pathophysiological, immunological and clinical aspects. Springer, Wien, pp 91-105
6. Wissler JH (1980) Habilitationsschrift, Ruhr-Universität Bochum

7. Wissler JH (1982) Z Physiol Chem 363:131
8. Burdach StEG, Evers KG, Wissler JH (1982) Monatsschr. Kinderheilkd 130:789-791
9. Wissler JH, Logemann E (1979) Z Physiol Chem 360:1204-1205
10. Wissler JH (1979) Z Physiol Chem 360:402-403
11. Wissler JH (1979) Immunobiology 157:253-254
12. Wissler JH (1971) Experientia 27:1447-1448
13. Wissler JH (1972) Eur J Immunol 2:73-83
14. Wissler JH (1972) Eur J Immunol 2:84-89
15. Wissler JH, Stecher VJ, Sorkin E (1972) Eur J Immunol 2:90-96
16. Wissler JH, Stecher VJ, Sorkin E (1972) Int Arch Allergy 42:722-747
17. Wissler JH, Stecher VJ, Sorkin E (1972, 1973) In: Peeters H (ed) Proc 20th Coll Protides of the Biological Fluids, Brugge. Pergamon Press, Oxford, pp 411-416
18. Wissler JH (1981) Eur. Pat. Publ EP 0 042 560 A2 DOS DE 30 22 914 A1, Max-Planck-Gesellschaft zur Förderung der Wissenschaften, München, pp 1-53
19. Wissler JH, Pfefferkorn B, Rother K, Rother U, Schramm L, Renner H, Renker H, Wissler AM, Schaper W (1980) Z Physiol Chem 361:1358
20. Wissler JH (1982) German Patent DE 30 34 529 C2/1982, Max-Planck-Gesellschaft zur Förderung der Wissenschaften, München, pp 1-44
21. Wissler JH (1982) Z Physiol Chem 363:941
22. Gottwik M, Renner H, Wissler JH (1982) Z Physiol Chem 363:938-939
23. Fritz H, Tschesche H, Greene LJ, Truscheit E (eds) (1974) Proc 2nd Int Res Conference: Proteinase inhibitors, Bayer-Symposium V. Springer, Berlin Heidelberg New York
24. Dumonde DC, Wolstencroft RA, Panayi GS, Matthew M, Morley J, Howson WT (1969) "Lymphokines": Non-antibody mediators of cellular immunity generated by lymphocyte activation. Nature 224:38-42
25. Karlson P (1982) Naturwissenschaften 69:3-14
26. Leber Th (1888) Fortschr Med 6:460-464
27. Lewis T (ed) (1927) The blood vessels of the human skin and their responses. Shaw, London
28. Dold H (1915) Dtsch Arch Klin Med 117:206-223
29. Erdös EG (ed) (1979) Bradykinin, kallidin and kallikrein (Handbook of experimental Pharmacology, vol 25 suppl). Springer, Berlin Heidelberg New York
30. Werle E, Götze W, Keppler A (1937) Biochem Z 289:217-225
31. Wissler JH (1982) Eur Pat Publ EP 0 061 140 A2 DOS DE 31 10 561 A1, Max-Planck-Gesellschaft zur Förderung der Wissenschaften, München, pp 1-72
32. Wissler JH, Renner H, Gerlach U, Wissler AM (1981) Z Physiol Chem 362:244-245
33. Wissler JH, Renner H, Bodzian F, Gerlach U, Wissler AM (1981) Exp Hematol 9, suppl 9:35
34. Wissler JH, Renner H, Gerlach U, Gottwik M, Wissler AM (1981) Immunobiology 159:121-122
35. Wissler JH, Renner H, Pfefferkorn B, Gottwik M, Wissler AM (1981) Biochem Soc Trans 9:257P
36. Selye, H (ed) (1965) The mast cell. Butterworth, Washington DC
37. Houck JC (ed) (1976) Chalones. Elsevier-North-Holland, Amsterdam
38. Lobue J, Gordon AS (eds) (1973) Humoral control of growth and differentiation, vol 1: Vertebrate regulatory factors. Academic Press, New York
39. Rother K (1972) Eur J Immunol 2:550-558
40. Ghebrehiwet B, Müller-Eberhard HJ (1979) J Immunol 123:616-621
41. Whipple HE, Spitzer MI, Bierman HR (eds) (1964) Ann NY Acad Sci 113:511-1092
42. Van Waarde D, Hulsing-Hesselink E, Sandkuyl LA, van Furth R (1977) Blood 50:141-154
43. Van Waarde D, Hulsing-Hesselink E, van Furth R (1977) Blood 50:727-742
44. Grippa L, Sebastiani A (1978) Il Progr Medico 34:1-5
45. Wissler JH (1982) U.S. Patent 4,343,793/1982, Max-Planck-Gesellschaft zur Förderung der Wissenschaften, Washington pp 1-10
46. Wissler JH, Arnold M, Gerlach U, Schaper W (1980) Z Physiol Chem 361:351-352
47. Sauer E, Sebening H (eds) (1980) Myocard- und Ventrikelszintigraphie. Grundlagen und Anwendung. Mannheimer Morgen Verlag, Mannheim
48. Böyum A (1968) Scand J Clin Lab Invest 21, suppl 97:1-109

49. Wissler JH, Renner H (1981) Z Physiol Chem 362:244
50. Wissler JH, Schaper W (1982) Eur Pat Publ EP 0 061 138 A2 DOS DE 31 10 560 A1 Max-Planck-Gesellschaft zur Förderung der Wissenschaften, München, pp 1-64
51. Wissler JH, Gottwik M, Renner H, Klein HH, Gerlach U, Wissler AM, Schuurmans R, Schaper W (1981) Fed Proc 40:1638
52. Wissler JH (1981) Kurzreferateband 19. Hauptversammlung Gesellschaft Deutscher Chemiker, Verlag Chemie, Weinheim, p 186
53. Wissler JH (1981) Immunobiology 160:131-132
54. Wissler JH (1982) Eur Pat Publ EP 0 060 565 A2 DOS DE 31 10 559 A1, Max-Planck-Gesellschaft zur Förderung der Wissenschaften, München, pp 1-27
55. Wissler JH (1982) In: Brunner G, Fischer G (eds) Proc First Eur Conf on Serum-Free Cell Culture Advances in Hormone-Supplemented Media, Heidelberg 1982. Springer, Berlin Heidelberg New York
56. Wissler JH (1982) In: Bundesministerium für Forschung und Technologie BMFT (ed) Proc. BMFT-Statusseminar "Tierische Zellkulturen" Jülich 1981. Projektträger Biotechnologie Verlag KFA Jülich, pp 293-303
57. Wissler JH (1982) Eur Pat Publ EP 0 061 139 A2 DOS DE 31 10 611 A1, Max-Planck-Gesellschaft zur Förderung der Wissenschaften, München, pp 1-64
58. Wissler JH (1982) Eur Pat Publ EP 0 061 141 A2 DOS DE 31 10 610 A1, Max-Planck-Gesellschaft zur Förderung der Wissenschaften, München, pp 1-82
59. Bradshaw RA, Rubin JS (1980) J Supramol Struct 14:183-199
60. Maurer HR (1980) Apoth Z 120:839-844
61. Wissler JH, Maurer HR, Kastner M, Maschler R, Neumeier R, Arnold M, Gerlach U, Pfefferkorn B, Tschesche H, Schaper W (1980) Eur J Cell Biol 22:387
62. Maurer HR, Kastner M, Maschler R, Neumeier R, Arnold M, Gerlach U, Glendinning K, Pfefferkorn B, Wissler JH (1981) Z Physiol Chem 362:221
63. Maurer HR, Kastner M, Maschler R, Wissler JH (1981) Exp Hematol 9 suppl 9:34
64. Neumeier R, Maurer HR, Arnold M, Gerlach U, Glendinning K, Renner H, Wissler JH (1982) Z Physiol Chem 363:193-195
65. Pfeffer W (1884) Unters Botan Inst Tübingen 1:524-533
66. Pfeffer W (1904) In: Pfeffer W (ed) Pflanzenphysiologie, 2. Auflage, Leipzig pp 561-598
67. Rothert W (1901) Flora 88:371-421
68. Rosen WG (1962) Quart Rev Biol 37:242-259
69. Delbrück M (1972) Angew Chem 84:1-7
70. Baserga R (ed) (1981) Tissue growth factors (Handbook of experimental Pharmacology, vol 57). Springer, Berlin Heidelberg New York
71. Liebow AA (1965) In: Hamilton WF, Dow P (eds) Handbook of physiology, section 2, circulation, vol 2, Amer Physiol Soc, Washington DC, pp 1251-1276
72. Fromer CH, Klintworth GK (1975) Amer J Pathol 79:537-550
73. Fromer CH, Klintworth GK (1975) Amer J Pathol 81:531-554
74. Fromer CH, Klintworth GK (1976) Amer J Pathol 82:157-170
75. Peterson HI (ed) (1979) Tumor blood circulation: Angiogenesis, vascular morphology and blood flow of experimental and human tumors. CRC Press, Boca Raton
76. Schwartz CJ, Gerrity RG, Lewis LJ (1978) Atherosclerosis Res 3:109-124
77. Algire GH, Chalkley HW, Legallais FY, Park HD (1945) J Natl Cancer Inst 6:73-85
78. Folkman J (1974) Cancer Res 34:2109-2113
79. Abell RG (1946) Amer J Physiol 147:237-241
80. Folkman J, Cotran R (1976) Int Rev Exp Pathol 16:207-248
81. Klagsbrun M, Knighton D, Folkman J (1976) Cancer Res 36:110-114
82. Mostafa LK, Jones DB, Wright DH (1980) J Pathol 132:191-205
83. Mostafa LK, Jones DB, Wright DH (1980) J Pathol 132:207-216
84. Gospodarowicz D, Thakral KK (1978) Proc Natl Acad Sci USA 75:847-851
85. Gospodarowicz D, Brown KD, Birdwell CR, Zetter BR (1978) J Cell Biol 77:774-788
86. Polverini PJ, Cotran RS, Gimbrone jr MA, Unanue ER (1977) Nature 269:804-806
87. Sidky YA, Auerbach R (1975) J Exp Med 141:1084-1100
88. Fenselau A, Watt S, Mello RJ (1981) J Biol Chem 256:9605-9611
89. Weiss JB, Brown RA, Kumar S, Phillips P (1979) Brit J Cancer 40:493-496
90. McAuslan BR, Hoffman H (1979) Exp Cell Res 119:181-190
91. Ben Ezra D (1978) Amer J Ophthalmol 86:455-461
92. Folkman J, Hochberg M (1973) J Exp Med 138:745-753

93. Thom R (ed) (1975) Structural stability and morphogenesis. An outline of a general theory of models. WA Benjamin, Reading Massachusetts
94. Doerr W (ed) (1963) Perfusionstheorie der Arteriosklerose. G. Thieme, Stuttgart
95. Joris J, Majno G (1979) Adv Inflammation Res 1:71-85
96. Ross R, Glomset J, Harker L (1978) Atherosclerosis Rev 3:69-76
97. Keller HU, Wissler JH, Hess MW, Cottier H (1978) Eur J Immunol 8:1-7
98. Wissler JH, Jennissen HP, Keller HU (1978) Z Physiol Chem 359:1462
99. Wissler JH (1977) Z Physiol Chem 358:1301
100. Wissler JH (1979) J Mol Cell Cardiol 11 suppl 2:67

Signals of Cell Determination in Embryogenesis

H. Tiedemann[1]

Amphibian eggs and embryos have been widely used to study the development of vertebrates. Early stages which can be handled with relative ease, are endowed with a high regulatory capacity. To make us familiar with amphibian embryogenesis I will briefly outline the development of the early stages.

Development of the Amphibian Embryo

The amphibian oocyte is enlarged about 30,000 times during oogenesis and develops thereby an animal-vegetal polarity which is especially evident in the distribution of pigment (Fig. 1) and yolk. After fertilization a contraction wave moves towards the side of the sperm entry and the cytoplasmic streaming establishes a dorso-ventral polarity, whereby the dorsal side is located opposite to the random point of sperm entry [1]. Yolk and pigment do not have a direct function in the control of morphogenesis. But a polar distribution of molecules which control cell differentiation exists already in the fertilized egg and early embryo [2,3].

After a certain number of cell divisions, a cavity, the blastocoel, appears in the middle of the embryo, which separates the presumptive

Fig. 1. Egg of *Xenopus laevis*. The dark pigmented animal half develops to epidermis and the neural system. Endoderm and mesoderm derived organs develop from the unpigmented vegetal half. (Courtesy Dr. Grundmann)

1 Institut für Molekularbiologie und Biochemie der Freien Universität Berlin, Arnimallee 22, D-1000 Berlin 33

33. Colloquium - Mosbach 1982
Biochemistry of Differentiation and Morphogenesis

ectoderm in the animal half from the presumptive endoderm and mesoderm in the vegetal half. During gastrulation the blastopore is developing in the endodermal region and through it endoderm and mesoderm invaginate (Fig. 2). After invagination the mesoderm induces in the overlying dorsal ectoderm the neural plate, the primordium of the neural anlage. In the late neurula and the tail bud stage the differentiation of organs begins and finally the hatching larva is formed. The dorsal mesoderm separates into notochord and myomeres, whereas pronephric tubules, blood cells and the endothelium develop from the ventral mesoderm.

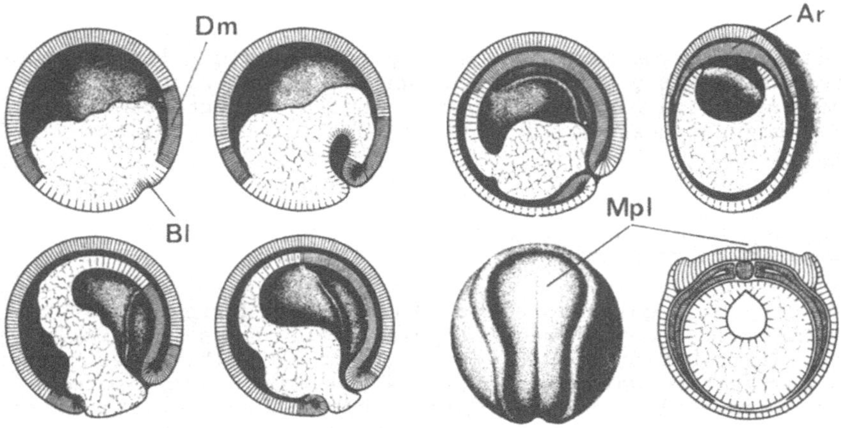

Fig. 2. Gastrulation and neurulation in amphibia. *Bl* blastopore, *Dm* dorsal mesoderm, *Ar* archenteron roof (anlage for notochord and myomeres), *Mpl* Medullary plate, primordium of the neural system. (Designed by V. Hamburger and B. Meyer from H. Spemann: Experimentelle Beiträge zu einer Theorie der Entwicklung, Springer, 1936)

About 50 years ago it was already discovered by Spemann, Mangold, Bautzmann, and Holtfreter [4-7] that the ectoderm until the early gastrula stage, in contrast to endoderm and mesoderm, is not irreversibly determined to its later fate. Isolated ectoderm forms epidermis-like cells [8] in irregular order. It can however, differentiate into other tissues such as muscle or brain, when appropriate inducers are added, i.e. its determination can be changed. Determination has been defined by Hadorn [9] as a process which initiates a specific pathway of development by singling it out from among various possibilities for which a cellular system is competent. This definition does not imply particular mechanisms, which can indeed widely differ.

Gastrula Ectoderm as a Test System for Inducing Factors

Because ectoderm of early gastrula stages is still totipotent and its differentiation can be channelled into other pathways gastrula ectoderm is very suitable for testing inducing factors. Different test methods have been developed (Fig. 3). The most convenient is the implantation method, whereby whole gastrulae are used [10]. A pellet of the material to be tested is implanted into the blastocoel so that during the gastrulation movements it is brought into contact with the ventral ectoderm, which can be induced. The substances can be tested

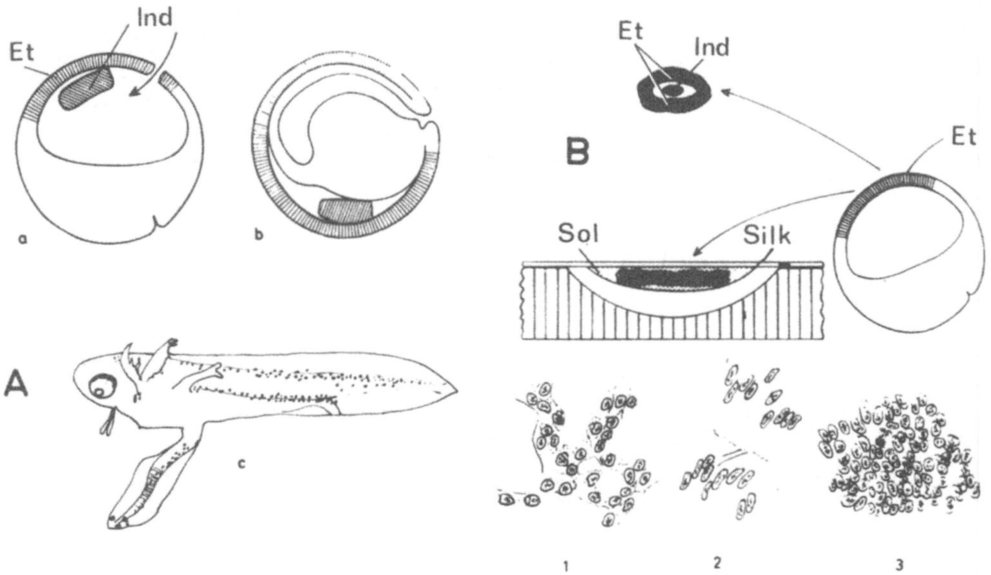

Fig. 3. Test methods for inducing factors. Ectoderm of an early gastrula (*Et*) is brought into contact with the inducer (*Ind*) either by the implantation method or by the sandwich method. In this method the inducer is wrapped into two pieces of isolated gastrula ectoderm (*Et*). In the third method [11] ectoderm is explanted in salt solution (*Sol*) and covered with silk [5] to prevent curling off. The results are shown schematically in the lower part of the figure. [1] Strands of cuboidal cells, if no inducing factor is added. [2] Differentiation of muscle after addition of crude vegetalizing factor. [3] Differentiation of neural tissue after addition of neural inducing factor

in different dilutions when during the preparation of the pellet an inert, i.e. non-inducing protein is added in proportions up to 1:250,000. The results are highly reproducible. Another method adopted from conventional tissue culture methods [12,13] allows a test of water soluble substances of low molecular weight, besides macromolecules. The inducing activity is judged from the percentage of positive cases and the size of the resulting inductions. The different test methods give similar results.

I will now briefly discuss: (1) the chemical nature of the substances which can change the determination of gastrula ectoderm, (2) the problem whether the formation of organ patterns depends on different threshold concentrations of one and the same factor or on the sequential interaction of several fators and (3) experiments to reveal the mechanism of action of inducing factors.

The Chemical Nature of Inducing Factors

A protein was isolated from chicken embryos, which induced primarily endoderm and, when secondary interactions are involved, also mesodermal organs as notochord, myomeres, renal tubules, blood cells, and endothelium. Because in normal development these tissues are derived

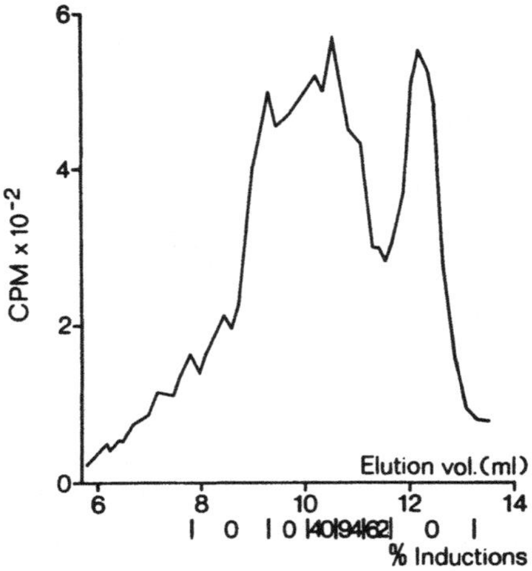

Fig. 4. Size exclusion chromatography of the vegetalizing factor from chicken embryos at 20°C. Four columns (250 × 4.6 mm) with glycerol-propyl derivatized LiChrospher Si 300 (10 µm) were connected. Eluent: formic acid/water/1-propanol (5:4:1; v/v). Flow rate: 0.15 ml/min. Pressure: 34 bar. Fraction volume: 0.15 ml

from the vegetal half of the egg, the factor has been called "vegetalizing factor". It can be separated from ribonucleic acids and polysaccharides by extraction with phenol. After several steps including isoelectric focusing, the factor was finally isolated by high pressure size exclusion [14,15] and high pressure reversed phase chromatography [16]. Figure 4 shows the size exclusion chromatography with 50% formic acid as eluent. Derivatized silica particles (Diol-LiChrospher Si 300) were used as column packing. The fraction with the highest inducing activity was then applied to a reversed phase column (RP8-LiChrosorb Si 100) and eluted with a linear formic acid-1-propanol gradient (Fig. 5). To facilitate the detection of very small amounts the factor was labelled to high specific activity by reductive methylation of the amino groups with formaldehyde and $[^3H]NaBH_4$. Up

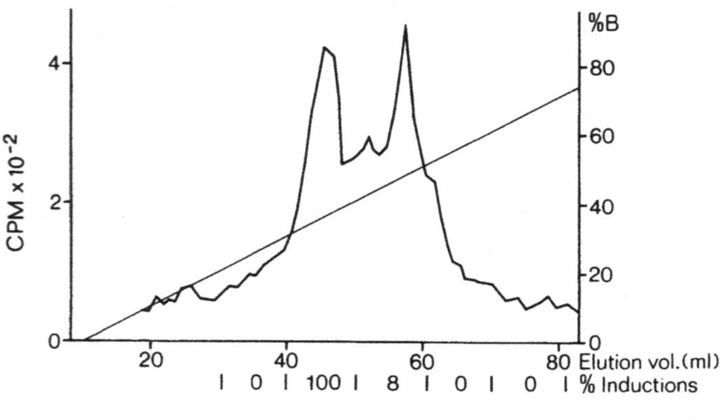

Fig. 5. Reversed phase chromatography of the vegetalizing factor on LiChrosorb-RP-8 (7 µm; 250 × 4.6 mm) at 20°C. The fraction with the highest inducing activity from size exclusion chromatography was diluted to 15% formic acid and applied to the RP-8 column. Solvent A: 15% formic acid. Solvent B: 30% formic acid/1-propanol (1:1; v/v). Slope of linear gradient: 1% B/min. Flow rate: 1 ml/min. Pressure: 100-180 bar

to 20% of the amino groups can be methylated without a decrease of the inducing activity [17]. The yield is very low, about 10μg from 1 kg of chick embryo trunks. About 1×10^{-4} μg/per gastrula is needed to evocate inductions in about 60% of the cases. Still a large number of molecules, more than 10^5, is applied per competent cell under these conditions. But only a relative small percentage of the molecules is taken up by the cells as the autoradiography with labelled factor has shown. The factor can be inactivated by proteolytic enzymes [18]. In 6 M urea the isoelectric point is at pH 8 and the molecular weight is about 26,000-30,000 Dalton. The factor is alkali labile, but very resistant to treatment with acids. The factor dissociates in 50% formic acid into subunits of a molecular weight of 13,000-14,000 Dalton [14]. The inducing activity is not diminished after dissolution in formic acid when the factor is transferred back to neutral aqueous solution before being tested. Chicken embryos contain a "neuralizing factor" which is also protein in nature. The factor induces forebrain with eyes (Fig. 6). It has only partially been purified [19].

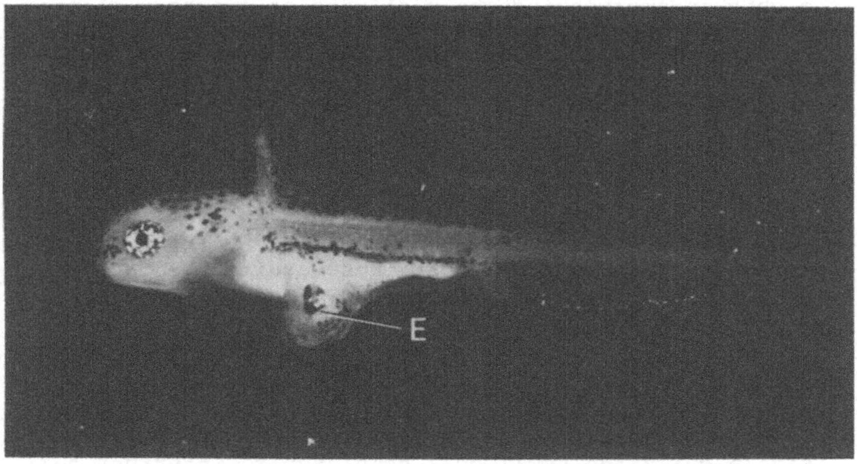

Fig. 6. Induction of forehead with forebrain and eye (*E*) by a neuralizing factor

The isolation of inducing factors from amphibian embryos is much more difficult because these embryos are only available in g-quantities but not in kg quantities like chicken embryos. Extracts from *Xenopus* gastrula stages induce preferentially neural tissues but contain also some vegetalizing factor. Both factors are inactivated by treatment with trypsin and must therefore be protein in nature like the inducing factors from chicken embryo [20]. The factors are soluble in phenol as the chicken factors and many other proteins. Other molecular properties differ however. The molecular weight of the "vegetalizing factor" from gastrulae of *Xenopus laevis* or *Triturus alpestris* is much higher (about 60,000-100,000 Dalton) than the molecular weight of the chicken factor [21]. It is possible that the amphibian factor is a precursor from which active factor with a short half life is generated. It is not unlikely that the inducing activity depends on similar domains of protein structure in the amphibian and in the chicken factor. The chicken factor can then be regarded a master-key for the factor acting in normal development. It should be emphasized that the chicken factor can induce a similar chain of reactions as those which take place in normal development. The factor is therefore a valuable tool to study these interactions.

The "neuralizing factor" in the 100,000 *g* supernatant from *Xenopus* gastrulae has been partially purified under non-denaturing as well denaturing conditions by Janeczek and John [22]. Size exclusion chromatography on sephacryl gels revealed a wide range of molecular weight from 500,000 to 50,000 with a maximum at about 100,000. When this fraction was subjected to isoelectric focusing the factor banded at about pH 5-6.

Secondary Interactions in Pattern Formation

Several experimental approaches suggest that endoderm and mesoderm derived organs as well the different regions of the nervous system are induced by secondary interactions. The highly purified vegetalizing factor from chicken embryos induces undiluted or slightly diluted with γ-globulin or another non-inducing protein preferentially endoderm (Fig. 7). When the vegetalizing factor was however mixed with crude proteins which were separated from the factor in the course of its purification quite different organs and tissues were induced [23]. The mixed factors induced little endoderm, but in a high percentage of the cases large masses of notochord, muscle and renal tubules (Table 1, Fig. 8) whereas the crude fraction itself did not induce such organs. This suggests that the state of determination of the primarily induced cells is labile for a certain period of time and that their pathway of differentiation can be changed by interaction with additional factors. It was shown that these additional factors are inactivated by trypsin or immobilized pronase and are therefore also protein in nature [23].

That secondary interactions play a major role is also shown by the cell dissociation experiments of Grunz and Minuth [24]. Gastrula ectoderm was isolated and induced with highly diluted vegetalizing factor, so that some ectoderm remains uninduced. Then the ectoderm was dissociated into single cells and small cell complexes to interrupt cell interaction, and after 20 h allowed to reaggregate. In a high percentage of cases endoderm derived tissues (liver, intestine) and organs

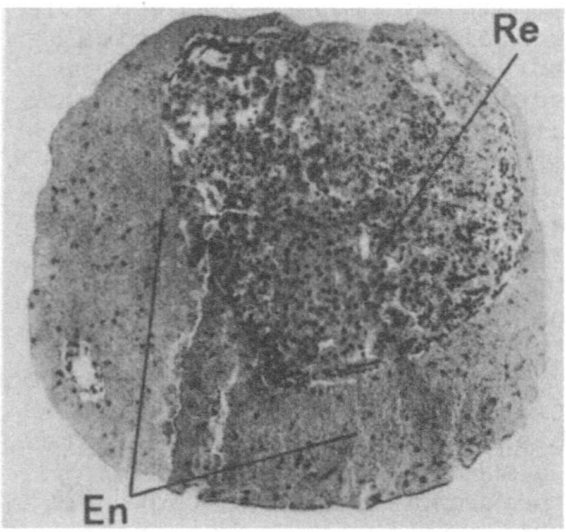

Fig. 7. Induction of endoderm by the vegetalizing factor from chicken embryos tested by the implantation method. *En* Endoderm, *Re* Renal tubules

Table 1. Inducing activity of a crude fraction: the vegetalizing factor (V.F.) and combinations of these fractions

	No	Positive %	Induced Region (%)			
			Endoderm	Mesoderm (Trunk/Tail)	Neural Hindhead	Neural Forehead
V.F. mixed with γ-Globulin (1:9)	27	100	93	7	0	0
Crude Fraction ($(NH_4)_2SO_4$ Sup.)	72	34	0	0	33	1
V.F. mixed with Crude Fract. (1:9)	30	100	3	97	0	0

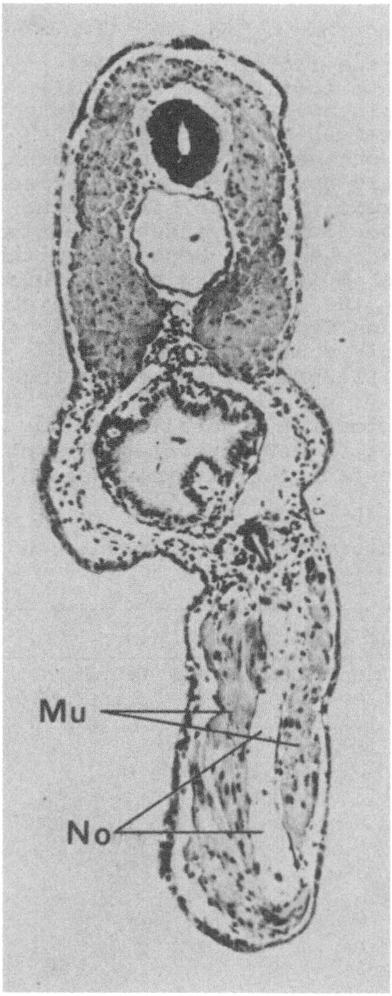

Fig. 8. Induction of muscle (*Mu*) and notochord (*No*) by a combination of vegetalizing factor and a crude protein fraction

derived from ventral mesoderm (blood cells, coelomic epithelium) were induced. In the induced but undissociated control ectoderm where the cell interactions were not interrupted, in addition muscle and notochord (i.e. organs derived from dorsal mesoderm) were induced by secondary interactions between the primarily induced endoderm and the ectoderm. All these experiments favor the idea that secondary cell interactions and additional factors are involved in the formation of the mesodermal organ pattern, and not different threshold concentrations of one and the same factor.

The segregation of the neural anlage into forebrain, hindbrain and spinal cord depends likewise on secondary interactions. The neuralizing factor induces forebrain. Hindbrain and spinal cord appear when neural inducers are combined in different proportions with vegetalizing inducers [25,26]. It was shown by Toivonen [27] that the induced mesoderm transformed the neural anlage into hindbrain or neural tube. Such interactions certainly play also a role in normal development.

Possible Mechanism of Action

The third topic, the mechanism of action of the factor is the most difficult one, because at present few concrete facts are known. Different experimental approaches have been employed. I will begin with the question whether the factors must be taken up by the competent cells or whether they act via binding sites on the cell surface. When the chicken vegetalizing factor was covalently bound to BrCN-Sepharose particles or BrCN-Sephadex particles to prevent the factor from being taken up by the ectoderm cells, the factor had lost its inducing activity ([28,29], Table 2). The result was similar when BrCN-Sepharose with reduced binding capacity was used, to prevent multipoint binding. This suggests that the factor must be taken up by the ectoderm cells, which easily incorporate foreign proteins [30]. When Sephadex, a derivative of dextran, to which the factor was covalently bound, was degraded with the enzyme dextranase, the fully active factor could be recovered (Table 3).

Asashima [2] has shown that in normal development a vegetalizing factor (or its mRNA) is already localized in the vegetal cytoplasm of the fertilized egg, i.e. in the region where the factor exerts its action.

Table 2. Inducing activity of a crude vegetalizing fraction mixed with Sepharose or covalently bound to Sepharose

	No	Positive (%)	Induced Region (%)			
			Trunk/Tail	Hindhead	Forehead	Neural (Not specified)
Crude Veg. Fraction (Inducer Control)	83	87	84	10	0	0
Veg. Fraction (2.5 mg) mixed with Sepharose (40 mg)	23	61	56	13	0	0
Veg. Fraction (2 mg) bound to Sepharose (40 mg)	75	55	0	5	26	24

Table 3. Inducing activity of vegetalizing factor after binding to Sephadex G 200 and of factor recovered after enzymatic degradation of the Sephadex matrix

	No	Positive	Size of Induction (%)			Trunk/Tail Induced (%)	Not Specified (%)
			Large	Medium	Small		
V.F. + γ-Globulin (1:49) (Inducer Control)	24	96	41	17	38	96	0
V.F. (0.7 mg) Coupled to Sephadex (20 mg)	30	33	0	15	18	30[a]	3
V.F. Recovered after enzymatic degradation of Sephadex + γ-Globulin (1:49)	28	100	31	36	33	100	0

[a]Includes 14% small mesenchyme tails

On the other hand, the neuralizing factor in the crude fraction remains fully active after covalent binding to BrCN-Sepharose or BrCN-Sephadex particles (Table 2). This suggests that the neural factor interacts with binding sites on the plasma membrane of ectoderm cells. This is probably also true for other secondary factors. It has recently been found out by Japanese colleagues [31,32] and in our laboratory [33] that concanavalin A and concanavalin A which was covalently bound to Sepharose induces neural tissues. This could suggest that the binding

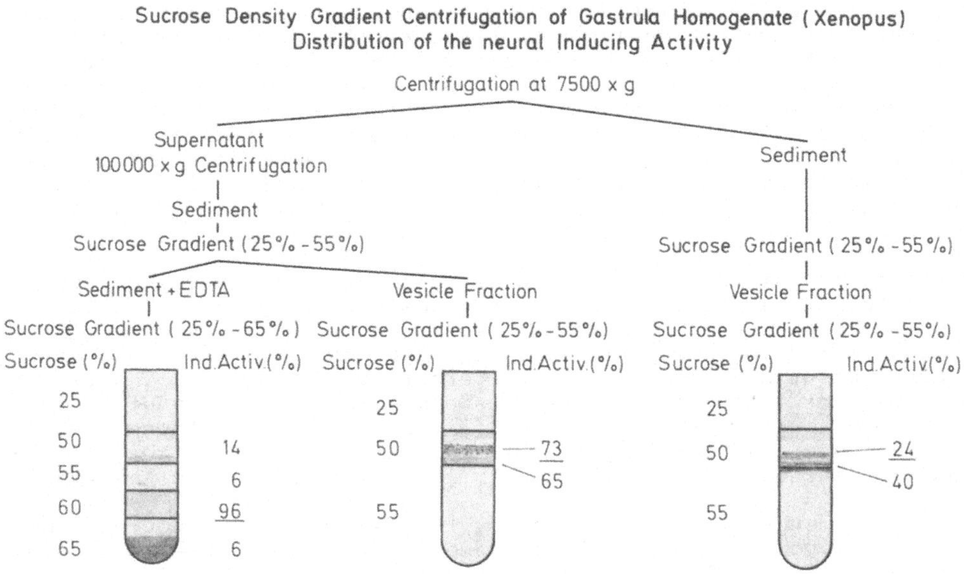

Fig. 9. Neural inducing activity of subcellular fractions after repeated isopygnic sucrose gradient centrifugation. The inducing activity of the fractions which are presented in the electronmicrographs (see Figs. 10-12) is underlined

site for the neuralizing factor contains mannose. Concanavalin A modulates however the distribution of cell surface receptors [34] and such modulation could perhaps trigger signals which are more specifically generated by the neuralizing factor.

In a second approach the subcellular distribution of the neuralizing factor in early gastrulae was investigated. After repeated separation by density in stepwise sucrose gradients a high neural inducing activity is found in a fraction released by EDTA treatment from "microsomes", which consists of RNP-particles and a few cytoskeletal proteins and in another fraction which consists mostly of a heterogeneous population of small vesicles. The fraction in which the plasma membranes are enriched shows only little neural inducing activity (Fig. 9). The electronmicrographs of the subcellular fractions are shown in Figs. 10-12.

Further experiments revealed that the neuralization depends on the activation of a factor. Intact gastrula ectoderm has no inducing activity. It contains however neuralizing factor in a masked particulate form. The factor is activated after homogenization in a process which is not energy dependent. The very small neuralizing activity of the 100,000 *g* supernatant increases when the homogenate was incubated for 5 h at 22°C. Proteases are activated after homogenization, but whether proteolysis or other enzymatic processes are involved needs further investigation. The total inducing activity in the ectoderm does not increase after neural induction. This strongly indicates that a masked factor is activated, a notion which is also supported by experiments with actinomycin D. The distribution of the factor in different subcellular fractions changes however somewhat after induction. The neuralizing activity in the 100,000 *g* supernatant from neural plates increases as compared to non-induced ectoderm. This could be correlated to the well-known acquisition of neural inducing activity by the induced neural plate [35,36] and an export of the factor or on the other hand to an intracellular translocation of active factor The target for such a translocation could be the nuclear chromatin of the cells. It was shown by Bretzel [37] that the neural inducing activity in highly purified nuclei increases somewhat from the neurula stage on. A part of the neuralizing factor and the chicken vegetalizing factor bind tightly to DNA, but binding to specific sequences has not yet been proven [38]. It is also not known whether the inducer molecules which are exported and those which act within the induced cells are structurally related, for instance whether they are derived from a common precursor or whether they are unrelated.

The subcellular distribution of the neuralizing factor may indicate that the factor is stored after its synthesis on polysomes in ribonucleoprotein particles perhaps in association with the cytoskeleton, then transferred to exocytotic vesicles, exported from the mesoderm and bound to surface receptors of competent ectoderm cells. The factor is probably not inserted into the plasma membrane as an integral protein. We are now preparing monoclonal antibodies against those particles which contain neuralizing factor to separate them from the heterogeneous population and to follow their localization in eggs and embryos.

The proposed mechanism is supported by the isolation of a neuralizing protein or glycoprotein from the extracellular space between mesoderm and dorsal ectoderm. The protein was separated from proteoglycans by extraction with phenol. The proteoglycan could modulate the inducing activity or could be involved in the tansport of the neuralizing factor. Several years ago it was shown that a proteoglycan which was

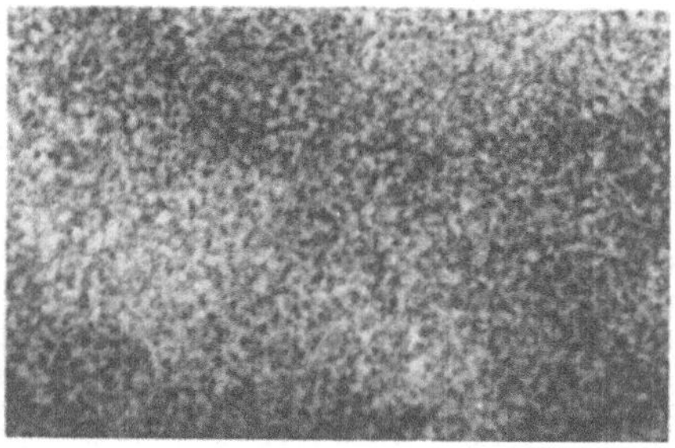

Fig. 10. Electronmicrograph of a neural inducing fraction (96% inducing activity; left side gradient in Fig. 9) which consists of heterogenous ribonucleoprotein particles and polysaccharides which could be separated from the ribonucleoprotein particles in a 10%-30% sucrose gradient. The fraction contains only a few ribosomal subunits. The most part of the ribosomal subunits is found in the sediment of the 25%-65% sucrose gradient (left side gradient in Fig. 9) which has only a very low inducing activity. x 90,000

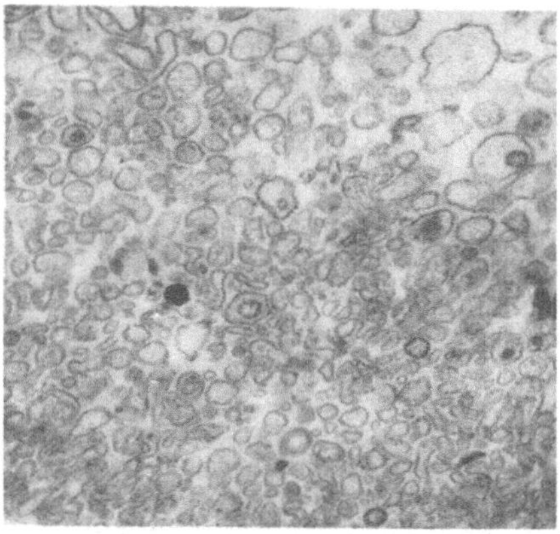

Fig. 11. Electronmicrograph of a heterogeneous fraction of mostly small vesicles (73% neural inducing activity; middle gradient in Fig. 9). x 30,000

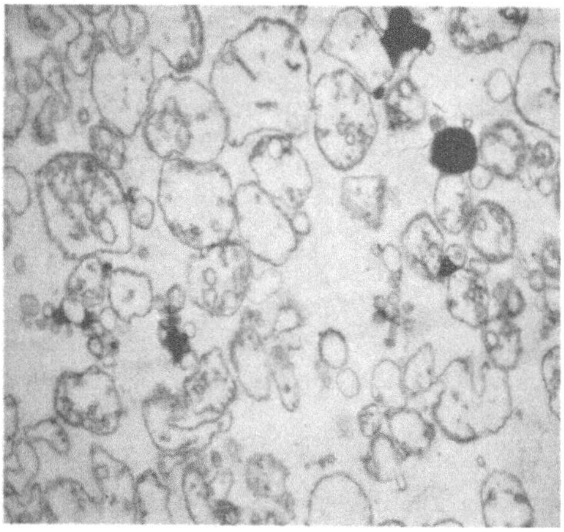

Fig. 12. Electronmicrograph of a fraction of larger vesicles. This fraction contains the most part of the plasma membranes as labeling experiments with diazotized ^{125}I-sulfanilic acid has shown. The inducing activity is low (24% neural inducing activity). Right side gradient in Fig. 9. x 24,000

isolated from chicken embryos, forms a complex with inducing factors and inhibits their inducing action when applied in larger excess [39,40].

The neuralizing factor as we have seen probably binds to a cell surface receptor of ectodermal cells. How the plasma membrane does communicate with constituents within these cells to activate the neuralizing factor is unknown. The possible involvement of enzymatic processes has already been mentioned. Cyclic nucleotides or their mono- and dibutyrylderivatives [41], change in the Na^+/K^+ ratio or treatment with the Ca^{++} ionophore A 23187 did not evocate inductions in the ectoderm of *Triturus alpestris* [42].

The membrane mediated activation of a neuralizing factor can explain the so-called "autoneuralization effect". Neuralization can easily be brought about by unspecific means in ectoderm of some amphibian species especially *Ambystoma mexicanum*, but not ectoderm of *Triturus alpestris*. Ectoderm of *Triturus alpestris* was therefore used in all test experiments.

When the factor is stored in RNP particles several questions to the mechanism of action could be asked. (1) How is the expression of mRNA in these RNP particles regulated? (2) Has the factor an affinity to RNA as well to a DNA sequence complementary to this RNA? (3) Are repetitive sequences involved which are inserted in the gene sequence, which codes for the factor as well, in regulatory sequences upstream or downstream from other genes which are regulated by the factor? (4) What is the mechanism of activation of the factor and which enzymatic processes are involved? We have no answer to these questions, but at least we are now in the position to formulate mechanisms which can be experimentally tested.

The experiments have definitely shown, that proteins are involved in in the sequential determination of cells which finally leads to the formation of organ patterns. This is an example for the evolution of a system in which proteins or peptides act over short distances. Systems in which proteins or peptides partake in short range cell communication are the maturation and interaction of immune cells [43,44] and the transfer of signals in the nervous system [45]. In 1936 Hans Spemann [5] closed the Silliman lectures with the words "these processes of development are comparable, in the way they are connected, to nothing we know in such a degree as those vital processes of which we have the most intimate knowledge, the psychical ones". At that time it was not known that they have, indeed, a common biochemical basis.

Acknowledgements. This investigation was supported by the Deutsche Forschungsgemeinschaft (Sonderforschungsbereich Embryonale Entwicklung und Differenzierung; Embryonalpharmakologie).

References

1. Gerhart J, Ubbels G, Black S, Hara K Kirschner M (1981) Nature (London) 292: 511-516
2. Asashima M (1975) Wilhelm Roux's Arch 177:301-308
3. Boterenbrood EC, Nieuwkoop PD (1973) Wilhelm Roux's Arch 173:319-332
4. Spemann H, Mangold H (1924) Arch Mikrosk Anat Entw Mech 100:599-638
5. Spemann H (1936) Experimentelle Beiträge zu einer Theorie der Entwicklung. Springer, Berlin
6. Mangold O (1936) Naturwissenschaften 24:753-760
7. Holtfreter J, Hamburger V (1955) In: Willier BH, Weiss PA, Hamburger V (eds) Analysis of development. Saunders, Philadelphia London
8. Grunz H (1977) Wilhelm Roux's Arch 181:267-277
9. Hadorn E (1965) Brookhaven Symp Biol 18:148-161
10. Mangold O (1924) Wilhelm Roux's Arch 100:198-301
11. Holtfreter J (1933) Wilhelm Roux's Arch 128:584-633
12. Becker U, Tiedemann H, Tiedemann H (1959) Z Naturforsch 14b:608-609
13. Yamada T, Takata K (1961) Dev Biol 3:411-423

14. Geithe HP, Asashima M, Asahi KI, Born J, Tiedemann H, Tiedemann H (1981) Biochim Biophys Acta 676:350-356
15. Schwarz W, Tiedemann H, Tiedemann H (1981) Mol Biol Rep 8:7-10
16. Schwarz W, Hoppe P, Tiedemann H, Tiedemann H (1982) (in preparation)
17. Asahi KI, Asashima M, Geithe HP, Born J, Tiedemann H, Tiedemann H (1982) (in press)
18. Tiedemann H, Tiedemann H, Kesselring K (1960) Z Naturforsch 15b:312-319
19. Tiedemann H, Kocher-Becker U, Tiedemann H (1963) Biochim Biphys Acta 74:557-560
20. Tiedemann H, Becker U. Tiedemann H (1961) Embryologia 6:204-218
21. Faulhaber I (1970) Hoppe-Seyler's Z Physiol Chem 351:588-594
22. Janeczek J, John M (1982) (in press)
23. Asahi KI, Born J, Tiedemann H, Tiedemann H (1979) Wilhelm Roux's Arch 187:231-244
24. Minuth M, Grunz H (1980) Cell Differ 9:229-238
25. Saxén L, Toivonen S, Vainio T (1964) J Embryol Exp Morphol 12:333-338
26. Tiedemann H, Tiedemann H (1964) Rev Suisse Zool 71:117-137
27. Toivonen S (1972) In: Harris R, Allen P, Viza D (eds) Proc 1st Int Conf Cell Differ. Munksgaard, Copenhagen
28. Tiedemann H, Born J (1978) Wilhelm Roux's Arch 184:285-299
29. Born J, Grunz H, Tiedemann H, Tiedemann H (1980) Wilhelm Roux's Arch 189:47-56
30. Vainio T, Saxén L, Toivonen S, Rapola J (1962) Exp Cell Res 27:527-538
31. Kawakami I, Oseko H (1979) Zool Mag 88:457
32. Takata K, Yamamoto Y, Ozawa R (1981) Wilhelm Roux's Arch 190:92-96
33. John M, Tiedemann H (1982) (in press)
34. Edelman GM (1976) Science 192:218-226
35. Mangold O, Spemann H (1927) Wilhelm Roux's Arch 111:341
36. Tiedemann-Waechter H (1960) Wilhelm Roux's Arch 152:303-338
37. Bretzel G (1982) (in preparation)
38. Tiedemann H, Born J, Tiedemann H (1972) Wilhelm Roux's Arch 171:160-169
39. Born J, Tiedemann H, Tiedemann H (1972) Biochim Biophys Acta 279:175-183
40. Niebel J, Tiedemann H, Tiedemann H (1973) Eur J Biochem 32:242-246
41. Grunz H, Tiedemann H. (1977) Wilhelm Roux's Arch 181:261-265
42. Siegel G, Grunz H, Tiedemann H (1982) (in press)
43. Schimpl A, Wecker E (1973) Exp Med 137:547
44. Hadden JW, Stewart WE (eds) (1981) In: The lymphokines: biochemistry and biological activity. John Wiley, New York (review)
45. Snyder SH (1980) Science 976-983 (review)

Differentiation and Embryogenesis in *Volvox carteri*

L. Jaenicke and R. Gilles[1]

Introduction

In many freshwater algae sexual processes have a much more immediate advantage than that of genetic recombination, namely survival under temporary adverse conditions. The thick-walled zygotes are able to endure long periods of drought and heat and germinate on re-wetting. From zygote-containing mud samples collected in Japan R.C. Starr isolated in this way the female (HK-10) and male (HK-9) strains of *Volvox carteri f. nagariensis* [1]. These algae can be grown under laboratory conditions in a synthetic glycylglycine/glycerophosphate buffered medium containing the essential trace substances [2]. During the past 15 years the sexual and asexual reproduction of *Volvox* species was intensively investigated in different laboratories [3-5].

In *Volvox carteri* it is impossible to achieve sexual development directly by conversion of existing cells to gametes (as e.g. in *Chlamydomonas*). Sexual spheroids with eggs or sperm-packets are formed by differentiation of reproductive cells (gonidia) which otherwise would give rise to asexual spheroids (Fig. 1). The switch from the asexual to the sexual embryogenetic pathway is triggered by a glycoproteinaceous pheromone, secreted into the medium by the male strain [6]. Thus, *Volvox carteri* provides us with an ideal system to study hormonal control of differentiation.

One would hardly expect that a thallophyte exhibits such a complex embryogenesis resembling in many aspects that of higher organized forms of life. However, matters are not too complicated, since the *Volvox* spheroid consists of only two kinds of cells, somatic and reproductive ones, which are determined by a single unequal cleavage early in embryogenesis. In the following we summarize what is known in *Volvox* about the mechanism of sex induction and differentiation; two themes which are intensively intertwined.

Sex Inducer

In 1965, Darden demonstrated that a substance in the culture medium from male spheroids of *Volvox aureus* is able to induce the formation of sexual individuals [7]. Since then sex inducers were described in several other species of *Volvox* [6,8-11]. In most cases - as in *Volvox carteri f. nagariensis* - sex inducers are high molecular. We succeeded in

1 Institut für Biochemie, Universität Köln, An der Bottmühle 2, D-5000 Köln 1, FRG

33. Colloquium - Mosbach 1982
Biochemistry of Differentiation and Morphogenesis

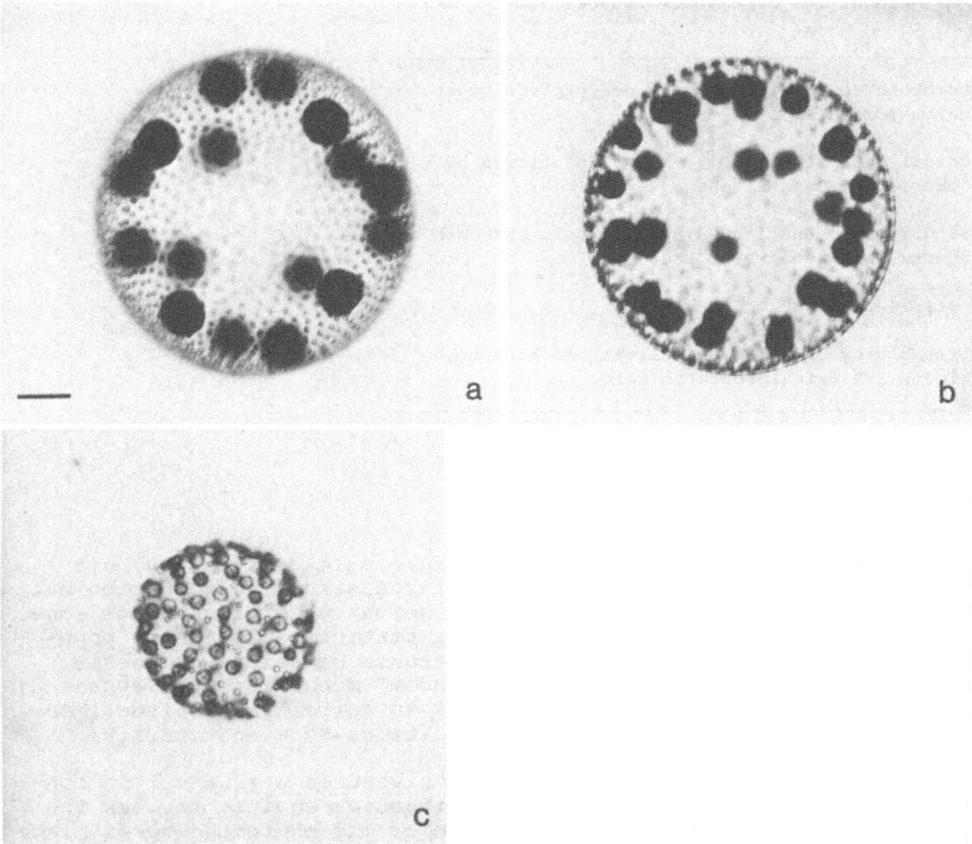

Fig. 1a-c. *Volvox carteri f. nagariensis*. (a) Asexual spheroid of the HK-10 (female) strain; the corresponding one of the 69-1b (male) strain looks almost identical. The individual consists of a peripheral layer of *Chlamydomonas*-like somatic cells separated by thick envelopes of sheath material. Within the hollow sphere 16 large reproductive cells (gonidia) are embedded in a mucilaginous matrix. Gonidia form a reproducible pattern of high symmetry (details in 14). (b) Sexual spheroid of the female strain with 32 eggs which are smaller and darker than gonidia. (c) Male spheroid showing the typical 1:1 ratio of small somatic cells and large androgonidia lying both in the same peripheral layer. Androgonidia divide to sperm-packets containing 64 biflagellated sperm. Bar 50 µm

purifying this glycoprotein to electrophoretic homogenity. The specific biological activity surpasses that of previous preparations [6]: 10^{-17} M give 20% induction in the bioassay [12]. From this we may calculate that less than 100 molecules per gonidium suffice for pheromone action. Complete chemical characterization of the sex inducer is not yet achieved, since the starting concentration in the medium is only approximately 20 µg/l and the losses during purification are high due to irreversible polymerization and inactivation. Some of the properties of the molecule listed in Table 1 show that the sex inducer does not behave like a "normal" protein. Some of its unusual characteristics may result from the large carbohydrate moiety of the molecule. Our current interest is therefore focused on the pheromone's carbohydrate portion which also seems to be involved in biological action.

Table 1. Properties of the sex inducer

Glycoprotein (M.W. 27,000), one polypeptide chain, 40% carbohydrate
Biological activity at 10^{-17} mol/l, destroyed by periodate and α-mannosidase
Self-aggregation unaffected by pH, ionic strength and detergents, prevented by 6 M urea
Heat-stable up to 80°C
Precipitable by 3M ammonium sulfate and acetone, soluble in 12% trichloroacetic acid

Formation of the Pheromone

The first appearance of sexual colonies in a male culture may be a "hen-and-egg" problem. The sex inducer starts sexual reproduction but the pheromone is exclusively produced by sexual spheroids. Hence, one speaks of "spontaneous" males which occur statistically in the population [4]. If we realize the vital importance of sexual processes stressed in the introduction, such a "random" mechanism will appear unlikely. However, so far all our efforts to correlate the induction of the first male with environmental conditions were unsuccessful.

The site and time of pheromone formation have been worked out in considerable detail [13]. Using specific antibodies against the sex inducer we were able to show that excretion of the pheromone takes place in the sperm-packets within 6 h. The biologically inactive polypeptide chain is already translated in the uncleaved androgonidium. During cell divisions which give rise to the 64-128 sperms bundled to one packet, translation is interrupted. The excretion is blocked by tunicamycin indicating that the final processing involves glycosylation. Isolation of mRNAs and analyses of the products of cell-free translation should elucidate whether precursors of sex inducer exist.

Mode of Action

The most important questions concerning the processing of a chemical signal are: (1) Where is the binding site of the molecule, (2) how does it elicit its specific response? The gonidia are obviously the target cells for the pheromone. However, neither binding studies with iodinated sex inducer nor fluorescence microscopy with anti-inducer antibodies revealed a specific binding site on the gonidial surface. Using the bioassay we were able to demonstrate indirectly that approximately 80 molecules per gonidum are bound. As mentioned above the sensitivity of the system is extremely high (10^{-17} M). We therefore assumed a high affinity and almost irreversible pheromone-receptor-interaction. But to our surprise, we found the response to be reversible upon diluting out the sex inducer even after long periods of incubation; only at a defined developmental stage (4 h before the first cleavage) the cells are ultimately induced. Thus, the pheromone's action may be biphasic; in the first step the molecule is reversibly bound without consequences for the determination of the cell; then the state of the pheromone-receptor-complex is modified by a conformational change, degradation or internalisation, and simultaneously the gonidia become competent for induction.

Sexual induction requires synthesis of new membrane glycoproteins. Therefore, a transmission of the signal into the cells is a prerequisite. We have investigated the mode of transmission by means of agents interfering with cyclic nucleotide or calcium mediated processes, but we found no indication for involvement of these second messengers in sexual induction.

Embryogenesis of *Volvox carteri*

While we know almost nothing about the mode of pheromone action the developmental consequences of sex induction are clear [4,14]. On asexual reproduction each gonidium will cleave 11-12 times resulting in a daughter spheroid of 2000-4000 cells (see Fig. 1). These cells are unable to cleave further and they will eventually die. Only the 16 large reproductive cells (gonidia) lying in the posterior half of the spheroid can form the next generation. This makes *Volvox carteri* such a promising object for the study of differentiation. As in all metazoa, we have somatic and germ cell lines. Figure 2 shows the cleavage pattern in asexual development: Differentiation of the gonidial initials is carried out by unequal division of distinct cells at the sixth cleavage. Whereas in the female strain only the cells 1 and 2 of the 16-celled embryo (see Fig. 2) are differentiated in both, the asexual as well as the sexual pathway, in the sexual development of the male strain *all* cells are affected by an unequal division which is the last one in embryogenesis. Thus, the sexual male spheroids consist of an equal number of somatic and reproductive cells (Fig. 1).

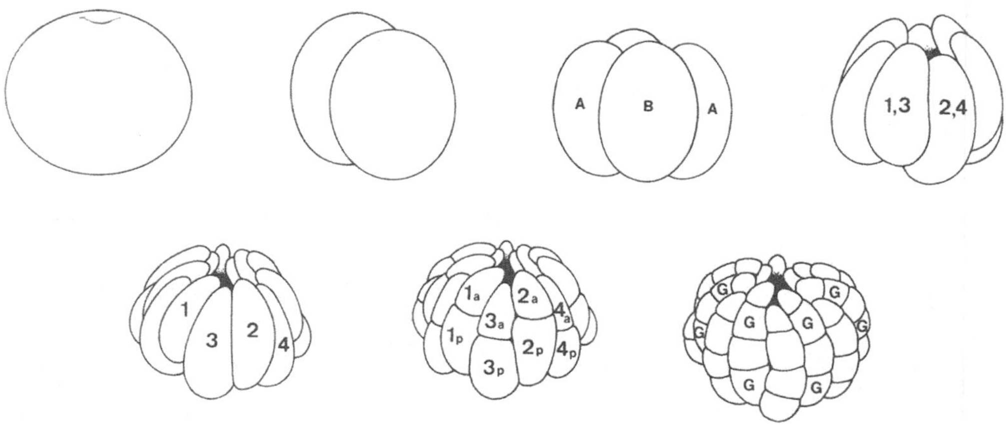

Fig. 2. Schematic representation of the gonidial cleavage pattern in asexual development up to the 64-celled stage. The first five cleavages are equal, then the 16 cells which derive from cells *1* and *2* of 16-celled embryo cleave unequally. The gonidia are formed from the large initials, whereas the small initials as well as the other cells give rise to the somatic cells of the spheroid. Nomenclature according to [17]

Under the influence of the inducer this unequal cleavage is delayed to the 64-celled or to the 128-celled stage for the female or male strains respectively.

Model of Differentiation

A working model of differentiation able to explain the different embryogenetic events during sexual and asexual reproduction is based on two principal parameters of differentiation. (1) The potency of the cells to undergo differentiation, (2) the point in embryogenesis at which unequal differentiating cleavage occurs. We assume that both parameters are mediated by membrane proteins. Before discussing the plausibility of such an assumption, let us see which characteristics these proteins should exhibit. The two basic hypotheses are:

1. A differentiation protein (D-protein) exists on the gonidial membrane which is synthesized in the uncleaved gonidium. It is extruded exclusively at the anterior pole, whence it diffuses down the membrane, generating a radially symmetric distribution. During cell cleavages diffusion will be limited by cleavage planes. All cells carrying the D-protein on their membranes at a distinct stage will cleave unequally.
2. A counting protein (C-protein) exists in the gonidial membrane which is also synthesized before onset of cleavages. Its distribution is homogeneous. The concentration of the C-protein decreases during cleavage by modification or by diluting-out. (The surface area grows exponentially). If its concentration (i.e. active molecules per cell) reaches a distinct level (which may or may not be zero), all cells which carry the D-protein on their membranes cleave unequally.

Figure 3 shows how the different embryogenetic pathways could be governed by the C- and D-protein. Sexual induction is carried out by increasing the starting concentration of the C-protein and altering the distribution of the D-protein. The latter could be achieved, if the uncleaved gonidia start earlier to produce D-protein. Hence the covered area at the beginning of the cleavages will already be larger and at the critical 16-celled stage all cells carry the D-protein. The total number of cleavages which varies as well, especially in the sexual

→

Fig. 3. Schematic diagram of the three developmental pathways guided by the hypothetical (protein) compounds, *C* (counting) and *D* (differentiating). Cell cleavages are indicated in one quarter only and in side view with the anterior pole at the top and just before onset of the next cleavage. In the asexual cycle embryogenesis of both, the male and female strain, is identical. Patterns are common in all of the three pathways up to the 32-celled stage when, in asexual development, the first unequal cleavage occurs. In the sexual cycles it is delayed to the 64- (female) or to the 128-celled stage (male; the final, unequal cleavage at the 256-celled stage is not shown for clarity). Cells marked *S* (stem cells) will cleave unequally and eventually form reproductive cells. After unequal cleavage the larger derivative located more to the posterior pole will continue to cleave unequally at least two or three times; the smaller one will divide equally. *Shaded areas* on the embryos represent distribution of the D-protein by membrane diffusion (keep in mind that the distribution is radially symmetrical!). *The triangular bars* represent the respective concentration of C-protein assuming that the decrease is linear and triggering occurs at zero level

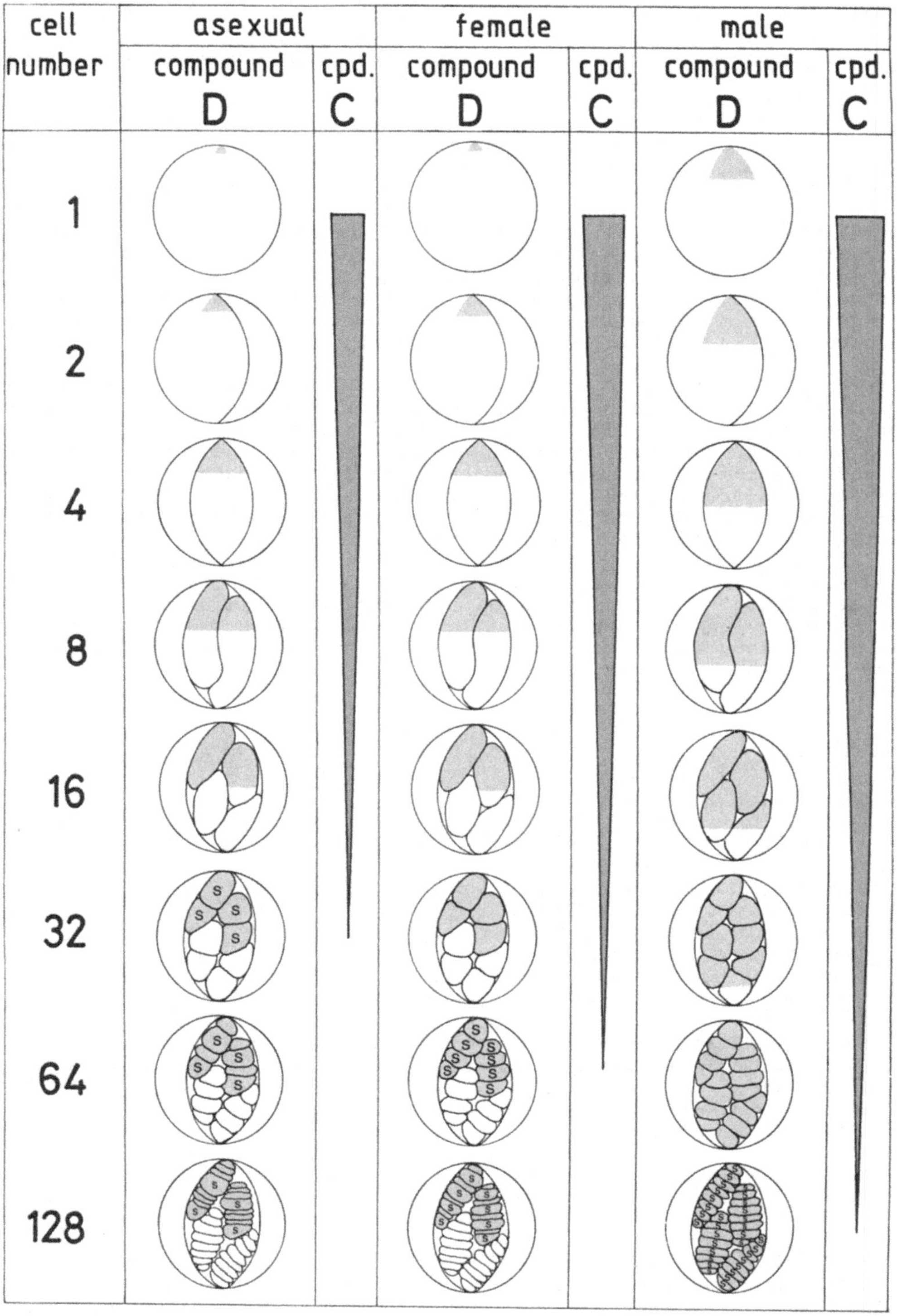

cell number
asexual
female
male
compound D
cpd. C
compound D
cpd. C
compound D
cpd. C
1
2
4
8
16
32
64
128

development of the males, could be determined by a similar counting mechanism; perhaps this is also done by the C-protein, triggering unequal cleavage at a higher and cessation of cleavages at a lower level[2].

This differentiation model is still hypothetical, but there are experimental results which completely agree with our assumptions. Sumper postulated a counting mechanism on the basis of contact sites built up by a sulfated glycoprotein. In this model the number of contact sites grows exponentially during cell divisions, thus the contact protein is consumed, and the first cells which are devoid of it are triggered to differentiate [15,16]. It was meanwhile demonstrated by electron microscopy that *Volvox carteri* has no contact sites at all; and the actual cell lineages are different from his assumptions [17]. Nevertheless, the postulation of a counting mechanism seems a reasonable idea and we have integrated it in a modified form to our model. The sulfated glycoprotein could be a candidate for the C-protein.

A further support may be the observation of a glycoprotein, the distribution of which fulfills our statements about the D-protein. We study gonidial membrane proteins by fluorescence microscopy using FITC-Con A as a probe. At the very beginning of the cleavages the anterior pole of the gonidium becomes flattened and a small fluorescent spot appears. This spot enlarges until at the 8-celled stage it will be undistinguishable from the background staining. Studies are in progress to verify whether this glycoprotein is the D-protein.

Thus, indeed, *Volvox carteri* is a promising object for investigating differentiation at a molecular level. Differentiation is assumed to be controlled by spatial and temporal gradients of membrane glycoproteins, and sexual induction could be explained by systematic variations of these chemical gradients.

References

1. Starr RC (1969) Arch Protistenk 111:204-222
2. Provasoli L, Pintner IJ (1959) Spec Publications No. 2, Pymatuning Laboratory of Field Biology, University of Pittsburgh, pp 84-96
3. Jaenicke L (1982) TIBS: 61-64
4. Starr RC (1970) Dev Biol Suppl 4:59-100
5. Kochert G (1975) Symp Soc Dev Biol 13:55-90
6. Starr RC, Jaenicke L (1972) Proc Natl Acad Sci USA 71:1050-1054
7. Darden WH (1966) J Protozool 13:239-255
8. Kochert G, Yates I (1974) Proc Natl Acad Sci USA 71:1211-1214
9. Vande Berg WJ, Starr RC (1971) Arch Protistenk 113:195-219
10. Starr RC, O'Neil RM, Miller CE (1980) Proc Natl Acad Sci USA 77:1025-1028
11. Karn RC, Starr RC, Hudock GA (1974) Arch Protistenk 116:142-148
12. Gilles R, Bittner C, Cramer M, Mierau R, Jaenicke L (1980) FEBS-Letters 116:102-106
13. Gilles R, Bittner C, Jaenicke L (1981) FEBS-Letters 124:57-61
14. Gilles R, Jaenicke L (1982) Z Naturf (in press)
15. Sumper M (1979) FEBS-Letters 107:241-246
16. Wenzel S, Sumper M (1981) Proc Natl Acad Sci USA 78:3716-3720
17. Green KL, Kirk DL (1981) J Cell Biol 91:743-755

---

2 Control of the two parameters of differentiation - potency and point - by two proteins is an obvious concept, one could also envisage both parameters to be coded by a single protein. Its distribution would determine the potency of cells and its concentration would trigger the point of differentiation. But this modification would not touch the basic principles of our model

Subject Index

GPSR Compliance

The European Union's (EU) General Product Safety Regulation (GPSR) is a set of rules that requires consumer products to be safe and our obligations to ensure this.

If you have any concerns about our products, you can contact us on ProductSafety@springernature.com

In case Publisher is established outside the EU, the EU authorized representative is:

Springer Nature Customer Service Center GmbH
Europaplatz 3
69115 Heidelberg, Germany

Batch number: 10640288

Printed by Printforce, the Netherlands